ALMOST HUMAN

BOOK SEVEN OF THE WAVE

Other books by Laura Knight-Jadczyk

The Secret History of the World

9/11: The Ultimate Truth
(with Joe Quinn)

High Strangeness

The Horns of Moses
(Coming Soon)

Laura Knight-Jadczyk

ALMOST HUMAN

Book Seven of The Wave

Red Pill Press
2009

Red Pill Press
10020–100 Ave.
Grande Prairie, AB
T8V 0V3

Printed in Canada.

Table of Contents

Introduction
Forget About Global Warming, We're One Step from Extinction!

Often, some of the most important news comes from local papers, stories that don't make it up the feeding chain and onto the news wires or major newspapers or nightly network news. It can be news that at first glance wouldn't appear to have a national or international impact. Second glance, and a good memory, can reveal that the impact may well be quite significant.

The other day (mid-March 2007) a reader of our alternative news website *Signs of the Times* (sott.net) sent us an article link that he found on *The Guardian* website: "Bad news — we are way past our 'extinct by' date" (Robin McKie, *The Observer*, March 13, 2005).

Since we had run the same story back when it first came out, we thought, "Yeah! Flashback!" After all, with all the talk about "Global Warming", it served to remind readers that human-caused CO_2 levels are not all there is to what is going on here on the Big Blue Marble today.

Later in the evening, we had a conversation here at SOTT Central about this article and how it relates to a book that is currently being passed around in the house here: *The Cycle of Cosmic Catastrophes: Flood, Fire, and Famine in the History of Civilization*, by Richard Firestone, Allen West and Simon Warwick-Smith. This book is about the "event" that took place about 12,000 years ago that is recorded in myth and legend variously as the Fall of Atlantis and Noah's Flood.

Plato describes a destruction that occurred in a day and a night, and the Bible recounts the story of torrential rains and an immense flood in which most of the life on Earth perished. There is also a rich body of Native American literature about a worldwide cataclysm of fires, followed by floods and death raining down from the skies. As many as fifty different cultures around the globe record versions of this story, and physicist Firestone, along with his geologist co-authors, have put together a book, based on hard scientific evidence, describing a cosmic chain of events that they believe culminated in the global catastrophe of 12,000 years ago. They believe that the event was triggered by a nearby supernova that occurred 41,000 years ago.

Regular readers of *Signs of the Times* are familiar with my Cassiopaea website and the experiment in superluminal communication that I began in 1992 and which finally bore fruit in 1994 on the day that the fragments of Comet Shoemaker Levy began impacting the planet Jupiter. We find it amusingly synchronous that one of the themes of the Cassiopaean information is planetary destruction via a Comet Cluster which cycles through the solar system every 3,600 years as a consequence of the orbit of our Sun's solar companion, a smaller, dark, Twin Sun. As it happens, Firestone, West and Warwick-Smith also talk about a bombardment of planet Earth by literally thousands of asteroids, comets or other debris, though they attribute it to the supernova 28,000 or 29,000 years earlier; it took that long for the ejecta from the supernova — along with debris it kicked out of the Oort Cloud — to reach Earth.[1]

With the idea that there is a Cometary Bombardment Cycle, we have naturally been alert to the fact that the last few years have brought increasing evidence that this theory may very well be the correct one. This evidence includes the fantastic increase in the number of "moons" attached to Jupiter that have so recently been "discovered," as well as the increase in frequency of comets over the past few years, along with the astonishing increase in meteorites and fireballs entering Earth's atmosphere and falling to Earth. In some cases, these events have resulted in damage to human beings and property, and one recent case even resulted in deaths, as we will see further on.

Getting back to our conversation about humanity being past its "extinct by" date, I mused that anybody with eyes and ears and a bit of scientific knowledge can look around and see that something is going on "out there." It's in the news everyday, you just have to search for it (or read SOTT: we do the searching for you). The problem is, of course, that the masses of humanity are so distracted by all the concerns of everyday life — many of which are quite serious nowadays,

---

[1] Recently (2008) we came across the work of Victor Clube and Bill Napier, authors of *The Cosmic Serpent* and *The Cosmic Winter*. Both works present a wealth of historical and scientific evidence for the cyclical bombardment of the Earth by a cluster of cometary debris. While we speculate that the cause of this cycle is the result of a binary star, Clube and Napier disagree. Regardless of the cause, their work presents the hard data showing that "something wicked this way comes."

especially the threat of nuclear war brought to us by George W. Bush and the Ziocons — that most of them haven't got a clue that they probably don't have to worry about Global Warming. (Just because I say that people don't have to worry about Global Warming doesn't mean they don't have to worry!) The evidence that is all around us nowadays even helps us to realize that there was nothing really magical or mysterious about the story of Noah. The Bible tells us that God told Noah that something was up, something was coming, and that he should build an ark and that would enable him and his family and a few critters to survive. But obviously, in this day and time, we really don't need God to tell us that "Something Wicked This Way Comes." Noah probably didn't either.

Then, of course, it was pointed out to me that it was the Cassiopaeans that told me about the cycle of cometary disasters. I thought about that for a moment and said, "Well, partly." I did write about all of this in my book *The Noah Syndrome* back in 1985, long before the Cassiopaeans proper ever introduced themselves. (That book was never published, but much of it is incorporated into *Secret History*.) Of course, back then, I had started from a purely metaphysical question: "Is there going to be an end of the world as described in the Bible, and if so, what does it really mean?" It was that question that lead to a deep study of the Bible, which then led to a realization that the destruction described in the Book of Revelation was almost identical to what was described in the story of the Exodus, so whatever happened then was being predicted to happen again. It wasn't until I read Velikovsky's *Worlds in Collision* that I realized that this was very likely talking about a bombardment of the Earth by rocks and bolides from space. Velikovsky, of course, attributed it to an errant planet, Venus, that came careening into the solar system, just as Firestone *et al* attribute it to a supernova 41,000 years ago. The Cyclic Comet Cluster related to a Companion Sun explanation is a better fit to all the data, though a supernova could also be involved, as well as a "newcomer" to the solar system. Reading Velikovsky changed the flavor of my research from metaphysical to scientific, and the result was *The Noah Syndrome*.

In any event, what is perfectly clear is that the story of Noah and the story of Atlantis are apocryphal: Many small groups of people around the world survived the event of 12,000 years ago here and there (not to mention similar, but less severe events since then until the present day!), and very likely many of them survived because they realized what was coming — they "read the signs." Afterward, in their stories and legends their descendants ascribed their survival to the intervention of their particular deity to give that deity more authority. Bottom line is: Anybody can be a Noah today if they are informed and pay close attention to what is going on.

Getting back to the article that started the discussion ("Bad news — we are way past our 'extinct by' date"), we are told:

> Some say the world will end in fire, some say in ice, wrote Robert Frost. But whatever is to be our fate, it is now overdue.
>
> After analysing the eradication of millions of ancient species, scientists have found that a mass extinction is due any moment now.
>
> Their research has shown that every 62-million years — plus or minus 3-million years — creatures are wiped from the planet's surface in massive numbers. Even worse, scientists have no idea about its source.
>
> "There is no doubting the existence of this cycle of mass extinctions every 62-million years. It is very, very clear from analysis of fossil records," said Professor James Kirchner, of the University of California, Berkeley. "Unfortunately, we are all completely baffled about the cause."

This part of the article is actually quite disingenuous. It is well known that there have been other major extinctions and the cycle is not *only* every 62-million years! There is also a very strong signal for a 26-million-year extinction cycle. The different estimates of the number of major mass extinctions in the last 540-million years are due mainly to what the individual researcher chooses as the threshold for naming an extinction event as "major," as well as what set of data he selects as the determinant measure of past diversity. As it happens, the 62-million event data stems mainly from *marine fossil* evidence.

> The report, published in the current issue of *Nature*, was carried out by Professor Richard Muller and Robert Rohde, also from the Berkeley campus. They studied the disappearances of thousands of different *marine* species (whose fossils are better preserved than terrestrial species) over the past 500m years.
>
> Their results were completely unexpected. It was known that mass extinctions have occurred in the past. During the Permian extinction, 250m years ago, more than 70 per cent of all species were wiped out, for example. But most research suggested that these were linked to asteroid collisions and other random events.
>
> But Muller and Rohde found that, far from being unpredictable, *mass extinctions occur every 62m years*, a pattern that is "striking and compelling," according to Kirchner.
>
> But what is responsible? Here, researchers ran into problems. They considered the passage of the solar system through gas clouds that permeate the galaxy. These clouds could trigger climatic mayhem. However, there is no known mechanism to explain why the passage might occur only every 62m years.

Alternatively, the Sun may possess an undiscovered companion star. It could approach the Sun every 62m years, dislodging comets from the outer solar system and propelling them towards Earth. Such a companion star has never been observed, however, and in any case *such a lengthy orbit would be unstable*, Muller says.

Or perhaps some internal geophysical cycle triggers massive volcanic activity every 62m years, Muller and Rohde wondered. Plumes from these would surround the planet and lead to a devastating drop in temperature that would freeze most creatures to death.

Unfortunately, scientists know of no such geological cycle.

"We have tried everything we can think of to find an explanation for these weird cycles of biodiversity and extinction," Muller said. "So far we have failed. And, yes, we are due one soon, but I would not panic yet." (This author's emphases)

Well, they have a problem, don't they? They think it's coming and, based on the ancient legends, it happens very fast and almost without warning.

The classical "Big Five" mass extinctions identified by Raup and Sepkoski in 1982 (interestingly, right about the same time I started asking questions about the "End of the World" as prophesied in Revelations, culminating in *The Noah Syndrome* in 1985!) are widely agreed upon as some of the most significant. They are:

The late Ordovician period (about 438-million years ago) — 100 families extinct — more than half of the bryozoan and brachiopod species extinct.

78-million years later: The late Devonian (about 360-million years ago) — 30% of animal families extinct.

106-million years later: At the end of the Permian period (about 245-million years ago) — Trilobites go extinct, 50% of all animal families, 95% of all marine species, and many trees die out.

37-million years later: The late Triassic (208-million years ago) — 35% of all animal families die out. Most early dinosaur families went extinct, and most synapsids died out (except for the mammals).

143-million years later: At the Cretaceous-Tertiary (K-T) boundary (about 65-million years ago) — about half of all life forms died out, including the dinosaurs, pterosaurs, plesiosaurs, mosasaurs, ammonites, many families of fishes, clams, snails, sponges, sea urchins and many others.

As you can see from the above, using the number "62-million years" and building a theory on it is really a bit misleading.

Raup and Sepkoski are mentioned as identifying the "Big Five," but the fact is that Sepkoski, a University of Chicago paleontologist, actually suggested that the extinction of the dinosaurs 65-million years ago was part of a *26-million-year* cycle. However, I would like to mention that if you multiply 26 there are interesting results, such as "3 x 26 is 78," which just happens to be the time between the Ordovician and Devonian extinctions; "4 x 26 is 104," which is very close to the 106-million years between the Devonian and Permian extinctions; and "5 x 26 is 130," which (when dealing with these kinds of numbers) is close enough to the gap between the Triassic and K-T extinction to be in the ballpark. So, maybe there is something to this 26-million-year thing after all, only each "return" has varying effects based on many other solar system variables. A Companion Star with a 26-million-year orbit might be more stable, since Muller has suggested that a 62-million-year orbit is too great to be stable.

As it happens, if we postulate the 26-million-year orbit of a Companion Star, using the "Big Five" extinctions as our jumping off point, we would find that there ought to have been a return about 39-million years ago, and then another 13-million years ago, which would put us half-way into the Companion Star orbit cycle now. Question: Is that half-way, as in aphelion or perihelion? Wherever the theorized Twin Sun might be at the moment, what we know is that there are other extinction events of great magnitude that seem to have little to do with just a 62-million-year cycle and a great deal to do with some other cycle. Another thing that is a fact is that extinction events occur far more frequently than the general public is aware of and, yes, we are way overdue for one.

Leaving the possible triggers for mass extinctions for the moment, let's look at some ways extinctions might occur. Here is a handy little run-down of the problem:

Ways to Destroy Life
Of all the ways scientists have proposed to cause a mass extinction, here are a few of the front runners. Conceptually there are four main ways:
1. Freeze it (Snowball Earth)
2. Boil it (Greenhouse Earth)
3. Drop a meteor on it (Meteor Impact)
4. Cover it with Ash and Lava (Giant Volcanic Eruption)
It is important to realise that numbers 3 and 4 are essentially ways which have been put forward to explain 1 and 2, though Snowball Earth and Greenhouse Earth are still theories in their own right, as they are to some extent self-replicating, i.e. we rather get stuck in vicious circles as the more we warm the Earth, the more greenhouse gasses we can potentially release (at least in theory).
Now that's all very well and good, but how do these physical effects (1 and 2 alone) cause mass extinctions?

The simplest answer to this is climate. Every species alive on earth today, and so presumably in the past is adapted to a certain range of conditions. In the same way that if we were suddenly whisked off to the North Pole and expected to live there for a year with only a woolly jumper and a Mars Bar, we would surely die. It is the same in an extinction event; climatic zones essentially shift around the globe (so us being placed on the North Pole is not as far fetched as you first thought), or ecosystems are starved of light or nutrients. This results in plants and animals being out of equilibrium with their surroundings, which not only causes the death of individuals but whole species. With the death of a species there is a gap in the food chain, and so even animals which have adapted to the new climate find themselves with no food soon die out and so on.

EXTINCTION PROBLEMS IN A STRANGE ECOSYSTEM

If we consider the example of us at the North Pole, but this time with a whole box of Mars Bars, assuming we did not freeze to death in the first day, we would slowly become more accustomed to the cold, maybe we would find shelter from the cold, but either way as soon as our prey (the Mars Bars) ran out, unless we found an alternative source of food, we would starve. To create a more dynamic ecosystem let us assume we provide a food source for another organism such as a polar bear; now, if we died, what would the polar bear eat?

Unfortunately my example falls down here as no one is going to believe that without us eating Mars Bars they're going to reproduce uncontrollably. However, in our North Pole ecosystem of Mars Bars, us and polar bears, let us assume we find extra clothes and shelter, so the cold no longer controls our numbers, and we manage to find the recipe and ingredients for Mars Bars (and by some freak coincidence they provide us with all the essential nutrients for life); in this case it would be the polar bears controlling our numbers since they are our direct predators. So if polar bears were wiped out, then our numbers would no longer be controlled, so our population would over many generations grow, until eventually we could no longer supply ourselves with Mars Bars, or we may have even eaten all the Mars Bars in the world. This would cause a huge population crash, or maybe even extinction of the human race (or at least those dependent on Mars Bars). This rather abstract example helps illustrate the point that all the trophic levels of a food chain (or web) need to be in place, otherwise there will be instability in all other populations.[2]

The above description of how Global Warming or Global Cooling can cause mass extinctions is clever, but it does not take into account the creativity of human beings. Certainly, there could be massive reductions in the human population as a consequence of Global

---

[2] "Ways to Destroy Life," http://palaeo.gly.bris.ac.uk/Palaeofiles/Triassic/exttheory.htm

Warming or Cooling, but it would be unlikely to produce a mass extinction such as those for which we have evidence in the past.

In *Secret History*, I provide some of the evidence for a sudden Global Cooling and mass extinction 12,000 years ago. Some estimate that tens of millions of animals may be buried along the rivers of northern Siberia as a result of this cataclysm. The evidence suggests these animals were flash-frozen in an enormous tsunami. Mass extinctions took place in the Americas at the same time, killing off woolly rhinoceros, giant armadillos, giant beavers, giant jaguars, ground sloths, antelopes and scores of other entire species. This is the "event" that Firestone, West and Warwick-Smith discuss in their book, *The Cycle of Cosmic Catastrophes*, mentioned above.

But if the above accounts are the result of such a catastrophe, what might the catastrophe itself be like? The following is condensed and adapted from Chapter 11 of Firestone, West and Warwick-Smith's book:

It begins with meteors falling like raindrops, a few here and there. Perhaps a few hit the Sun, provoking large solar flares. The solar flares provoke colourful auroras even in the daytime sky. Then the day of the comets arrive. From horizon to horizon, growing larger every second, they streaked into the atmosphere, lighting up brighter than the Sun.

Heated to immense temperatures by its passage through the atmosphere, the lethal swarm exploded into thousands of mountain-sized chunks and clouds of streaming icy dust. The smaller pieces blew up high in the atmosphere, creating multiple detonations that turned the sky orange and red.

Then the largest comet smashed through the sheet of ice covering part of the northern hemisphere in what is now Hudson Bay. Other comets struck in Lake Michigan, Canada, Siberia and Europe. Then the ground shock-waves hit, shaking the Earth violently for ten minutes in great rolling waves and shudders. Fissures opened, trees shook and fell, and rivers and streams disappeared into the cracked earth.

Within seconds of the impact, the blast of superheated air expanded outward at more than 1,000 miles an hour, racing across the landscape, tearing trees from the ground and tossing them into the air, ripping rocks from mountainsides, and flash-scorching plants, animals, the earth, as well as any humans in its way. The only living things to survive would have been those who had sought shelter underground or underwater.

Across the upper part of North America and Europe, the immense energy from the multiple impacts blew a series of ever-widening, giant, overlapping bubbles that pushed aside the atmosphere to create a near vacuum inside. As the bubble passed by, the air pressure dropped making it difficult to breathe. Behind the expanding edge of the bubble, the Earth was stripped of the protective shield of the atmosphere. The blast had ejected tiny, fast-moving grains in all directions through the thin air. Some lodged into trees, plants and animals, while others went up only to fall back again at incredible speeds as there was no atmosphere to break their fall. At the same instant, high speed cosmic rays bombarded the area with radiation. Animals and humans dropped dead on the spot from the bombardment. Inanimate objects appeared to come to life, and shiver and quake on the ground from the barrage.

When the outward push of the shock-wave ceased, the vacuum began to draw back the air. As the expanded atmosphere rushed back toward the impact site, the bubbles collapsed, sucking white-hot gases and dust inwards at tornado speeds and then channelling them up and away from the ground. Some of the dust escaped from the Earth's atmosphere while the rest flowed out as a red mushroom cloud that flattened out for thousands of miles across the upper atmosphere, blocking the Sun and engulfing the Earth in darkness.

The dust and debris that was too heavy began crashing back down to Earth. Still super hot from the blast, it gave off a powerful lava-like glow. The pieces landed on the continental ice sheet, instantly melting untold gallons of water that coursed off the ice sheet in all directions causing flooding.

The raging updraft through the hollow bubbles created an equally powerful downdraft of frigid, high-altitude air, travelling at hundreds of miles per hour. With temperatures exceeding 150 degrees F below zero, the downward stream of air hit the ground and radiated out from the many blast sites in all directions, flash-freezing within seconds everything it touched. The howling, frigid blast turned trees and plants into brittle ice statues and flash-froze mastodons and mammoths with food in their mouths that we have uncovered still frozen in Siberia.

The rapid temperature fluctuations meant the end of millions of plants and animals.... but the destruction was only beginning.

The impacts and shock waves triggered enormous earthquakes along existing fault lines from the Carolinas to California while shaking awake dormant volcanoes from Iceland across to the Pacific. Erupting with furious activity, they spewed hot lava across the landscape and noxious chemicals into the air, adding to the already heavy cloud cover.

The impacts, the blast waves, and the eruptions started thousands of ground fires wherever there was fuel to feed them, some of which continued to burn for days. Fast-moving, wind-driven wildfires formed spiralling tongues of raging flames that twisted for thousands of feet into the air and the inferno raced through forests faster than birds and animals could flee. The roar of the fire shook the ground, and the fierce heat blew apart trees like bombs, exploded rocks like shrapnel grenades, and set off steam explosions wherever the fast-moving fire-front jumped across frozen ponds and streams. When the fires had finally burned themselves out, there was little left besides smoldering stumps and tell-tale charcoal strewn across the continents.

The noxious chemicals in the atmosphere fell back to Earth as poisoned rain. In some places, the air was too toxic and oxygen-depleted to support life.

(Fairchild Aerial Surveys for the Ocean Forest Company; Aerial view of some of the Carolina Bays taken in 1930)

The impact in Hudson Bay sent up 200,000 cubic miles of the glacier, throwing off the icy debris that followed the pieces of the comet out across the continent. A rain of incandescent debris and chunks of steaming ice showered down across most of North America, Europe and Asia. Within minutes, the massive, low-flying clumps crashed into the Carolinas and the eastern seaboard, exploding into fireballs and gouging out the Carolina Bays, over 500,000 of them. Other lumps exploded across the plains from Nebraska and Kansas to Arizona.

Pieces of flying ice and debris, large and small, fell from the Atlantic to the Pacific, from the Gulf of Mexico to the Arctic, from Europe over to Asia, and even down to Africa. More than one-quarter of the planet was under siege.

But even that was not all.

The impact through the glacier at Hudson Bay sent high velocity melt-water surging under the ice sheet. The surges lifted and floated large sections of ice, causing monolithic ice blocks to slide southward along hundreds of miles of the ice front. *Moving nearly as quickly as a horse is able to run*, the blocks plowed over forests, shearing off the trees.

The oceans, too, were targets. Thousands of ice chunks and clouds of slushy water hit the Atlantic, exploding with colossal detonations. The multiple concussions triggered immense underwater landslides off the Carolinas and Virginia, releasing thousands of cubic miles of mud. In turn, the mud unleashed a 1,000-foot-high tidal wave that raced away towards Europe and Africa at 500 miles an hour.

Nine hours later the wave hit [Europe], 1,000 feet tall at 400 miles per hour, probably taking with it some of the survivors of the first explosions. The wave broke over hundreds of miles inland, devastating everything in its path. Anything living on the coast was killed instantly.

Its momentum spent, the churning water paused briefly and then began its rush backwards to the coast, pulling with it the battered remains of plants and animals under its tow. The surge provoked, in turn, offshore landslides in Europe and Africa, sending a second round of mega-waves back towards North and South America. Miles of coast land was hit by the 100-foot waves that triggered yet another wave of tsunamis that hit Europe and Africa once again. But little was left to damage.

Within minutes of the impacts, the sub-zero air and rising water vapour combined to produce heavy snow and sleet that reached as far south as Mexico, the Caribbean and Northern Africa. In the south, the snow turned to rain and the northern hemisphere was under a steady downpour for months, a downpour of noxious water, contaminated and deadly. Anyone lucky enough to survive was now a potential victim of acid, toxic metals, cyanide, formaldehyde and arsenic, a combination that would kill many and render the rest gravely ill.

The melted water of the glaciers had another effect: Flooding into the North Atlantic, it turned off the ocean conveyor that brought warm water to the northern climes. Once shut off, coupled with the clouds of dust blocking the Sun, the temperature fell drastically. Within days or weeks after the impacts, continental temperatures fell well below freezing, and a brutal ice-age chill once again spread across the land, remaining in place for another thousand years.

And all of this in an instant, in less time than it takes to cook a meal or write an email. [This author's emphases]

You will, of course, notice that "12,000 years ago" is just a rough estimate because some of the dates of their data come back as old as 14,000 years ago and as recent as 10,000 years ago. When considering a 3,600-year Comet Cluster Cycle, this range could cover more than one event. But what is important is that the main event did, apparently, happen *in a single day* and based on the scientific data collected by Firestone *et al*, it was one of the most horrifying events ever to happen on planet earth since modern Homo Sapiens appeared.

Why do I keep referring to a 3,600-year cycle? Well, in addition to having been explicated within the context of the Cassiopaean Experiment, it seems that this 3,600-year period was important enough to certain ancient peoples that it was the basis of their mathematics.

Around 3200 BCE, the Sumerians devised their numerical notation system, giving special graphical symbols to the units 1, 10, 60, 600 and 3,600. That is to say, we find that the Sumerians did not count in tens, hundreds, and thousands, but rather adopted a base 60, grouping things into sixties, and multiplying by powers of sixty. Our own civilization utilizes vestiges of a base 60 in the ways we count time in hours, minutes and seconds, and in the degrees of a circle.

Sixty is a large number to use as a base for a numbering system. It is taxing to the memory because it necessitates knowing sixty different signs (words) that stand for the numbers from 1 to 60. The Sumerians handled this by using 10 as an intermediary between the different sexagesimal orders of magnitude: 1, 60, 60^2, 60^3, etc. The word for "60", *geš*, is the same as the word for "unity." The number "60" represented a certain level, above which multiples of 60 up to 600 were expressed by using 60 as a new unit. When they reached 600, the next level was treated as still another unit, with multiples up to 3,000. The number "3,600," or sixty sixties, was given a new name, *šàr*, and this in turn became yet another new unit.

So, the mystery is: Why did the Sumerians enshrine the number 60 — and its multiple "60 x 60" — in their numbering system?

Zecariah Sitchin believed that it was because there was a 10th Planet in the solar system that had an orbit 3,600 years long that the Sumerians based their numbering system on the cycle of this event. But the evidence for the 10th Planet — as a planet — and his related ideas, is rather skimpy, while the evidence for bombardment of the Earth by masses of cometary debris is growing every day. Examining the hard data, it doesn't take a genius to figure out that if there is something that returns every 3,600 years, it is more likely to be a cluster of cosmic bodies than a 10th Planet.

That is bad news.

In his article, "Comets and Disaster in the Bronze Age," published in *The Journal of the Council for British Archaeology* (December 1997, No 30, pp. 6-7),[3] Benny Peiser concludes, "The extent to which past cometary impacts were responsible for civilisation collapse, cultural change, even the development of religion, must remain a hypothesis. But in view of the astronomical, geological and archaeological evidence, this 'giant comet' hypothesis should no longer be dismissed by archaeologists out of hand."

> Occasionally … cosmic debris measuring between one and several hundred metres in diameter strike the Earth and these can have catastrophic effects on our ecological system, through multimegaton explosions of fireballs which destroy natural and cultural features on the

---

[3] Available online at: http://www.knowledge.co.uk/sis/ba9712bp.htm

surface of the Earth by means of tidal-wave floods (if the debris lands in the sea), fire blasts and seismic damage.

Depending on their physical properties, asteroids or comets that punctuate the atmosphere can strike the Earth's surface and leave and impact crater, such as the well-known Barringer Crater in Arizona caused by an asteroid made of iron some 50,000 years ago. At least ten impact craters around the world dating from after the last Ice Age, and no fewer than seven of these date from around the 3rd millennium BC — the date of the widespread Early Bronze Age collapses — although none occurred in the Near East.

Alternatively, comets and asteroids can explode in the air. A recent example — known as the Tunguska Event — occurred in 1908 over Siberia, when a bolide made of stone exploded about 5 km above ground and completely devastated an area of some 2,000 km through fireball blasts. The cosmic body, although thought to have measured only 60 m across, had an impact energy of about 20 to 40 megatons, up to three times as great as the Arizona example (about 15 megatons), and was equivalent to the explosion of about 2,000 Hiroshima-size nuclear bombs — even though there was no actual physical impact on the Earth. [...]

Until recently, the astronomical mainstream was highly critical of Clube and Napier's giant-comet hypothesis. However, the crash of comet Shoemaker-Levy 9 on Jupiter in 1994 has led to a change in attitudes. The comet, watched by the world's observatories, was seen split into twenty pieces and slam into different parts of the planet over a period of several days. A similar impact on Earth, it hardly needs saying, would have been devastating.

According to current knowledge, Tunguska-like impacts occur every 100 years or so. It is, therefore, not far fetched to hypothesize that a super-Tunguska may occur every 2,000, 3,000 or 5,000 years, and would be capable of triggering ecological crises on a continental or even global scale. In the past, skeptics have demanded the evidence of a crater before they would accept an argument of cosmic impact, but it is now become understood that no crater is necessary for disastrous consequences to ensue. The difficulty this leaves scholarship, however, is that in a Tunguska event no direct evidence is left behind. It may be impossible to prove that one ever took place in the distant past.

Among the many side-effects of cometary bombardments are earthquakes, tsunamis and volcanic eruptions. As it happens, there was a significant volcanic event at the time of the collapse of Bronze Age civilizations that gives us a firmly fixed date: The island of Thera/Santorini. Recent developments published in the April 2006 issue of *Science* fix the date of the eruption between 1627 and 1600 BCE with 95% certainty. This, of course, is rejected by many archaeologists because they have spent their entire careers trying to date things according to the Bible, and it really upsets the apple cart to realize that they've been chasing an illusion.

Yoshiyuki Fujii and Okitsugu Watanabe (1988) demonstrate that "large-scale environmental changes possibly occurred in the Southern Hemisphere in the middle of the Holocene" (within the last 10,000 years). Their depth profiles of microparticle concentration, electrical conductivity and Oxygen-18 at circa 1600 BCE indicates a spike in readings for all of these elements. The evidence shows that this disturbance covered the designated period, but with a huge "spike" at around 1600 BCE.

Similar evidence exists at 5200 BCE. This period shows less severe but similar climatological stress. The Oxygen-18 profile is close to normal, but there is a visible volcanic dirt band. The dating of this segment is less close because it is clear that nobody is really looking for this cycle, but it appears to correspond to the ash band from the Byrd Station core (Schellhorn, 1991). In an article in *Nature*, November 1980, C.U. Hammer, H. B. Clausen and Dansgaard date a disturbance from the Camp Century ice-core to 5470 BCE +/- 120 years. This compares to the proposed Hekla eruption which was radiocarbon-dated to 5450 BCE +/- 190 years. There is an appreciably high acidity signal at these sections of the core which indicates a high level of volcanic activity — again, right at the 3,600-year cycle mark.

It is conjectured that the cycle goes unnoticed because of long-term after-effects, such as cooling climate, as well as the fact that each cycle has greater or lesser effects on the Earth depending on the particular dynamical interactions within the solar system at any disruption.

What is clear is that something happens at 3,600-year intervals, as shown by the ice cores, and is capable of setting off prolonged periods of Earth changes that are above the levels of ordinary uniformitarian geologic and climatological changes.

Looking further: Michel R. Legrand and Robert J. Delmas, of Laboratoire de Glaciologie et Geophysique de l'Environment, published an article (1988) in which they graphed the Oxygen-18 variations and the ionic components Na, NH_4 and Ca_2, and H and Cl, and NO_3 and SO_4. The time scale for each ionic-component level as well as the Oxygen-18 levels stretches back 30,000 years. The graph shows correlations to spikes at 5200 BCE, 8800 BCE, 12400 BCE, c. 16000 BCE and c. 19600 BCE. All of these were times of great geologic stress.

When looking at the data and taking into account the acknowledged dating inaccuracies for the more recent dates (some of the ranges of dates can go 100 years in either direction of the spike, even though the spiking is regular and rhythmic), and 300 to 600 years variance for the older dates (especially when one considers that these are broad analyses and nobody was really looking for anything specific, they just said "Wow! Look at that wavy line!"), we find that the southern ice-cores do not always register the same as the northern ones. The 1628 BCE event that really slammed the tree rings in the northern hemisphere shows almost no registration in the Antarctic cores in terms of volcanic activity. But the northern cores show the activity beginning 1644 BC.

The evidence for the 5200 BCE event is strong in the Dome-C core. The 8800 BCE event is well marked — in fact, seems to be the strongest of them all. Keep in mind that this was 10,800 years ago — exactly within the range of dates reported by Herodotus and Plato. The Oxygen-18 isotope variation is noticeable, the rise in sea-salt, elevated levels of Cl and Cl/Na. There is an extreme spike in SO_4 and H readings, suggesting widespread volcanic activity — great Earth changes were happening at that time, and they registered in the climate and the oceans, and were preserved in ice.

The 12,400 BCE event is also extremely pronounced in the cores. The graphs show a quick, vast change including the end of the Wisconsin Ice Age. There is a great Oxygen-18 isotope variation as well as peaks of Na and very pronounced spikes in Ca, SO_4 and H.

There is absolutely no question that the Thera/Santorini event occurred. The acid signal in the ice core is very strong. Which means that there is very little question about when it occurred. Something very unusual and specific happened then, starting in 1644 BCE, and culminating in a major cataclysm, and it seems that it walked all over the Aegean and Anatolian area, leaving tracks that are impossible to miss. Impossible for anyone, that is, except Egyptologists and their kin.

All over the Mediterranean there were kingdoms and cultures that communicated and traded with one another. Reading the many books on each region, produced by the various experts on the different cultures, again and again one encounters the fact that a period of severe disruption was noted in the historical and archaeological record. Somehow, such an event in one region is not necessarily connected to a similar event in another region. The idea that all of the disruptions in a given general time period may be simultaneous cannot be considered because it would disrupt the carefully constructed chronology that is based on endless acts of tetraphyloctomy (i.e., "splitting a hair four ways," coined by Umberto Eco in *Foucault's Pendulum*).

In his book *Stratigraphic Comparée et Chronologie de l'Asie Occidentale*, Claude Shaeffer's life-long archaeological investigations led him to propose that a great natural catastrophe brought about the end of the Middle Kingdom in Egypt, and also devastated by fire and earthquake almost every other populated region of Crete, Cyprus, the Caucasus, Syria, Palestine, Persia and Asia Minor in general. That's one heck of a "local event." It is only logical to conclude that the Thera/Santorini event and the end of the Middle Bronze Age are one and the same event.

It happened a little over 3,600 years ago.

In other words, we are overdue.

Now, let's come back to what I mentioned above about being half-way through a 26-million-year Companion Star orbit. We really have no idea where the theorized critter is or what it is up to, but we do have some clues. But the first question we want to ask is what is the relation between this Companion Star — Nemesis — and extinction? How can a star, way out beyond the solar system, have an effect on the third rock from the Sun?

Far beyond the orbit of Pluto lies the Oort Cloud.

The Oort Cloud, alternatively termed the Öpik-Oort Cloud, is a postulated spherical cloud of comets situated about 50,000 to 100,000 AU (astronomical units) from the Sun. This is approximately 2,000 times the distance from the Sun to Pluto or roughly one light year, almost a quarter of the distance from the Sun to Proxima Centauri, the star nearest the Sun.

The solar system is engulfed by this cloud comprised of billions of comets. Imagine what would happen if a star passed through that cloud, knocking the comets in the same way a bowling ball sends bowling pins scattering in all directions. Imagine then a certain number of those comets heading towards the centre of the cloud, our Sun and its solar system. The Sun, being the largest object in the neighborhood, would be the attraction point. The comets in the incoming cloud would be pulled into an orbit around the Sun.

Although no direct observations have been made of such a cloud, it is believed to be the source of most or all comets entering the inner solar system (some short-period comets may come from the Kuiper Belt, based on observations of the orbits of comets).

So far, only one potential Oort Cloud object has been discovered: 90377 Sedna. With an orbit that ranges from roughly 76 to 928 AU, it is much closer than originally expected and may belong to an "inner" Oort Cloud. If Sedna indeed belongs to the Oort Cloud, this may mean that the Oort Cloud is both denser and closer to the Sun than previously thought.

So, we have a mechanism which can hypothetically trigger the launching of a swarm of comets into the solar system. The orbit of the proposed binary twin of our Sun conforms to the cycles of major extinctions on Earth. But it is still a hypothesis. More importantly, you might ask, even if we assign a high probability to the truth of the hypothesis, if these cycles happen every 26-million years, what evidence do we have that we are living during one of the, shall we say, "unlucky" periods?

One of the corollaries of the Nemesis theory is that the Dark Companion might well become visible as a Second Sun in the sky when it is closest to the Sun. Is there any evidence that might suggest that people have ever seen a "Second Sun"?

In her book *Comets and Popular Culture and the Birth of Modern Cosmology*, Sara J. Schechner writes:

> The sunny disposition of the weather during the coronation (of Charles II) was seen as the fulfillment of a prophecy. In 1630, at the time of Charles' birth, a *noonday star* or *rival sun* allegedly had appeared in the sky... Aurelian Cook in *Titus Britannicus* explained its import: "As soon as Born, Heaven took notice of him, and eyed him with a *star*, appearing in defiance of the Sun at Noonday..."
>
> For Cook, the extra sun announced that Charles ruled by divine right. Moreover, the timing of Charles' entry into London on his birthday was politically calculated to fulfill what had been portended at his birth. Abraham Cowley, poet, diplomat and spy for the court wrote: "No Star amongst ye all did, I believe, Such Vigorous assistance give, As that which thirty years ago, At Charls his Birth, did in despight of the proud Suns' Meridian Light, His future Glories, this Year foreshow."
>
> Edward Matthew devoted an entire book to the fulfillment of the prophecy, declaring Charles "ordained to be the most Mighty Monarch in the Universe..."
>
> Charles' return was seen as a rebirth of England and duly recorded by a special act in the statute book, which proclaimed that 29 May was the most memorable Birth day not only of his Majesty both as a man and Prince, but likewise as an actual King...

Well, that certainly sounds like it fits the bill: 377 years ago a Second Sun appeared and no one, so far as I know, has ever linked this to either a specific comet or a supernova. Interestingly, it was followed thirty years later by the sighting of several comets.[4]

But the 17th century was interesting for another anomaly involving our Sun: the Maunder Minimum. Between the years 1645 and 1715, our Sun stayed in a period of solar minimum.

> During one 30-year period within the Maunder Minimum, for example, astronomers observed only about 50 sun spots, as opposed to a more

---

[4] See appendix A for details.

typical 40,000-50,000 spots. The Maunder Minimum coincided with the middle — and coldest part — of the so-called Little Ice Age, during which Europe and North America, and perhaps much of the rest of the world, were subjected to bitterly cold winters. Recently published research suggests that the Sun's rotation slowed in the deep Maunder Minimum (1666-1700).[1] At our current level of understanding of solar physics, a larger and slower Sun necessarily implies a cooler Sun that provides less heat to Earth.[5]

Perhaps a close approach, astronomically speaking, of the Dark Companion could cause this dampening? The lower solar activity during the Maunder Minimum also affected the amount of cosmic radiation reaching the Earth. The resulting change in the production of Carbon-14 during that period caused an inaccuracy in radiocarbon-dating until this effect was discovered.

However, standing against the idea that the close approach of a companion star could be the cause of such events, we note that, in total, Carbon-14 analysis as well as tree rings and ice-core studies indicate there seem to have been eighteen periods of sunspot minima in the last 8,000 years, and studies indicate that the Sun currently spends up to a quarter of its time in these minima. It is hardly likely that there is a companion star with that short an orbit so we can only speculate at the moment whether these cycles have a relationship to our Dark Companion and/or its cometary children. The work of Victor Clube more directly addresses cometary cycles in relation to the break-up of a giant comet and the fact that the earth regularly passes through the debris stream of this comet and this could be the cause of the "second sun" seen at the time of the birth of Charles II, as well as a contributor to solar minima.[6]

Returning to our hypothesis, if there is a Dark Companion, Nemesis, then there might be other ways to detect it early, before it is seen. If it is there, even if invisible at present, there should be comets heading to the inner solar system. Depending on their locations and orbits in the Oort Cloud, there would be variable groups — variable both in size and arrival time. Is there any evidence for this?

The third edition of the university textbook *Exploration of the Universe*, by George O. Abell, published in 1975, informs us that Jupiter has nine moons as of 1974. It says:

> The outer seven, however, have rather eccentric orbits, some of which have a large inclination to Jupiter's equator. The four most distant satellites revolve from east to west, contrary to the motions of most of the

---

[5] "Maunder Minimum", *wikipedia.org*.

[6] See Victor Clube's and Bill Napier's *The Cosmic Serpent* and *The Cosmic Winter*.

other objects in the solar system. They may be former minor planets captured by Jupiter. (324)

Please note that Abell is suggesting that some of Jupiter's moons have been captured by Jupiter's gravity.

Now let's time-travel back to the future, and see what the latest information tells us about Jupiter's moons: "Jupiter is now given 63 satellites. Forty-seven of those satellites have been discovered since 1999".[7] What if they weren't there before?

Planet	1975	2005
Jupiter	9	63
Saturn	10	62
Uranus	5	28
Neptune	2	13

Table 1. Number of moons

What about Saturn? Our 1975 text tells us that Saturn has ten satellites. In 2007? Well, there are so many that one source declines to give a precise number![8]

However, counting the named satellites on the "Timeline of discovery of solar-system planets and their natural satellites"[9] gives us a count of 62, with 41 being discovered since 2000 and another ten in the 1980s and 1990s.

Moving outward, we come to Uranus, given five satellites in 1975, it now has 28, with ten being discovered in the 1980s, six in the 1990s, and seven since 2000. Neptune had two satellites in 1975; now it has 13.

The explanation given most often to explain this surge in the numbers of satellites for these planets is that telescopes have gotten better. That is, we can see further, with greater detail, and can therefore find things that we couldn't see before. It is an explanation that makes sense. One small problem with this theory is that *the "new" moons of Neptune and Uranus showed up (were seen) before the "new" moons*

---

[7] "Jupiter's Natural Satellites," *wikipedia.org*

[8] "Saturn," *wikipedia.org*

[9] See: *wikipedia.org*

of Jupiter and Saturn. One would think that powerful telescopes capable of finding moons as far away as the seventh and eighth planets wouldn't have found it hard to see moons of the fifth and sixth first.

Another possible explanation, and one which fits with new moons appearing around Neptune and Uranus prior to appearing around Jupiter and Saturn, is that these new moons, or some of them, are objects that have been trapped into orbits around these planets only recently, that they were captured by the gravity of these planets and removed from the incoming comet cloud. Passing the orbits of the outer planets first, they would arrive at the inner planets afterward.

We also note that the much derided Immanuel Velikovsky, in his book *Worlds in Collision*, gives a time frame of *nine years* as the time it would take for a comet to cover the distance between Jupiter and Earth. The new Jovian moons were discovered beginning in the late1990s.

Do the math.

Which brings us to a series of local stories that give "impact" a less than metaphoric meaning. On February 14, 2007, people throughout Ohio and New Jersey heard a large bang, sparking rumors of a meteor strike or sonic boom. Granville resident Jeff Gill saw a meteor headed east to west that lasted three seconds. It had a long red, green and gold tail, and produced a sonic boom as it faded. "I saw it first. It was the most eerie, cool, scary, wonderful thing. You just see this dragon tail going across the sky. All of a sudden, everything goes boom".[10]

Notice the reference to the "dragon tail." Could reports of meteorites be the basis for some of the ancient myths about dragons fighting in the sky? Check out Mike Baillie's book *Exodus to Arthur* for more on that subject.

Some months ago, while much of the United States was watching the XLI Super Bowl, some people in the Midwest were being treated to a different type of spectacle. Reports of a flaming-orange meteor with a blue-green tail reached Wisconsin, Illinois and Iowa. There were reports of numerous objects exploding on impact with the ground.

On February 6, 2007, a meteor was spotted over Hangleton, Hove (UK), lighting up the sky.[11] A week earlier, police in Turkey were flooded with calls from Didim to Bodrum, telling of a meteor and a large boom. It narrowly missed a man, landing ten meters from him, as reported in *Voices* on February 2, 2007. Fortunately, no one was hurt with these meteors, but people in India weren't so lucky. Three were

[10] Jim Sabin, "Strange noise might have been meteor," *Newark Advocate*, February 16, 2007

[11] Rachel Pegg, "Meteor lights up the sky," *The Argus*, February 7, 2007

killed by a meteor in Ragasthan just a couple days after the incident reported above by Rachel Pegg.[12]

Then there was the meteorite that crashed into a house in New Jersey recently and embedded itself in the wall,[13] and the cottage destroyed by a meteorite in Germany in October 2006, injuring the 77-year-old owner.[14] Also, the falling ice that hit a car in Florida.[15]

So what are we to make of this sudden appearance of so many "once in a lifetime" meteors across the globe? Anything? Coincidence? Or something else?

Let's put together an overview of the data we have been collecting and outline the working hypothesis that we have to explain it. It isn't very hopeful, we'll tell you that right at the start. It may well make Tunguska look like a firecracker in comparison. The blast at Tunguska has been described thusly:

> The explosion was probably caused by the air burst of a meteorite or comet 6 to 10 kilometers (4-6 miles) above the Earth's surface. The energy of the blast was later estimated to be between 10 and 15 megatons of TNT, which would be equivalent to Castle Bravo, the most powerful nuclear bomb ever detonated by the US. It felled an estimated 60-million trees over 2,150 square kilometers (830 square miles). ("Tunguska Event," *wikiedia.org*)

The Binary System

We look out at our sky and we see only one Sun. We naturally conclude that our star system includes only the Sun. However, binary star systems are very frequent.

There is an hypothesis that argues that our Sun is part of such a binary system. The Sun's hypothetical companion has been named, as mentioned above, "Nemesis." The projected orbit of Nemesis is 26-million years, give or take the time necessary for the rising and falling of several civilizations.

Studies of the fossil record by Dave Raup and Jack Sepkoski have shown that there is a cyclic repetition to periods of extinction. The Nemesis theory was drawn up to explain the extinction cycle.[16]

---

[12] *playfuls.com*, February 8, 2007
[13] Chris Newmarker, "Possible Meteorite Crashes Through New Jersey Roof," *AP*, January 4, 2007
[14] "German cottage destroyed by meteor," *Reuters*, October 20, 2006
[15] "Mysterious, Large Ice Chunk Falls on Tampa Man's Car," *First Coast News*, January 28, 2007
[16] http://muller.lbl.gov/pages/lbl-nem.htm

We can offer no proof for the working hypothesis outlined above. We are working on a limited data set. The most that we can say is that an argument can be made suggesting the following scenario:

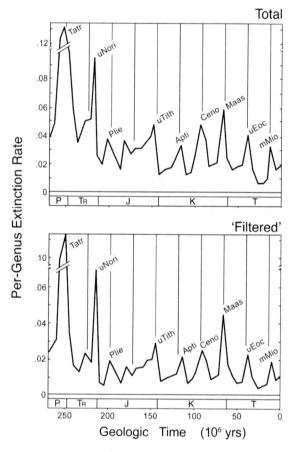

(© Raup & Sepkoski. Chart of the 26-million-year extinction cycle)

The Sun's Dark Companion, on its 26-million-year orbit, came close to the solar system 377 years ago, showing itself and pulling with it comets, a few of which appeared thirty years after the 1630 showing of the Dark Star itself. The passage of the companion through the Oort Cloud, dragging hundreds of thousands of other bodies in its wake, swung around in its orbit, and flung a swarm of them our way, traveling now for nearly four hundred years. Members of that swarm have been picked off by the gravitational fields of the outer planets, increasing the numbers of moons in recent years. If the swarm has passed by Jupiter, then it may be arriving here very soon.

Are the articles pulled from papers around the world over the last few months the announcement of their arrival?

One final point. There have been reports that Earth is not the only planet being hit by "Global Warming." Might it be possible that this apparently widespread change of "climate" in the solar system is linked to an incoming comet cloud? We do not know and are sorely lacking in the means to acquire data to refine or reject that working hypothesis. Perhaps someone else out there does have the means. Whatever the explanation for a generalized warming of several planets, it is clear that we know very little about the fundamental mechanisms behind it. We are a speck in the universe, a drop in an ocean more vast, more complex, and more mysterious than we can imagine.

To close, I want to quote from the final pages of *The Cycle of Cosmic Catastrophes: Flood, Fire, and Famine in the History of Civilization*, because the words there should have meaning for all of us:

If you want more evidence for what happened to the mammoths, you need only to look up at the clear night sky. In almost any month, you can see shooting stars from one of many meteor showers. Nearly every fiery streak you see is the tiny remnant of some giant comet that broke up into smaller pieces. Of course, most of those pieces are microscopic, but their parent comet was not — it was enormous. Astronomers know that, even today, hidden in those cosmic clouds of tiny remnants, there are some huge chunks of comet pieces. We pass through their clouds every year like clockwork, so eventually we will collide with some of bigger pieces.

In 1990, Victor Clube, an astrophysicist, and Bill Napier, an astronomer, published *The Cosmic Winter*, a book in which they describe performing orbital analyses of several of the meteor showers that hit Earth every year. Using sophisticated computer software, they carefully looked backward for thousands of years, tracing the orbits of comets, asteroids and meteor showers until they uncovered something astounding. Many meteor showers are related to one another, such as the Taurids, Perseids, Piscids, and Orionids. In addition, some very large cosmic objects are related: the comets Encke and Rudnicki, the asteroids Oljato, Hephaistos, and about 100 others. Every one of those 100-plus cosmic bodies is at least a half-mile in diameter and some are miles wide. And what do they have in common? According to those scientists, every one is the offspring of the same massive comet that first entered our system less than 20,000 years ago! Clube and Napier calculated that, to account for all the debris they found strewn throughout our solar system, the original comet had to have been enormous.

So was this our megafauna killer? All the known facts fit. The comet may have ridden in on the supernova wave [or was knocked into the solar system by the Companion Star], then gone into orbit around the Sun less than 20,000 years ago; or, if it was already here, the supernova debris wave may have knocked it into an Earth-crossing orbit. Either way, any time we look up into the night sky at a beautiful, dazzling display of shooting stars, there is an ominous side to that beauty. We are very likely seeing the leftover debris from a monster comet that finished off 40-million animals 12,000 to 13,000 years ago.

Clube and Napier also calculated that, because of subtle changes in the orbits of Earth and the remaining cosmic debris, *Earth crosses through the densest part of the giant comet clouds about every 2,000 to 4,000 years* [or 3,600 years?]. When we look at climate and ice-core records, we can see that pattern. For example the iridium, helium-3, nitrate, ammonium, and other key measurements seem to rise and fall in tandem, producing noticeable peaks around 18,000, 16,000, 13,000, 9,000, 5,000 and 2,000 years ago. In that pattern of *peaks every 2,000 to 4,000 years*, we may be seeing the "calling cards" of the returning mega-comet.

Fortunately, the oldest peaks were the heaviest bombardments, and things have been getting quieter since then, as the remains of the comet break up into even smaller pieces. The danger is not past, however. Some of the remaining miles-wide pieces are big enough to do serious damage to our cities, climate and global economy. Clube and Napier (1984) predicted that in the year 2000 and continuing for 400 years, Earth would enter another dangerous time in which the planet's changing orbit would bring us into a potential collision course with the densest parts of the clouds containing some very large debris. Twenty years after their prediction, we have just now moved into the danger zone. It is a widely accepted fact that some of those large objects are in Earth-crossing orbits at this very moment, and the only uncertainty is whether they will miss us, as is most likely, or whether they will crash into some part of our planet.

That may seem like bad news, but there is a glimmer of good news too. For the first time in human kind's known history, we have ways to detect those objects and prevent them from hitting us again. One such effort is Project Spaceguard, a multinational cooperative attempting to locate those Earth-threatening objects, and other similar programs include the Near-Earth Asteroid Tracking (NEAT) telescope and the Spacewatch Project at the University of Arizona. *Unfortunately, not one of them is funded nearly well enough to complete the job for many years*, but they are working at it steadily.

No one knows exactly how many dangerous comets and asteroids are out there, but astronomers are certain that hundreds to thousands of them remain undiscovered. The worst part is that many of those space objects are so dark and difficult to see that they are nearly invisible until they come very close, and by then it is too late. It is certain that one of these monsters is on a collision course with Earth — we just do not know the details. Is it days from now or hundreds of years from now? Even if we were sure one was coming, there is just very little that we can do about it currently.

We are years away from being able to control our own destiny as it relates to supernovae and giant comets and asteroids, but scientists are working on solutions. *This is not a high priority with the world's governments*, however, which typically prefer to confront terrestrial threats rather than cosmic ones. *To prevent one of those giant objects from smashing into us, collectively, we spend about $10 to $20-million annually, an amount less than the cost of one or two sophisticated fighter jets.* Almost no money is spent trying to detect imminent supernovae [or comets].

Our politicians are seriously underestimating these severe threats, which are capable of ending our species, just as they snuffed out the mammoths a mere 13,000 years ago, only an eye blink in cosmic terms. There are few threats of that magnitude facing us today. *The survival of the human race is not seriously threatened by the avian flu, Al Qaeda attacks, the end of the Age of Oil, monster hurricanes, giant earthquakes, or enormous tsunamis;* if any of those occur, most of us will continue with our lives. Furthermore, nothing on that list is broadly accepted as having caused worldwide extinctions in the past. *The same cannot be said about supernovae and massive [cometary] impacts.* Those two cosmic events are implicated in many of the largest extinctions on our planet over the last millions of years. Fortunately, we survived them, but many of our fellow species did not. Humankind might not survive the next one. *It seems reasonable to forgo several of our military fighter jets each year to decrease our chances of being "nuked" from space by a supernova or a comet.* [This author's emphases]

There is a climate scientist in our research group who recently analyzed the data I covered above and wrote:

Looking at the longer time series from the NGRIP data we again see the Precessional Cycle more resolved at 25,000 years and 20,500 years. The larger signal at 41,000 years is another Milankovitch Cycle, the obliquity or axial tilt of the earth. The 61,500-year signal is most likely picking up the 100,000-year Eccentricity Cycle, given that the data set is only 112,900 years long and the error so large.

But what we're concerned with here are the shorter cycles. Because of the longer time series of the data (roughly a factor of three), the errors are reduced by roughly 1.7 (sqrt(3)).

As can be seen in the plot, there is an approximate 6,000-year cycle which has been recognized in the Greenland ice-core data.

"[S]ignificant ice growth did, in fact, occur at around 6,000 years Before Present. Studies of some of the largest of the 3,000 small glaciers that occur in Fennoscandia today indicate that the majority of these started to grow shortly after 6,000 years ago." The biblical flood of Noah is believed to date back to this time period (Olsen and Hammer, "A 6-ka climatic cycle during at least the last 50,000 years," *NGU Bulletin* 445, 2005, 89-100).

According to recent archaeological research the flooding of the Black Sea dates to this period. [...]

The next largest signal below 6,000 is approximately 4,300 years, and then below that, a 3,600-year cycle. A similar spectral analysis using the Lomb-Scargle periodogram (not shown) shows that the 4,300 and 6,000 periodicity are significant, but not the 3,600-year cycle.

So from the longer time series data we see a stronger, 4,300-year cycle and an even stronger 6,000-year cycle. *These two (possibly three) cycles are unexplained.* We do not know what causes them.

Or do we?

The answer is important, for not only was there a global event *4,200 years ago* (and again at 8,200 years Before Present), *there was one 6,000 years ago* (and again at approximately 12,000 years Before Present). On top of that, there was the rapid rise in sea level 19,000 years ago and the associated warming of the North Atlantic region that began 21,000 to 22,000 years Before Present, coinciding with the Precessional Cycle (see Clark, *et al.* "Rapid Rise of Sea Level 19,000 Years Ago and Its Global Implications," *Science* 304, 1141, 2004). Then there was an event 41,000 years ago, possibly a supernova (Firestone, *et al.*), aligning with the Obliquity Cycle.

So we have four cycles, possibly five if you include the eruption of Thera/Santorini as part of the 3,600-year cycle, that are converging at the present moment. I believe my theory that it's a 4,300-year cycle that we need to be most concerned about is the correct one, not a 3,600-year cycle. Arguably, the 4,300-year cycle could fit into the 41,000-year cycle, but that wouldn't fit with the known solar forcing due to the Obliquity Cycle.

Just a quick look at the plot shows that if recent warm periods (or interglacials) are a guide, then we may soon slip into another glacial period and that *whatever brings it about is due right about now as at least four cycles coincide* to bring down global temperatures a few notches. The hype about "Global Warming," (which preceded every glacial cycle for the past million years), is just mere distraction by comparison, because the bigger surprise is yet to come.

There was another article recently by Alexander Cockburn ("Sources and Authorities: Dissidents Against Dogma," *Counterpunch.com*, June 9/10, 2007), perhaps the only liberal writer out there speaking against "Global Warming" or the idea that man is the cause of it. He writes:

> We should never be more vigilant than at the moment a new dogma is being installed. The claque endorsing what is now dignified as "the mainstream theory" of Global Warming stretches all the way from radical greens through Al Gore to George W. Bush, who

signed on at the end of May. The left has been swept along, entranced by the allure of weather as revolutionary agent, naïvely conceiving of Global Warming as a crisis that will force radical social changes on capitalism by the weight of the global emergency.

Amid the collapse of genuinely radical politics, they have seen it as the alarm clock prompting a new Great New Spiritual Awakening.

Alas for their illusions.

Capitalism is ingesting Global Warming as happily as a python swallowing a piglet. The press, which thrives on fear-mongering, promotes the nonexistent threat as vigorously as it did the imminence of Soviet attack during the cold war, in concert with the arms industry. There's money to be made, and so, as Talleyrand said, "Enrich yourselves!"

Now, some of the facts Mr. Cockburn confuses, but the gist of his article is correct (see comments in above link). What is happening is that the controllers are using Global Warming as a distraction to consolidate their hold on the public, the economy and now even the military:

A secret report, suppressed by US defence chiefs and obtained by The Observer, *warns that major European cities will be sunk beneath rising seas as Britain is plunged into a "Siberian" climate by 2020.* Nuclear conflict, mega-droughts, famine and widespread rioting will erupt across the world.

The document predicts that abrupt climate change could bring the planet to the edge of anarchy as countries develop a nuclear threat to defend and secure dwindling food, water and energy supplies. The threat to global stability vastly eclipses that of terrorism, say the few experts privy to its contents.

"Disruption and conflict will be endemic features of life," concludes the Pentagon analysis. "Once again, warfare would define human life." The findings will prove humiliating to the Bush administration, which has repeatedly denied that climate change even exists. Experts said that they will also make unsettling reading for a President who has insisted national defence is a priority.

The report was commissioned by influential Pentagon defence adviser Andrew Marshall, who has held considerable sway on US military thinking over the past three decades. He was the man behind a sweeping recent review aimed at transforming the American military under Defence Secretary Donald Rumsfeld.

Climate change "should be elevated beyond a scientific debate to a US national security concern," say the authors, Peter Schwartz, CIA consultant and former head of planning at Royal Dutch/Shell Group, and Doug Randall of the California-based Global Business Network.

An imminent scenario of catastrophic climate change is "plausible and would challenge United States national security in

33

ways that should be considered immediately," they conclude. As early as next year widespread flooding by a rise in sea levels will create major upheaval for millions (Townsend and Harris, "Now the Pentagon tells Bush: climate change will destroy us," *The Observer*, February 22, 2004).

Of course, it didn't prove humiliating at all to Bush. In fact, it played right into his hands as he subsumed even more power like the python swallowing the piglet.

As I wrote at the beginning of this article, "If we can't stop an asteroid or comet from hitting us, do you think they'll panic the public by announcing an impending collision beforehand?" Nevertheless, panicking the public is *de rigeuer* for the Powers That Be, but *only for those things they can manage.* Global Warming that would slowly play out over decades is the ideal problem for those seeking power; the ever more dire warnings feed into the controllers' hands as they use it to justify ever higher military spending, dismantling of civil liberties, more control for corporations, and "pre-emptive" invasions in the quest for dwindling resources. What the public doesn't realize is that this may very well be all in preparation for when that first comet swarm panics us little earthlings who have no underground bases, no stockpiles of supplies, no protocol for surviving such an event, to emerge on the other side still in control.

Whether it's a 3,600, 4,200, 6,000, 20,000 or 41,000-year cycle, and that it was 3,600, 4,200, 6,000, 12,000, 20,000 or 41,000 years ago, doesn't make any difference to our "extinct by" date. As I noted above, the cycles are converging and the time is now.[17] [This author's emphases]

So, indeed, humanity has passed its "extinct by" date, and as it was in the days of Noah...

They did eat, they drank, they married wives, they were given in marriage, until the day that Noah entered into the ark, and the flood came, and destroyed them all.

Likewise also as it was in the days of Lot; they did eat, they drank, they bought, they sold, they planted, they builded. But the same day that Lot went out of Sodom it rained fire and brimstone from heaven, and destroyed them all.

---

[17] The full article is available at http://www.signs-of-the-times.org/articles/show/134637-+The+Younger+Dryas+Impact+Event+and+the+Cycles+of+Cosmic+Catastrophes+-+Climate+Scientists+Awakening

Chapter 57
It's Just Economics

Until recently, impacts by extraterrestrial bodies were regarded as, perhaps, an interesting but certainly not important phenomenon in the spectrum of geological processes affecting the Earth. This has only been the case since Lyell, Laplace and Newton put a period to such speculations. What seems to have happened is that, through repeated cataclysms, man has been brought low, relegated to darkness regarding his history, and at the very point when he began to study and analyze his environment objectively, *religion stepped in and put a period to such ideas*. Velikovsky's work was vilified by the scientific community, and shortly after, there came an onslaught of ideas promoting extraterrestrials as the source of civilization anomalies; and then, of course, Zecharia Sitchin's infamous "10th Planet" hypothesis to explain away planetary disruptions that could not be adequately concealed.

The question is, why? What, in the name of all things reasonable, would prompt anyone to wish to hide these matters? What kind of sick mind would divert the attention of humanity away from what is evident all over the planet to those with open eyes, and promote so assiduously ideas that mislead, misguide, and generally placate the populace with an assurance that either nothing is going to happen, or if it does, it will be preceded by a long period of approach by a body that is well organized and clearly seen, and that the government can probably "fix it?"

Well, the clue is right there: "Placate the populace." Control.

But, heavenly days! What kind of lunatics would want to keep everything under control in that sense if they have some idea that they, themselves, might be destroyed by the very processes they are concealing?

Obviously, they don't think so. Obviously, they think they have a plan. That suggests that, obviously, they know a lot more about what's going on, what the possibilities and probabilities are, than the rest of us. One of the "official views" of the subject that is available to the public tells us:

Our concept of the importance of impact processes, however, has been changed radically through planetary exploration, which has shown that virtually all planetary surfaces are cratered from the impact of interplanetary bodies. [...] The Earth, as part of the solar system, experienced the same bombardment as the other planetary bodies.

Most of the terrestrial impact craters that ever formed, however, have been obliterated by other terrestrial geological processes. Some examples, however, remain. To date, approximately 150 impact craters have been identified on Earth. Almost all known craters have been recognized since 1950 and several new structures are found each year.

Meteorite fragments are found only at the smallest craters and they are quickly destroyed in the terrestrial environment. [...] Although the number of known impact craters on Earth is relatively small, the preserved sample is an extremely important resource for understanding impact phenomena. They provide *the only ground-truth data currently available and are amenable to extensive geological, geophysical and geochemical study*. [...] In some cases, the large size of terrestrial impact craters, up to approximately 300 km in diameter, requires orbital imagery and observation to provide an overall view of their structure and large-scale context. [...]

The tendency to discount impact processes as a factor in the Earth's more recent geologic history was severely challenged by the interpretation in 1980 that Cretaceous-Tertiary (K-T) boundary sediments worldwide were due to a major impact event, and that impact was the causal agent for a mass extinction event. The acceptance of the K-T impact hypothesis by the more general terrestrial geoscience community *was not instantaneous* and considerable controversy and debate was generated. Today, *there are few workers who would deny that there is abundant diagnostic evidence that a major impact event occurred at the K-T boundary*. It is fair to say, however, there is less consensus on the role of impact in the associated mass extinction event, with *some workers still having difficulty in accepting impact-related processes* as the cause.

The impact signal of the K-T event is recognizable globally, because *large impact events have the capacity to blow out a hole in the atmosphere above the impact site*, permitting some impact materials to be dispersed globally by the impact fireball, which rises above the atmosphere. These materials do not require atmospheric winds for dispersal and have the capacity to encircle the globe in relatively short time periods, before eventually returning to the surface. Model calculations indicate that it does not require a K-T-sized event, which produced the buried 180 km diameter Chicxulub impact structure in the Yucatan, Mexico, to result in atmospheric blowout. Relatively small impact events, resulting in impact structures in the 20 km size-range can produce atmospheric blowout. [...]

From estimates of the terrestrial cratering rate, the frequency of K-T-sized events on Earth is of the order of one every 50 to 100-million years. Smaller, but still significant impact events, occur on shorter time scales and will affect the terrestrial climate and biosphere to varying degrees. The formation of impact craters as small as 20 km could produce light reductions and temperature disruptions similar to a nuclear winter. [...]

The most fragile component of the present environment, however, is modern human civilization, which is highly dependent on an organized and technologically complex infrastructure for its survival. [...] There is little doubt that if civilization lasts long enough, it will suffer severely or may even be destroyed by an impact event. [...]

[A]n impact anywhere in the Atlantic Ocean by a body 400 m in diameter *would devastate the coasts on both sides of the ocean with wave run-ups of over 60 miles.* The 1960 tsunami, generated by a magnitude 8.6 Chilean earthquake, is thought to have been the largest this century. *An impact-generated tsunami ten times more powerful will occur with a typical recurrence time of a few thousand years.*

Small impacting bodies release their energy in the atmosphere as an air burst. The threshold size at which this is exceeded depends on the strength of the impacting body. For example, iron impacting bodies up to 20 miles will deposit their energy in the atmosphere and not reach the surface; whereas, comets as large as 200 miles will deposit their energy in the atmosphere. Such air-burst explosions, fortunately, are not efficient at delivering their energy to the ground, because some of initial energy is blown into space. The Tunguska event in 1908 was due to the atmospheric explosion of a relatively small, approximately *few tens of meters*, body at an altitude of 10 km. The energy released, has been estimated to be 10-100 megatons TNT equivalent. Although *the air blast resulted in the devastation of 2000 sq. km of Siberian forest*, there was no loss of human life due to the very sparse population. *Events such as Tunguska occur on a time-scale of hundreds of years.* [...]

The next large impact with the Earth could be an "impact-winter"-producing event or even a K-T-sized event. To emphasize this point, in March 1989 an asteroidal body named "1989-FC" passed within 700,000 km of the Earth. This Earth-crossing body was not discovered until it had passed the Earth. It is estimated to be in the 0.5 km size-range, capable of producing a Zhamanshin-sized crater or a devastating tsunami. *Although 700,000 km is a considerable distance, it translates to a miss of the Earth by only a few hours, when orbital velocities are considered.* At present, no systems or procedures are in place, specifically for mitigating the effects of an impact.[18] [This author's emphases]

---

[18] C.R. Chapman and D. Morrison, *Cosmic Catastrophes* (New York: Plenum Press, 1989). T. Gehrels (ed.), *Hazards Due to Comets and Asteroids* (Tucson: Univ. Arizona Press, 1994). R.A.F. Grieve, "Impact cratering on the Earth," *Scientific American* (1990),

The conclusion that I have reached after the research that went into *Secret History* is that the inner solar system experiences swarms of comets/asteroids on a fairly regular schedule. One major swarm comes at 3,600-year intervals, and there are minor swarms at other intervals; and these cycles are sub-cycles of even larger swarms, at intervals of hundreds of thousands of years, and even millions of years. The massive bombardment of the East Coast of the U.S. around 12,000 to 12,500 years ago, resulting in the Carolina Bays, is evidence of the great number and relatively small sizes of many of these bodies, indicating that they released their energy into the air above the ground, similar to the Tunguska event. Keep in mind, of course, that *the air blast of a single small body over Tunguska resulted in the devastation of 2000 sq. km of Siberian forest.* This pretty much confirms the Cassiopaeans' remark about this, made on August 22, 1998:

Q: (L) Okay, we would like to know, what was this famous explosion in Tunguska?
A: Comet fragment.

This would also explain many other discontinuities in history that have destroyed centers of civilization without necessarily leaving any impact craters. Such events would also be viewed by uneducated peoples as "the wars of the gods." Of course, larger asteroid type bodies are also well represented in the geological record with impact craters being discovered every day.

At present, the Carolina Bays (discussed in *Secret History*) are numbered at somewhere around 500,000. I want the reader to just stop and think about that for a minute. Think about 500,000 Tunguska-like events. Forget movies like *Armageddon* or *Deep Impact* where there is a single "big one" and the government can save us. Instead, go rent the movie *Asteroid*, and watch the rain of rocks on New York City for a much better idea of what we are talking about. The effect will be similar, only it may occur over the entire globe. Then just think about the fact that the Carolina Bay phenomenon *covers several states.* Put that together with the evidence of "nuclear" explosions in the great lakes regions, as well as other areas, *dated to about the same time*, and then consider a most interesting fact reported in the news quite recently: That one of the largest asteroids known to have approached the Earth zipped past about 450,000 km away on March 8, 2002 — but nobody recorded it until four days later. Yes, a miss is as good as a

---

v. 262, 66-73. A.R. Hildebrand, "The Cretaceous/Tertiary boundary impact (or the dinosaurs didn't have a chance)" *Journal of the Royal Astronomical Society of Canada* (1993), v. 87, 77-118.

mile, but these "near misses" have been occurring with greater and greater frequency in recent years. This could be seen as a result of increased observational ability, or it could be because we are already experiencing the first influx of a swarm of such bodies entering the inner solar system.

On March 15, 2002, *New Scientist* magazine said that the object was "hard to spot because it was moving outward from the innermost point of its orbit, 87-million km from the Sun. When it passed closest to the Earth — just 1.2 times the distance to the Moon — it was too close to the Sun to be visible. Asteroids approaching from this blind spot cannot be seen by astronomers. If a previously unknown object passed through this zone on a collision course with Earth, it would not be identified until it was too late for any intervention."

How many other "blind spots" are there?

How many other "objects" are out there zipping this way? How many of them are in "swarms" where any single one of them might not produce a "global" event, but taken together, could destroy our civilization and leave almost no trace of their activity other than the widespread destruction?

With just the brief review we have presented, it seems obvious by now that human civilization very likely has had a very different history than the standard teachings will admit. Here and in other volumes, we have presented only a tip of the iceberg of evidence. This evidence is available to the researcher who is diligent, thorough, and refuses to accept the material propagated in popular, mass-market books written by authors who either have not searched widely and deeply enough, or who have an agenda. Yes, it may be a bit harder to get, because it seems that the numerous government-sponsored scientific studies on the periodic rains of comets/asteroids are concealed behind obscure titles in technical journals, or on microfiche in university libraries. But with persistence, the material can be gathered together, examined, and extrapolated upon.

Again, we come back to the question as to why this topic is — at the very least — subjected to such extreme prejudice that it amounts to concealment? Why are there so many books about things like the 10th Planet, precessional alignments related to world ages, Stargates, Galactic Core Explosions, and so forth, that pass with only mild criticism; yet Velikovsky's proposal of rains of cometary bodies was so viciously vilified? This brings us back to that most interesting passage written by Wilhelm Reich:

> Why did man, through thousands of years, wherever he built scientific, philosophic, or religious systems, go astray with such persistence and

with such catastrophic consequences?" [...] Is human erring necessary? Is it rational? Is all error rationally explainable and necessary? If we examine the sources of human error, we find that they fall into several groups: Gaps in the knowledge of nature form a wide sector of human erring. Medical errors prior to the knowledge of anatomy and infectious diseases were necessary errors. But we must ask if the mortal threat to the first investigators of animal anatomy was a necessary error too.

The belief that the Earth was fixed in space was a necessary error, rooted in the ignorance of natural laws. But was it an equally necessary error to burn Giordano Bruno at the stake and to incarcerate Galileo? [...] We understand that human thinking can penetrate only to a given limit at a given time. What we fail to understand is why the human intellect does not stop at this point and say: "This is the present limit of my understanding. Let us wait until new vistas open up." This would be rational, comprehensible, purposeful thinking. (Reich, 1949)

The only problem here is that we have discovered that the material *is* available! There are *many* vistas of understanding that are open in our world to those with eyes to see! We have listed our sources. There are photographs and images of the proof that our planet has been repeatedly bombarded with showers of comets/asteroids; there exists stacks of *evidence* — recorded in books, papers, monographs, theories — that civilizations of greater advancement than our own have existed on the earth, and have been repeatedly destroyed. There are piles of *evidence* that attest to a far greater age of mankind than is presently accepted or taught by mainstream science. In every case, this evidence is ignored, marginalized, explained or argued away, and those who wish to make such information publicly available to a wider audience, are similarly marginalized and dealt with. What is even more interesting, as we noted previously, is the fact that the "mainstream occultists" have jumped on the uniformitarian bandwagon and are vigorously promoting reams of disinformation. Our own experience has been that our attempts to talk about this sort of thing have resulted in coordinated attacks from certain so-called "occultists," who have even sunk so low as to spend inordinate amounts of time roaming the internet, denouncing our research group as a "cult," writing libelous, defamatory public postings about us personally; and most importantly, attempting to destroy the Cassiopaeans' material as a source of clues to our reality. Reich notes this problem and writes:

What amazes us is *the sudden turn from the rational beginning to the irrational illusion.* Irrationality and illusion are revealed by the *intolerance and cruelty* with which they are expressed. We observe that human thought-systems show tolerance as long as they adhere to reality. The more the thought process is removed from reality, the more intolerance and cruelty are needed to guarantee its continued existence. (Reich, 1949; this author's emphasis)

Here is where we have to begin to really think about things in a reasonable and objective way. If there is so much evidence available — and there is — why is it so rabidly attacked and dismissed? Why is it necessary to defame, libel, crush, and destroy those who bring up these matters, and who also *produce the evidence*? Most especially, why is this so, when we know that there *are* groups — such as the government and certain academics — who *do* study these things, who do commission reports on them, who collect the data? Just what they heck is going on? Why do they permit — no, actually encourage — the crazy ideas about 10th Planets or galactic explosions or precessional clocks of world ages, when the simple truth is so evident?

Reich proposed that the "adherence to the surface of phenomena" was related to "a certain connection with the structure of the human animal." He thought that the function of seeking the truth must be somehow "buried" since the tendency to "evade the obvious" was so powerful. In this idea, we are, of course, reminded of Castaneda's "Predator."

"We have a predator that came from the depths of the cosmos and took over the rule of our lives. Human beings are its prisoners. [...] You have arrived, by your effort alone, to what the shamans of ancient Mexico called *the topic of topics*. I have been beating around the bush all this time, insinuating to you that *something is holding us prisoner*. Indeed we are held prisoner! This was an energetic fact for the sorcerers of ancient Mexico. [...] They took over because *we are food for them*, and they squeeze us mercilessly because we are their sustenance. Just as we rear chickens in chicken coops, the predators rear us in human coops. Therefore, their food is always available to them." [...]

"I want to appeal to your analytical mind," don Juan said. "Think for a moment, and tell me how you would explain the contradiction between the intelligence of man the engineer and the stupidity of his systems of beliefs, or the stupidity of his contradictory behavior. Sorcerers believe that the predators have given us our systems of beliefs, our ideas of good and evil, our social mores. They are the ones who set up our hopes and expectations and dreams of success or failure. They have given us covetousness, greed and cowardice. It is the predators who make us complacent, routinary, and egomaniacal." [...]

In order to keep us obedient and meek and weak, the predators engaged themselves in a stupendous maneuver — stupendous, of course, from the point of view of a fighting strategist. A horrendous maneuver from the point of view of those who suffer it. *They gave us their mind!* Do you hear me? The predators give us their mind, which becomes our mind. The predators' mind is *baroque, contradictory, morose, filled with the fear of being discovered any minute now.* [...] Through the mind, which, after all, is their mind, *the predators inject into the lives of human beings whatever is convenient for them.* (Castaneda, 1998, 213-220; this author's emphases)

Reich, of course, decided that the cause of "needless human erring" was due to the "pathological quality of human character." In fact, this fits quite well with what Castaneda has written: "The Predators gave us their mind." Reich also pointed out, quite reasonably, that religion, education, social mores, suppression of a true understanding of love, and so on, were merely *symptoms* of this fact. This is what led Reich to conclude: "The answer lies somewhere in that area of our existence which has been so heavily obscured by organized religion and put out of our reach. Hence, it probably lies in the relation of the human being to the cosmic energy that governs him."

As we have mentioned, Reich was moving dangerously close to describing the hyperdimensional reality. And, as a result, he, too, was subjected to overt intolerance and cruelty. He rightly noted: "Irrationality and illusion are revealed by the intolerance and cruelty with which they are expressed. [...] The more the thought process is removed from reality, the more intolerance and cruelty are needed to guarantee its continued existence."

It is in this last remark that we come to some clue about the matter: Intolerance and cruelty are *needed* to guarantee the "cover-up." *A certain kind of "human being" acts on behalf of this cover-up.* A certain kind of human being acts as the playing pieces in the "Secret Games of the Gods." The Cassiopaeans have said that this has something to do with economics.

This brings us to other strange matters, something called "Alternative 3" that is somehow mixed up with this cover-up of the ideas of the Holy Grail, the Ark of the Covenant, and the true history of mankind, including cyclic cometary destruction. Let me just insert here a couple of very strange little exchanges with the Cassiopaeans that I would like the reader to have in mind as we go along:

0-25-95

Q: OK, where is the Ark of the Covenant currently located?

A: "Alternative 3."

Q: (L) "Alternative 3" is the plan to take all the people, all the smart guys, all the elite, off the planet and leave everybody else here to blow-up, isn't it?

A: Maybe. Maybe not. [This "maybe not" was delivered in the midst of the next question and was originally included with the nest answer. But it rightly belongs here, and so I have moved it to retain the proper context.]

Q: (L) Where is it currently located?

A: Discover.

Q: (L) We're trying to discover, through our interaction with you. How else can we discover something as obscure as this? I mean, that's a pretty darned obscure question, I would think. (SV) Who's in charge of "Alternative 3," Laura? (L) That's too complicated.... (SV) Well, maybe they have it, who ever is in charge of it. (L) Well, are you going to tell us anything about it?

A: Study "Alternative 3" to find answer!

Curiously, as we will discover, the very next question relates to the subject of "Alternative 3," though I did not know it at the time:

Q: (L) OK, the Matrix material says that Henry Kissinger is the current head of MJ-12. Is this correct?

A: No.

Q: (L) Is he just a red herring, so to speak?

A: Yes. MJ-12 is no longer MJ-12.

Q: (L) What is MJ-12 now known as?

A: Institute of Higher Learning.

Q: (L) Are you talking about Brookings Lab, or Brookhaven?

A: Not really.

Q: (L) Is it a specific institute of higher learning?

A: Yes.

We will soon discover that "Alternative 3" has a lot to do with HAARP as well:

February 17, 1996

Q: (L) Some people on the net want me to ask about this HAARP thing... seems to be some sort of antennae thing....

A: Disguise for something else.

Q: (L) What is that something else?

A: Project to apply EM-wave theories to the transference of perimeters.

Q: (L) What does that mean?

A: If utilized as designed, will allow for controlled invisibility and easy movement between density levels on surface of planet as well as subterranially.

Q: (L) Who is in charge of building this thing?

A: More than one entity.

Q: (L) What groups?

A: INVELCO is one guise as well as UNICON and banking interest.

Q: (L) Who is in disguise as INVELCO and UNICON? Are they just dummy companies for cover?

A: Close.

Q: (L) Can you tell us if this is a human organization, or aliens, or a combination?

A: Human at surface level. [...]

Q: (L) Is there more you can tell us about this?

A: It has nothing to do with weather or climate. These things are emanating from 4th density, as we have told you before. [...]

Q: (L) So, HAARP has nothing to do with the weather?

A: And also, EM [electromagnetic] associated with same as reported.

Q: (L) So, when is this HAARP thing scheduled to go into operation?

A: Open.

Q: (L) Is it currently in operation?

A: Experimental.

Q: (L) How long have they been working on this thing?

A: Since the 1920s.

Q: (L) What?! The 1920s?

A: Yes.

Q: (L) Well, that certainly is strange.

So what is "Alternative 3"? On Monday, June 20, 1977, the UK's Anglia Television broadcast the last program in a series of serious science documentaries called *Science Report*. All of the shows were produced by highly-respected science reporters and it was simulcast in UK, Australia, New Zealand, Canada, Iceland, Norway, Sweden, Finland, Greece, and Yugoslavia. The title of the segment was "Alternative 3," and it was never broadcast in the United States. Some say that the "powers that be" prevented it from being screened here.

The documentary presented the following purported "facts":

There is a secret joint US/USSR space program that has gone far beyond what the public sees. Astronauts landed on Mars in 1962. It has been discovered that there is other intelligent life in the universe. The earth is dying. We have polluted it beyond repair. The increasing "greenhouse effect" will cause the polar ice caps and glaciers to melt and flood the Earth.

There are three possible solutions for mankind:

Alternative 1: Stop all pollution immediately and blow two huge holes in the ozone layer. This would allow excessive UV light to reach the earth and millions would die of skin cancer.

Alternative 2: Immediately begin digging underground cities for the *elite* [the "New World Order"?] — the lucky ones deemed worth saving — and let the teeming billions perish on the polluted surface.

Alternative 3: Build spaceships and get the *elite* off the planet — to the Moon and Mars. Kidnap and take along some "ordinary" people for use as slave labor. Use "mind control" techniques to control them. Leave the remainder of humanity to wallow in its own filth.

After the television show was aired, a book was published with three authors listed: Leslie Watkins, Christopher Miles and David Ambrose. Christopher Miles, who also directed and co-wrote the documentary, is the brother of actress Sara Miles. The actual author of the book version, Leslie Watkins, was a writer of thrillers even before writing *Alternative 3*. *Fortean Times* later published an article in which Nick Austin, then editorial director of Sphere Books, revealed that he commissioned Leslie Watkins to write the book version for literary agent Murray Pollinger.

The question is, however, whether there was any fact behind that fiction. Bill Cooper, killed in a shoot-out with the law, incorporated many of the "Alternative 3" elements into his conspiracy theories. The stories came fast and furious, and "Alternative 3" was grafted onto the mythos of the Philadelphia Experiment that became the Montauk Experiment as promulgated by Al Bielek and others. What is most interesting are the comments of the author of the book about the correspondence she received after its publication:

> In fact, the amazing mountains of letters from virtually all parts of the world — including vast numbers from highly intelligent people in positions of responsibility — convinced me that I had ACCIDENTALLY trespassed into a range of top-secret truths.
>
> Documentary evidence provided by many of these correspondents decided me [sic] to write a serious and COMPLETELY NON-FICTION sequel. Unfortunately, a chest containing the bulk of the letters was among the items which were mysteriously LOST IN TRANSIT some four years later when I moved from London, England, to Sydney, Australia, before I moved on to settle in New Zealand. For some time after *Alternative 3* was originally published, I had reason to suppose that my home telephone was being tapped, and my contacts who were experienced in such matters were convinced that certain intelligence agencies considered that I probably knew too much.
>
> It would be a mistake to file *Alternative 3* away too cozily with Panorama's spaghetti harvest and other hoaxes. Suppose it were fiendish double-bluff inspired by the very agencies identified in the program, and that the superpowers really are setting up an extraterrestrial colony of outstanding human beings to safeguard the species?[19]

I think that Leslie Watkins is onto something in saying that this story may have something to do with "top-secret truths," but when we consider the modes of disinformation, we have to realize that the "truth" may not be necessarily the story that is promoted, *as* it is

---

[19] "The Truth About Alternative 3," letter from Leslie Watkins to Ms. Dittrich of Windwords Bookstore, http://www.sacred-texts.com/ufo/watkins.htm

promoted. It may have little to do with the setting up of an extraterrestrial colony of "outstanding human beings in order to safeguard the species," as Watkins suggests.

Looking at the first "fact," and considering it as an item of disinformation, can we speculate about what it is designed to cover up or distract attention away from?

> There is a secret joint US/USSR space program that has gone far beyond what the public sees. Astronauts landed on Mars in 1962. It has been discovered that there is other intelligent life in the universe.

The first part is the "joint US/USSR space program" that is concealed by political posturing and maneuvers. Keeping in mind that disinformation is generally composed of "truth wrapping a lie," we can suppose that there is, indeed, a "One World Government" at some level, where the US and Russia are unified in some effort. But it seems sort of transparently obvious that the statement is promoting the idea that it is a *human* government. The statement seems designed to distract our attention from the highly advanced technological capabilities of hyperdimensional existence, and focuses it on a somewhat advanced human technology. The next part of the disinformation above seems to be that there is a space program that is far beyond what the public sees. How else to cover up UFO and extraterrestrial activity? It's too evident to deny, so give a different "reason" for its existence.

The next item: "Astronauts landed on Mars in 1962 and there is other intelligent life in the universe."

We see here the program of Richard Hoagland, Zechariah Sitchin, Courtney Brown, and many others. If it is suggested that this intelligent life is "Martian," or even related to a "10th Planet," it is a lot easier to accept its existence. It is, after all, more or less "like us" in the idea that it exists in our reality. Again, this distracts our attention away from the hyperdimensional nature of this "intelligent life," and conceals its true capabilities. We can all go back to bed and get some rest if "aliens" are more or less physical beings as we are, with the same limitations and the same reliance on human-oriented technology. What's more, it covers the fact that there may be anomalies on Mars or the Moon that are difficult to explain. If too many people start looking at the history of the Earth, they may discover something frightening, and we can't have that happening. Let's attribute everything to the "Martians" and then we won't have to deal with the ongoing, *multi-millennial manipulation of human beings* here on our own planet. After all, a tall blonde Martian or Annunaki is socially acceptable; hyperdimensional beings who manifest a form truer to their intrinsic nature are not invited to the best parties! As the Cassiopaeans pointed out when the existence of rocks from Mars on Earth was announced:

08-11-97

Q: (L) OK, what's the scoop on this Mars rock? [...] Is this rock from Mars?

A: Yes.

Q: (L) How long have they known that rocks from Mars are on the planet?

A: 12 years.

Q: (T) About the same length of time it took them to analyze the soil samples that they got from Mariner probes. (L) OK, Why all of a sudden are they revealing or releasing this information about this Mars rock in such a big and manipulative way? (T) You just said it!

A: You have already figured it out yourselves. [...]

Q: (L) Well, OK, but is this Mars rock, and is this opening of the doors concept, is this leading up to some definite, overt interaction with aliens?

A: Gradually. [...] Notice how you heard nothing about the Mars Probes until the rock announcement?

Q: (T) This is the new stuff? [...]

A: The excavation robot spacecraft. One Probe is already on its way, another to follow. No further explanation about "loss" of Mars Explorer.

Q: (L) What did happen to the Mars Explorer?

A: Blacked out. You see, "too risky." And too much too soon, due to pressure from Hoagland and others.

Q: (T) My own opinion is that they've already been there, and they know what's there.

A: No. Microbes are easier to swallow than humans in togas! [...]

Q: (T) OK, you just mentioned that somebody from this planet already launched a Mars Probe. A new Mars Probe, that no one in public knows about. Because it's never been talked about. So, it's a secret probe. Who does it belong to?

A: Was secret US government. [...]

Q: (T) What is the purpose of these probes?

A: Excavation to display living organisms.

Q: (T) Display? (L) Yes, for public consumption. In other words, not only do we have a rock now, that shows evidence that there was.... (T) Oh, "display," as when they find it and dig it up, they're going to show it on camera! (L) Yes! (T) Connie Couric will interview it! (L) Right! (F) First they said they found no evidence, then they said it was inconclusive.... Now, who the hell knows what they found! In revealing things, we'll start with fossilized life, and then move on.... (L) So, they're going to display the discovery of living organisms on Mars to take the next step to acclimate....

A: Yes.

Q: (L) So, in other words, this process is going to be something of an ongoing thing, and all of these people who are cranking around about, you know, alien landings....

A: No faces, though.

Q: (L) There's not going to be any "Faces On Mars?" They are not going to show us....

A: Won't be revealed, what do you think happened with Mars Explorer?

Q: (L) Well, what did happen with the Mars Explorer? (T) Now, now, now, let me....

A: Hoagland forced their hand.

Q: (T) What do we think happened to the Mars Explorer? I think they switched channels. They just moved it from one communication post to another, and it's doing exactly what it's supposed to be doing. And they did it in such a way, that the NASA people really didn't know what happened, so that when they were asked, they could say, "We don't know what happened to it!" Because they really don't know what happened! (L) When we're talking about attack, as we were before, as in plane crashes, the Olympics, all these different things — this dealing with these Mars Explorers — is all this stuff, or most of this stuff, coming from the 4th density manipulations of human minds, rather than....

A: Yes.

Q: (L) ...rather than actual, physical entry [into our plane of existence] and doing of deeds? Is that it?

A: Yes. [...]

Q: (T) I have a question. They're going to display live organisms, like... how did they put that...? "Living organisms"? How big are these living organisms going to be? How advanced?

A: Teeny-tiny.

Q: (T) So, we're still talking about microscopic organisms here?

A: Yes.

Q: (J) So, they won't wave at us!

A: But these will be alive. Can't you see the progression here? Don't want to scare Grandma Sally Bible Thumper/Stockmarket Investor!

Again there is a reference to economics as the means of control. But getting back to the next purported fact in "Alternative 3": "The earth is dying. We have polluted it beyond repair. The increasing 'greenhouse effect' will cause the polar ice-caps and glaciers to melt and flood the Earth."

It is an obvious fact that there is most definitely something going on in terms of weather. What we see here is the effort to divert attention away from hyperdimensional manipulations. At the same time, what this manipulates us to feel is that we are in deep doo-doo and we need Big Brother to haul us out of the soup.

Some time ago I wrote about Climate Change as being probably the most pressing problem facing humanity today. It is so pressing that I am convinced that possibly 90% of the human race — over 6-billion people — could be at risk of certain death in the very near future, perhaps within ten years, if this matter is not addressed adequately and appropriately very, very soon by our "glorious leaders," who seem to have little on their mind other than blowing up innocent people.

But then, that war-mongering has a hidden agenda behind it: To grab and hold resources. However, rest assured that the intent is not to grab and hold those resources for you and me; it is to get them for the "elite," that 6% of humanity that is on the top of the heap and intends to stay there, regardless of the fact that those genes should never be passed on.

Well, the Climate Change confusion factor is heating up.

The U.K.'s Channel 4 recently broadcast a special on the "Climate Change Swindle," that was intended to "expose the myths about climate change that have been promulgated in order to hoodwink the world into accepting the man-made theory of Global Warming."

As far as it went, this special wasn't too bad. However, it didn't really tell the whole story, which is that, yes, Climate Change is real and a serious threat, but not for the reasons given.

Keep in mind that this is really just a distraction, something to keep the masses busy so that they don't see the real agenda — that it is intended that they should be "left out in the cold," because they didn't act to get rid of corrupt leaders in time to do anything to prepare for what is coming.

How many of you have seen the movie "The Day After Tomorrow," based on the book by Art Bell and Whitley Strieber? If you haven't, the thesis of the movie is that Global Warming causes large areas of the Arctic to melt, so that the northern Atlantic Ocean is diluted by large amounts of fresh water, which changes the density of the water layers causing a disruption of the Thermohaline current. This then leads to a rapid and unnatural cooling of the northern hemisphere which triggers a series of anomalies, eventually leading to a massive "global superstorm" system consisting of three gigantic hurricane-like superstorms, which suck up heat and drop the super-cold upper atmospheric air down onto the planet, resulting in an "Instant Ice Age."

This idea is nothing new and it didn't really originate with Art Bell and Whitley Strieber. A NASA report[20] from 2004 tells us "Andrew Marshall, a veteran Defense Department planner, recently released an

---

[20] http://science.nasa.gov/headlines/y2004/05mar_arctic.htm

unclassified report detailing how a shift in ocean currents in the near future could compromise national security."

In a 2003 report,[21] Robert Gagosian cites "rapidly advancing evidence [e.g., from tree rings and ice cores] that Earth's climate has shifted abruptly and dramatically in the past." For example, as the world warmed at the end of the last ice-age about 12,500 years ago, melting ice sheets appear to have triggered a sudden halt in the Conveyor, throwing the world back into a 1,300-year period of ice-age-like conditions called the "Younger Dryas." It is also now known that the Gulf Stream weakened in that "little ice age."[22]

On December 6, 2005, Michael Schlesinger, a professor of atmospheric sciences at the University of Illinois at Urbana-Champaign, leading a research team, said: "The shutdown of the thermohaline circulation has been characterized as a high-consequence, low-probability event. Our analysis, including the uncertainties in the problem, indicates it is a high-consequence, high-probability event." See also: "Failing ocean current raises fears of mini ice age."[23]

There is another danger that comes with Global Warming: release of huge amount of methane from the methane clathrates buried in the arctic seabed, and even from other subterranean sources as the Earth struggles to shift around and balance itself. Methane can contribute to Global Warming, but it is a highly unstable gas. Mostly it just stinks, and can kill. Consider a September 2006 report ("Methane gas leak kills miners in Ukraine"[24]) which says: "Emergency Situations spokesman Ihor Krol said 'an unexpected eruption of a coal and gas mixture' — later identified as methane — 'occurred early this morning at a depth of 3,500ft'…"

We learn from further research that a high number of mining accidents are the result of methane pockets igniting or poisoning miners.

In short, if big bubbles of methane gas are released from the ground and a flock of birds happen to be in the area, they could very easily die and fall to the ground within a few minutes. Now, of course, methane itself is odorless, but it is a byproduct of organic decomposition and, as a consequence, is often associated with hydrogen sulfide, and a "rotten egg" smell. If you can smell it, the level is probably unsafe.

Of course, there is tremendous dispute about this, with a whole raft of critics (some of them scientists with questionable loyalties) pooh-

---

[21] http://www.whoi.edu/institutes/occi/currenttopics/climatechange_wef.html

[22] http://blogs.nature.com/news/blog/2006/11/gulf_stream_weakened_in_little.html

[23] http://www.newscientist.com/article.ns?id=dn8398

[24] http://www.breakingnews.ie/2006/09/20/story277612.html

poohing the idea and continuing along the line of "It's all Global Warming, and if we concentrate on cutting emissions, over time, things will stabilize."

Not very likely.

Why do we think so?

In December 2006, for almost a week, the Gulf Stream ceased to flow northward to Europe. Now, keeping those dates in mind, let's look at some headlines selected from the SOTT Weather Archive from the days during and after the temporary reversal of the Thermohaline current:

December 14, 2006: "Duck die-off in Idaho sparks fears"

December 19, 2006: "Lewiston residents unnerved by dead crows"

December 21, 2006: "Colorado reels under blizzard"

December 26, 2006: "Christmas storm brings devastation"

January 2, 2007: "Sections of Colorado Remain Buried in Snow"

January 3, 2007: "Cherry Blossoms Bloom In Brooklyn"

January 3, 2007: "Record snowfall buries Anchorage"

January 4, 2007: "Warm winter wreaks havoc"

January 4, 2007: "Scientists Say 2007 May Be Warmest Yet"

January 5, 2007: "2 dead after strong storms, tornadoes rip through southern Louisiana"

January 8, 2007: "Gas-like odor blankets Manhattan"

January 8, 2007: "Ducks die en mass in Vietnam's southern province"

January 8, 2007: "Dead birds shut down Austin"

January 8, 2007: "Outgassing: The environmental 'surge' you're not hearing anything about"

January 8, 2007: "NY gas smell shuts trains, forces evacuations"

January 8, 2007: "Wacky warm weather throws birds and bees off balance"

January 9, 2007: "Warm December Pushes 2006 to Record Year"

January 10, 2007: "Are the dead porpoises on Scottish beaches more evidence of Global Warming?"

January 10, 2007: "Freak tornado-like storm hits Barbados"

January 12, 2007: "Storm Warnings Across UK"

January 13, 2007: "Icy Weather Hits U.S. Midwest"

January 13, 2007: "Record Cold, Snow in Southern California!"

January 13, 2007: "Smelly Outgassing in Louisiana"

January 13, 2007: "Staten Island: More Bad Smells — Outgassing?"

January 14, 2007: "Powerful storm dumps ice and rain on central U.S."

January 14, 2007: "Ice storm lashes much of U.S. — 20 dead"

January 15, 2007: "Near Hurricane Force Storm Batters Baltic States, 2nd time in 2 years for 'once in a lifetime' event"

January 17, 2007: "Schwarzenegger seeks disaster aid for freeze ruined crops"

January 17, 2007: "Ice plays havoc with U.S. power grid"

January 17, 2007: "Thousands shiver as storm death toll hits 51"

January 17, 2007: "Big freeze hits $1bn crop"

January 17, 2007: "Wildfires burn in southern Australia"

January 17, 2007: "Scores killed, crops devastated in harsh U.S. winter weather"

January 17, 2007: "Storms forecast to batter UK"

January 17, 2007: "Warm spell in Russia wakes up the bears"

January 18, 2007: "Snow in Malibu!!!"

January 18, 2007: "Severe storms batter northwestern Europe"

January 18, 2007: "Travel in Europe disrupted by wild storm"

January 19, 2007: "Hurricane force winds rip into eastern Europe"

January 19, 2007: "Storm kills 27 in northern Europe"

January 19, 2007: "Germany limps back to life after storm claims 10 lives, wind gusts up to 202kph"

January 19, 2007: "Killer of 29, Kyrill Hurricane approaches Russia: 'The Day After Tomorrow'?"

January 19, 2007: "Experts can't find source of mysterious NYC odor"

January 19, 2007: "Okla., Mo., and Texas brace for storm"

January 19, 2007: "Icy storm blamed for 65 deaths in U.S."

January 19, 2007: "Germans told to stay indoors as hurricane nears"

January 20, 2007: "Europe counts cost of storms as stricken freighter is beached"

January 20, 2007: "New winter storm stalks southern plains"

January 21, 2007: "Snow storm rolls across plains; 8 dead"

January 25, 2007: "Getting colder in U.S. northeast, polar plunge underway"

January 25, 2007: "Anchorage hit with twice normal snowfall"

The $64,000 question: Is there a connection between the reversal of the Gulf Stream in December, the numerous reports of birds falling dead from the sky, and the wild and deadly weather during the first half of January?

The next question is: Does this small example suggest what *might* happen if the now highly unstable Gulf Stream finally and completely stops flowing North to Europe?

Meanwhile, on January 23, President Bush told the American Congress *"we must confront the serious challenge of global climate change."* However, the main thrust of his speech was to ask that Congress and the American People give his war escalation plan yet another chance.

Now, even though Senator Jim Webb (D-VA) responded to this nonsensical speech in a very appropriate way, telling Bush that "the majority of the nation no longer supports the way this war is being fought; nor does the majority of our military, nor does the majority of

Congress," Senator Webb himself is probably not aware of the truly great danger that may be looming over all of us, rich and poor, in every country of the world. Just as Nero fiddled while Rome burned, the leaders of our world are acting as though they have unlimited time to play their political games.

The question is, will any of us survive the threat to our civilization that is inherent in a Global Warming that can turn into an Ice Age in an instant while Bush and other world leaders engage in endless arguments about wars and economics? The real enemy is not "over there," it is "out there," in terms of changes to our environment that we all need to understand as fast as possible. We mean *all* people. When the sword falls, the billionaire oil tycoons will suffer just as much, if not more, than the Third World subsistence farmer.

Over the past few years, we have been saying repeatedly that there is something up in the Cosmos, and it isn't going to be just business as usual for the next few years. We have even speculated that whatever it is, the Powers That Be are somewhat more aware of it than the general public, and that, in fact, they are hiding a lot of things from the masses of humanity.

What seems to me to be true about this line of thinking is that it is a layer of "disinformation" that has been promulgated to even many of the outer circle of the "insiders" in order to induce them to put forth effort to deal with the human population issues as well as the "survival" of the elite. How better to induce scientists to work on such projects than to convince them that the future of the planet depends on it? This means that they will be busy building shelters, seeking technology to "transcend" the problem, and just generally creating an infrastructure of control that they *think* is designed to solve this problem, but which is actually intended for something else altogether.

One of the primary problems of dealing with a large human population in such terms seems to be economics and most particularly "Game Theory." This brings us to John Forbes Nash and his time at Princeton, which is where he came up with his ideas about Game Theory. Keep in mind, of course, what we discussed in a previous volume about the "economics" of Stockholm Syndrome — it's a lot more energy efficient to get the victim to like his captivity and to even help you to use and abuse him or her. It's all a matter of economics.

It was a curious set of "coincidences" that led me to the subject of John Nash, and the insights of his life story and how it relates to our subject of "economics" and the mysteries behind our apparent reality.

For me, the issue right after finishing Chapter 55 (in the previous *Wave* volume, *Facing the Unknown*) was how to move back into the past — the recapitulation — and I did as I usually do when preparing to write, I sort of pace around the house, looking at things, doing a little cleaning here and there, maybe cook a little, do a little gardening, or whatever attracts me in an instinctive way. After some time of allowing my mind to lie fallow, the flow begins again and I get an "urge" to do something that opens the door to the clues.

During this process, I remembered having read some things that would be useful in explicating this deep-level reality, since grasping 4th density is what seems to be the big chasm over which so many people simply cannot cross. The book was Nigel Pennick's *Secret Games of the Gods*.

I pulled it out and put it on my desk. A bit later, Ark and I were discussing several other individuals we were aware of who met unusual ends — either through death *or insanity* — while engaged in research on the boundary of physics and mathematics. One case in particular that always intrigued us was the story of Armand Wyler. Ark also mentioned that Hugh Everett (who first proposed the "many worlds" interpretation of quantum physics) mysteriously dropped out of academic physics, and later died rather young.

As I was thinking over Chapter 55, I kept coming back to this issue of M1 and G5, and the "economics" reference. This, in concert with the recent Enron debacle and several discussions on the e-group, kept popping up in my head. What the heck did "economics" have to do with anything? Was Enron involved? Are all of these things occurring at the present time simply the moving of the playing pieces in the 'Secret Games of the Gods"?

I mused over the fact that the guy who visited us, just before Vincent Bridges entered our lives in a direct way and Frank Scott exited, was an *economist*. He had done his Ph.D. under Milton Friedman, the 1976 Nobel Laureate in economics.[25]

One thing that bothered me was the fact that Bridges had pretended not to know this guy when I first wrote and told him about the impending visit. Bridges had, in fact, been planning to visit and attend a session, and after I wrote and suggested he might enjoy meeting this man, there was about ten days of no response. The guy came and went with no news from Bridges about whether he was coming, and if not, why. Then, what was more puzzling was the fact that it was just a few days after this visit that Bridges wrote to inform me that whatever I was doing, it was "hot," and he had been under severe attack from dark forces — both physical and ethereal — for even *talking* to me! But, not

---

[25] We now know a great deal more about Milton Friedman with the publication of Naomi Klein's *The Shock Doctrine*, a must read!

to worry! He was so determined to help me that he would manage somehow. Of course, it was obvious to him that I needed help — and *fast*! At this point, Bridges began "winding me up" to think that I needed his services as a hypnotherapist and specialist on government/Satanic mind control programs.

Seven months later, when I mentioned this economist again to Bridges, with the idea of identifying who was or was not part of a COINTELPRO operation — though I didn't call it that then — Bridges suddenly revealed that he had met him in Egypt in 1995 when they were in the same "tour group." He effectively deflected my suspicions by describing the economist as just a "harmless New Age groupie," and I was not to worry about him!

The reader should keep in mind that this was at a point in time when Bridges was working very hard to convince me that everybody in my environment was an "agent" out to get me, and the only one I could trust was Bridges himself. Meanwhile, there was a whole series of truly bizarre behaviors being manifested by people in our e-group, so that everything Bridges was saying seemed to be backed up by evidence in the environment.

So, the question becomes: Why did he wish to deflect my attention away from this economist whose presence at our house was immediately followed by serious maneuvers by Bridges to gain our confidence, control of our material, and *me*? What does this have to do with the almost simultaneous sudden defection of Frank Scott for the most absurd reason that anyone can imagine? Also, how was this connected to the sudden appearance of several people who wanted to join our e-group, to get inside the research group, and who later turned on us viciously and became, effectively, clappers for the Williams/Bridges gang? Was this entire scenario set up and scripted and controlled from higher levels in order to "herd" me into a corner where Bridges would be able to accomplish certain aims on behalf of these forces? Even if there was the possibility that he and the other players were unconscious of the maneuvers?

After the research team and e-group had investigated the matter, after we had assembled any private emails that any of us had received from the various parties involved, the "scripting" became even more evident. Each individual involved seemed to have been scripted to write and act in certain ways designed to "trigger" certain responses in myself and the group at large. The chronological assembling of the emails and comparison of one thread to another demonstrated clearly that there was a concerted, unified intent behind them. However, unless it could be determined that they all knew each other prior to joining our e-group, or that they had connections to some human COINTELPRO operation — and that did not seem likely — then the only conclusion that could be drawn was that each of them was, more or less, an "alien reaction machine," or a playing piece on the board of the "Secret Games of the Gods."

Most interesting of all, the one theme that they all had in common was "time travel," the search for the "deepest secrets," which they all sought to extract from me by emotional button pushing, manipulation, and later, intimidation and threat.

However, even when Bridges wrote about his connection to the economist, the lightbulb did not go on in my head. It was to be many more months before we even had the idea to investigate the many links between some of these people, coming to the realization of what we call the Cosmic COINTELPRO operation. Could it be a human-orchestrated operation? Well, sure; it's possible. We have certainly found enough links, nebulous though some of them are, to suggest that all the "players" know each other at the human level. But I am still inclined to think that most of the parties are being manipulated hyperdimensionally and that some of them are even unaware of the fact that their actions are "agent oriented." Perhaps that is my failing, that I cannot conceive of all of them being deliberately evil and manipulative to the extent that the evidence will show. But I will leave it to the reader to make their own assessment.

Among the things that this economist who visited us revealed about himself was a close relationship with Drunvalo Melchizedek, as well as certain members of what we now know of as "The Aviary" (discussed in Book Six, *Facing the Unknown*). He also claimed that he had worked under a security clearance at Wright Patterson Airforce Base, famous for the "Hanger 18" legend. He regaled us with stories about the underground storage facilities there — under a golf course, he said — and confirmed the rumor that there were "alien craft" stored there. We were suitably impressed that an individual with such credentials and obvious intelligence was giving such open credibility to some of the stranger stories of UFO mythology. He was also a past president of the American Cybernetics Society, a group with a very interesting membership.

At the time he was telling his stories, it never occurred to me to ask what in the world an economist was doing at Wright Patterson. What the heck did they want an *economist* for?

After posting Chapter 55 online, I brought up these questions to Ark. He agreed that there was something strange about all this that we ought to think about. Right after placing the Nigel Pennick book on my desk,

I had an overwhelming urge to go to Sam's Club and see if I couldn't find my son a new shirt to wear to a webmasters' conference he was going to attend with his employer. I could have gone the next day. But, for some reason, I wanted to go right then. I had that "antsy" feeling that I just had to *do* something.

We went, and after finding a shirt, we strolled through the book aisle and Ark noticed Sylvia Nasar's book *A Beautiful Mind*, about John Nash. (All subsequent quotations relating to Nash are taken from Nasar's book.) He picked it up and read the blurb on the back — which tells how the guy spent most of his life as a schizophrenic who thought that aliens were communicating with him, and how he had been awarded a Nobel Prize in *economics* for his contribution to Game Theory — and had the idea that perhaps I ought to read this book for clues. We bought the book. The history of the rise of Princeton as the "mathematical center of the universe," and Nash's interactions with Johnny von Neumann and Einstein was most revealing. CNN published an article on March 18, 2002, which began:

> The makers of *A Beautiful Mind* (which has just won a couple of academy awards), have objected to what they say is a whisper campaign to hurt the Oscar chances of their movie, which is up for eight Academy Awards including best picture.
>
> In a CBS *60 Minutes* interview, Nash and his wife, Alicia, denied allegations that he was gay, anti-Semitic or a poor father. And Sylvia Nasar, author of the 1998 biography, *A Beautiful Mind*, on which the film was based, wrote a commentary in the *Los Angeles Times* last week that accused many media outlets, including *The Associated Press*, of misstating details of Nash's life. [...] Both Nash and Nasar said his anti-Semitic remarks were made while he was suffering from paranoid schizophrenia.
>
> "I did have strange ideas during certain periods of time," Nash, 73, said on *60 Minutes*. "It's really my subconscious talking. It was really that. I know that now."
>
> Other aspects of his life not mentioned in the movie were a son he fathered by another woman before he married Alicia, and the fact that Nash and Alicia later divorced. The divorced couple lived together for many years and eventually remarried in 2001. [...] Nasar also criticized some reports that said Nash was a homosexual. Despite a 1954 indecency arrest and allusions in her book to his flirtatious behavior with men, she said he is an avowed heterosexual. The indecency charge was later dropped, she said.

I nearly choked when I read that both Nash and his wife denied that he was "gay, anti-Semitic or a poor father," when the book is full of evidence that Nash was all of those things, and his wife was certainly an enabler. Nash's quoted remarks above — "I did have strange ideas during certain periods of time. It's really my subconscious talking. It was really that. I know that now" — raises the important issue of how such things "arrived" in his subconscious mind. There is, of course, the ever popular "my parents damaged me as a child" routine that he could adopt to back this one up — after all, his parents are conveniently dead. And, if he is going to denigrate his "strange" subconscious ideas about some of the "politically incorrect" subjects, what are we going to do with the ideas he claims came from the same source for which he won a Nobel Prize? An interesting exchange between Nash and Harvard professor George Mackey is described in the prologue of the book:

> "How could you, how could *you*, a mathematician, a man devoted to reason and logical proof... how could you believe that extraterrestrials are sending you messages? How could you believe that you are being recruited by aliens from outer space to save the world? How could you...?"
>
> And Nash answered: "Because the ideas I had about supernatural beings came to me the same way that my mathematical ideas did. So I took them seriously." (Nasar, 1998)

The entire time I was reading *A Beautiful Mind*, I kept wondering when we were going to get to the part where the mind became beautiful. Everything in the book describes something repellent and cold and barely human. John Nash's brain was brilliant, oh, indeed! But at the end of the book there had not been a single thing of "beauty" about either his thinking or his life. In fact, as I read, my skin crawled with the realization that I was reading about a person whose life story was identical to that of Frank Scott.

John Forbes Nash was big and brainy, handsome and arrogant. He had virtually no social graces or redeeming qualities despite the fact that he was carefully brought up in an environment that one would have thought would have inculcated some human values. He was, indeed, a star of the mathematical scene that promoted human rationality as the supreme virtue, and for ten years he was viewed as a kind of wunderkind who was going to push the mathematical boundaries of Games of Strategy, economic rivalry, computer architecture, the shape of the inverse and geometric space, number theory and more. Some commentators suggested that Nash had that "extra-human spark." But reading his story, one comes to the idea that he had very little human about him at all. It wasn't a beautiful mind, it was a deadly efficient machine; unnatural and mysterious.

Then, curiously, at the age of thirty or thereabouts, he suddenly manifested "paranoid schizophrenia," psychotic delusions, and was in and out of mental hospitals for a period. After his wife divorced him and his mother died, and his sister could no longer cope with his psychosis, he became a "phantom," haunting the halls and corridors of Princeton for twenty years as the resident *idiot savant*. If any of the readers have watched the very funny movie *Sheer Genius*, they will remember the strange character of Laszlo, the "burned-out genius" who used a closet in a dorm room as an entry to a vast underground laboratory, a secret world hidden from the eyes of the university authorities. It's rather a somewhat sympathetic and idealized portrait of Nash during his psychotic years at Princeton.

In the 1990s, Nash's "illness" more or less went into "remission." The question has been raised: Did he really suffer from schizophrenia? Psychotic symptoms do not necessarily, as psychiatrists now agree, make a schizophrenic. And, absence of overt evidence of psychosis does not mean a person is cured of whatever afflicted them. They can most certainly still be suffering, but having learned to cope with it, able to conceal it. Nash himself described his long illness as a persistent dreamlike state with bizarre beliefs not unlike those of other people diagnosed with schizophrenia. Mostly, however, he noted that his illness consisted in being unable to reason. Despite claims of recovery, Nasar quotes him as telling several people that he is still having "paranoid thoughts" and still hears "voices," though the noise level is greatly modulated. He has compared his "recovery" to simply learning how to police his thoughts, to recognize paranoid ideas and to reject them.

> "Gradually I began to intellectually reject some of the delusionally influenced lines of thinking [...] the rejection of politically-oriented thinking as essentially a hopeless waste of intellectual effort. [...] A key step was a resolution not to concern myself in politics relative to my secret world because it was ineffectual. [...] This in turn led me to renounce anything relative to religious issues, or teaching or intending to teach." (Nasar, 1998)

Nash's son has also been diagnosed as a paranoid schizophrenic. His illness became apparent when he "disappeared" one day. When he reappeared, he had shaved his head and become a born-again Christian. He began to read the Bible obsessively and had fallen under the influence of a fundamentalist cult called "The Way." Not too long after, it was clear that he was hearing voices and believed that he was a great religious figure who had to save the world. Reportedly, he occasionally talked about extraterrestrials, and once threatened a history professor. But, somehow, in spite of his illness, he managed to get a Ph.D.

Despite his lack of a high school or college diploma, he was admitted to Rutgers on a full scholarship. That fact raises questions of its own.

Let's leave Nash for the moment, and come back to Armand Wyler. There is, in physics and mathematics, something called the "Fine Structure Constant." The Fine Structure Constant has a value very near 137, and many physicists think that this indicates fundamental characteristic of space, time and matter. Armand Wyler came along and suggested that it is *a geometrical property of a suitably defined seven-dimensional space-time*, and that the correct theoretical value is 1/137.03608. He then related this to proton-electron mass ratios.

Although the numerical values Wyler derived were close to experimental data, the physical reasons he gave for using the particular volumes he chose were not clear. Freeman Dyson invited Wyler to the *Institute for Advanced Study* in Princeton for a year to see if Wyler could explain what he was thinking in clearer terms. The general story told about what happened at Princeton was that, because Wyler was primarily a mathematician, he was unable to rise to the task. And, naturally, since he couldn't explain what he was doing and why, his results were dismissed as "numerology."

12-26-98

Q: (A) It came to my mind that perhaps Einstein, when you spoke about variable physicality, that Einstein was afraid when he understood that in his work. I thought about this and I think that Einstein determined that the future must be determined from the past and present, and when he found that he had a theory where the future was open, he dismissed it and was afraid. Is this a good guess that variable physicality, mathematically, means a theory where there is a freedom of choosing the future when past and present are given?

A: Yes.

Q: (A) Is it related to the fact that we should use higher-order differential equations, not second-order?

A: Yes. Einstein found that not only is the future open, but also the present and the past. Talk about scary!!

Q: (A) All you have said so far points to an idea by a Swiss guy named "Armand Wyler." This Wyler found a way to compute from geometry so-called Fine Structure Constant, which is a number and can be found experimentally. Then, of course, he was invited to Princeton to explain how he did it, and apparently he failed to explain himself, and *he ended in an asylum for the mentally deranged*. The question is: If I follow his way of thinking, can I succeed in deriving and understanding the nature of this Fine Structure Constant?

A: Yes.

Q: (A) Well, if I do it, should I keep it a secret so that I won't end up in an asylum?

A: The problem with Wyler was with the audience, not the speaker.

Q: (A) What does that mean? (L) I guess it means that the people he was talking to couldn't grasp it, not that he couldn't explain it. Did he really lose his mind, or was he sort of 'helped' to go crazy?

A: He suffered a "breakdown."

The fact that Wyler was locked away in an institution for the insane is not widely known. We only learned it directly from a fellow Swiss physicist who was in Geneva at the time and had direct knowledge of the event. He told us over lunch one day that Wyler had "lost it" *while at Princeton*, and was sent home and institutionalized.

The question that occurred to me at this point was: What was Nash working on when he went bonkers? In Nash's biography, we discover an interesting passage:

> He apparently had devoted what little time he spent at the Institute for Advanced Study that year talking with physicists and mathematicians about quantum theory. Whose brains he was picking is not clear; Freeman Dyson, Hans Lewy, and Abraham Pais were in residence at least one of the terms. [...] Nash made his own agenda quite clear. "To me one of the best things about the Heisenberg paper is its restriction to the observable quantities," he wrote, adding that "I want to find a different and more satisfying *under-picture of a non-observable reality.*"
>
> It was this attempt that Nash would blame, decades later in a lecture to psychiatrists, for triggering his mental illness — calling his attempt to resolve the contradictions in quantum theory, on which he embarked in the summer of 1957, "possibly overreaching and psychologically destabilizing." [This author's emphasis]

One might also conjecture that such a program would attract certain attention. We also notice the presence of Freeman Dyson mentioned in reference to both Wyler and Nash, and both men went mad upon probing too deeply into hyperdimensional physics.

After we had discussed Wyler, we moved on to Everett.

Everett's name may be familiar because of what is called the "Everett-Wheeler" interpretation of quantum mechanics, a rival of the orthodox "Copenhagen" interpretation of the mathematics of quantum mechanics. The Everett-Wheeler theory is also known as the "many worlds" interpretation.

Hugh Everett did his undergraduate study in chemical engineering at the Catholic University of America. Studying von Neumann's and Bohm's textbooks as part of his graduate studies under Wheeler, in mathematical physics at Princeton University in the 1950s (at the same time Nash was there), he became dissatisfied with the collapse of the wave function. *While he was at Princeton*, during discussions with

Charles Misner and Aage Peterson (Bohr's assistant, then visiting Princeton), he developed his "relative state" formulation. Wheeler encouraged his work and preprints were circulated in January 1956 to a number of physicists. A condensed version of his thesis was published as a paper for "The Role of Gravity in Physics" conference held at the University of North Carolina, Chapel Hill, in January 1957.

Not long afterward, Everett flew to Copenhagen to meet with Niels Bohr and discuss his ideas, *but Bohr gave him the bum's rush and brush off*, and this was the general response he received from physicists in general. Everett left physics after completing his Ph.D., going to work as a defense analyst at the Weapons Systems Evaluation Group at the Pentagon, and later becoming a private contractor. He was very successful, becoming a multimillionaire. In 1968 Everett worked for the Lambda Corporation, now a subsidiary of General Research Corporation in McLean, Virginia. His published papers during this period cover things like optimizing resource allocation and *maximizing kill rates during nuclear-weapon campaigns.*

With the steady growth of interest in the "many worlds" theory in the late 1970s, Everett began to make plans to return to academia in order to do more work on measurement in quantum theory. In the late 1970s, he visited Austin, Texas, at Wheeler's or DeWitt's invitation, to give some lectures on quantum mechanics. Not long afterward, he died of a heart attack in 1982. He was only 52 years old.

I was curious about Everett's work for Lambda. A recent search of the literature turns up a paper written by Joseph George Caldwell entitled "Optimal Attack and Defense for a Number of Targets in the Case of Imperfect Interceptors."[26]

This article is an extraction, with some amplification and minor notational changes, of portions of the report: Caldwell, J. G., T. S. Schreiber, and S. S. Dick, "Some Problems in Ballistic Missile Defense Involving Radar Attacks and Imperfect Interceptors," Report ACDA/ST-145 SR-4, Special Report 4, prepared for the US Arms Control and Disarmament Agency by the Lambda Corporation (subsidiary of General Research Corporation), McLean, Virginia, May 1969. [...]

These results were also published as Appendix F of the report, Caldwell, J. G., "Theater Tactical Air Warfare Methodologies: Automated Scenario Generation," Final Report Contract No. F33657-88-C-2156 prepared for the USAF Air Force Systems Command, Aeronautical System Division by Vista Research Corporation, July 1989.

---

[26] Note: Responding to the original online publication of this material, J. G. Caldwell updated his site, adding ample and interesting historical information re: Lambda Corporation, Hugh Everett, and several other subjects relevant to our discussion here (http://www.foundationwebsite.org/ HistoricalNote1.htm).

The summary of the paper tells us:

> This article describes the optimal defense of a set of targets against an optimal attack, in the case of imperfect interceptors. This solution is obtained from a mathematical representation of strategic nuclear warfare as a two-sided resource-constrained optimization problem. The attacker and defender are assumed to know each other's total force sizes, and it is assumed that the attacker "moves last," i.e., the attacker allocates his weapons to targets after observing the defender's allocation of interceptors to targets. A one-on-one (or fixed salvo-to-one) firing doctrine is assumed. The attacker's goal is to maximize the total damage to the targets, and the defender's goal is to minimize this damage. It is assumed that damage on any single target can be adequately described by the "exponential" damage function defined below.

Aside from the fact that we see evidence of the use of pure mathematics — Game Theory, in fact — in matters of warfare strategy, which includes source notes connecting this work to Wheeler, we find Joseph George Caldwell to be a bit interesting for other reasons. He has a website where he promotes the following idea:

> What is the sustainable human population for Earth? I propose that a long-term sustainable number is on the order of ten-million, consisting of a technologically advanced population of a single nation of about five-million people concentrated in one or a few centers, and a globally distributed primitive population of about five million. I arrived at this size by approaching the problem from the point of view of estimating the *minimum* number of human beings that would have a good chance of long-term survival, instead of approaching it from the (usual) point of view of attempting to estimate the *maximum* number of human beings that the planet might be able to support. The reason why I use the approach of minimizing the human population is to keep the damaging effects of human industrial activity on the biosphere to a minimum. Because mankind's industrial activity produces so much waste that cannot be metabolized by "nature," any attempt to maximize the size of the human population risks total destruction of the biosphere (*such as the "sixth extinction" now in progress*). [This author's emphasis]

Let's stop right here and ask the question: Is this "sixth extinction" something that is generally "known" in the circles that do this kind of research? Is this *why* they are doing it? What do they know that the rest of us don't? Or better, what do they think that they aren't telling us? Caldwell writes:

> The role of the technological population is "planetary management": *to ensure that the size of the primitive population does not expand.* The role of the primitive population is to reduce the likelihood that a localized catastrophe might wipe out the human population altogether. The reason

for choosing the number five-million for the primitive population size is that *this is approximately the number* (an estimated 2-20 million) *that Earth supported for millions of years*, i.e., it is proved to be a long-term sustainable number (in mathematical terminology, a "feasible" solution to the optimization problem). The reason for choosing the number five-million for the technological population size is that it is my opinion that that is about the minimum practical size for a technologically advanced population capable of managing a planet the size of Earth; also, it is my opinion that the "solar energy budget" of the planet can support a population of five-million primitive people and five-million "industrial" people indefinitely. (www.foundationwebsite.org; this author's emphases)

Mr. Caldwell's ideas are a techno-representation of synarchy, a clue to the *real* "Stargate Conspiracy." It seems that there is, indeed, something very mysterious going on all over the planet in terms of shaping the thinking of humanity via books, movies and cultural themes, but at this point we understand that most of what is promulgated is lies and disinformation. We hope to come to some idea of what the "insiders" know that they aren't telling us, and perhaps we will find some clues as we continue our investigation here.

Princeton is often referred to as the "mathematical center of the universe." But it wasn't always that way. Until the Rockefeller family endowed scientific research *in the mid-1920s* (keep in mind that the Cassiopaeans referenced the mid-1920s as the beginning of certain "projects" related to HAARP, that this was connected to "Alternative 3" and the "Ark of the Covenant"), a student could learn little more than what amounts to high-school math and science there. Here I am not going to go into any background on the Rockefellers because there are enough researchers already who are doing that. I urge the reader to do his or her own research in those areas. Again, whether or not there is a conscious conspiracy I cannot say.

Nevertheless, when thinking about conspiracies, it is extremely difficult to conceive of these activities in strictly human terms. Yes, humans carry them out, but the real question is "Why?" What drives them?

Practically next door to Princeton is the *Institute for Advanced Study*. This ivory tower of academia was the result of a "synchronous" act of philanthropy. The Institute was founded in 1930 with a gift from Mr. Louis Bamberger and his sister, Mrs. Felix Fuld, under the guidance of the famous educator Abraham Flexner, who originated the concept from which the Institute took its form. The Bambergers originally thought of endowing a dental school, which is probably why they consulted Flexner when they decided that they wanted to give away some of their fortune.

Flexner was a high-school principal who wrote a report on American Colleges. Based on this report and the recommendation of his brother (a pathologist with connections to existing medical schools), he was chosen by the Carnegie Foundation to do a study of American medical education *just after the turn of the century*. Flexner visited medical schools across the nation on a schedule that barely allowed him a whole day each for the evaluation of some schools. His efforts were closely linked with the American Medical Association, which provided resources. Although purporting to be objective, the report actually established guidelines that were designed to sanction orthodox medical schools and condemn homeopathic ones and alternative therapies. In short, it was biased toward allopathy and the A.M.A.

Is this evidence of a conspiracy to gain control of medical education, doctors, and therefore the entire population by means of *promotion of certain drugs designed to control human beings, or to modify their behavior?* I can't say. All I notice is that there seem to be so many "coincidental" threads that weave together, bringing the inhabitants of our planet to the present state, that it is truly difficult to not see some "grand design" behind it. But can it be a human design? It doesn't seem so. When you track such conspiracies back in time and space, you discover that there is so much historical evidence that the groundwork for the present conditions was laid long, long ago — back in the mists of prehistory even — that conceiving of it as just human activity, or the results of human nature, is extraordinarily difficult.

In any event, Flexner's report had a tremendous impact on American medicine, and it was to him that the Bambergers turned for advice regarding the founding of their proposed dental school. Who knows, maybe they suffered from frequent toothaches?

Flexner had a better idea: Why not found a research institution with no teachers, no students, no classes — only researchers, shielded from the cares of the real world so that they could just hang out and produce great thoughts?

Believe me, I think it was a great idea! The only problem is: Who's on first? Just who might be "influencing" those thoughts? Just what thoughts might be being encouraged and rewarded? This is not a rhetorical question, as we will see.

One of the more curious things about Flexner's idea was that he initially thought that *a school of economics ought to be at the core of the Institute*, but decided on mathematics since it was more "fundamental."

Once the Institute was set up, Flexner set about finding talent. This "just happened" to coincide with Hitler's takeover of the German government and the mass expulsion of Jews from German universities, which had until then been the seats of higher learning in mathematics and science. Negotiations were begun to get Einstein, who finally agreed to become the second member of the Institute's School of Mathematics. Kurt Godel came, followed by Hermann Weyl, who wanted Von Neumann. In short, overnight Princeton became the new Gottingen.

08-17-00

Q: (A) What I do not understand is why a few years later [Einstein] completely abandoned [UFT in terms of Kaluza-Klein theories] and started working very hard on a completely different solution. If he knew....

A: Was under control.

Q: (A) Can you control somebody and make him spend years.... Oh! Mind control! They got him!

A: Why do you think he emigrated to the United States in the first place?

Q: (A) Well, that is not a surprise. He was a Nobel Prize winner and America was getting together every possible Nobel Prize winner, and also there was the persecution of the Jews, so it was natural.

A: More to it than that. What about Freud?

Q: (L) I guess they didn't want Freud! He didn't know anything about UFT! (A) Now, apparently Von Neumann was also involved in application of UFT. But Von Neumann was, as far as we know, doing a completely different kind of mathematics. He didn't even really know geometry, differential geometry. He was doing completely different things [Game Theory, to be precise]. So how come the UFT that was discovered by Einstein involved Von Neumann? What did Von Neumann contribute to this project?

A: Von Neumann was one of three overseers at Princeton, with level-7 security clearance and a clear budget-request permittance.

Q: (L) My question is about Von Neumann; as I understand it, Von Neumann was supposed to have been involved with the creation of a time machine, right?

A: Yes.

Q: (L) Did he succeed in such a project?

A: Yes.

Q: (L) Well, why was it that, when he developed a brain tumor and realized he was going to die — and I read that he screamed and yelled like a baby when he knew there was no hope — if he was somebody who had access to a time machine, why wasn't he able to do something about it instead of carrying on like a madman? The stories about his screams echoing all over the place are horrible. He realized that his great mind was going to soon be still; if he had access to a time machine, one would think that he would have used it, would have pulled every string he could, to forestall his own death.

A: No Laura, it does not work that way. And besides, if you had a brain tumor, you could be forgiven for a few mental peculiarities too!

Q: (L) I just don't understand why, when he knew he was sick, that he didn't just use the time machine to go forward in time for a cure, or backward in time to correct something in his past....

A: The time machine was not his property.

Q: (L) So they got what they needed from him and let him die. (A) It is not clear. *He got this cancer so suddenly, it may even have been induced.* (L) Well, that's a thought.

What we are basically seeing in the above recitation of the endowment of Princeton as a mathematical center, and the luring of scientific talent to America, is *part of the truth of one of the ideas of "Alternative 3."* It was effectively a "brain drain" on Europe. All the geniuses who were capable of certain specific things were being brought to America and settled in Princeton. This produced an almost immediate scientific earthquake. Of course we see that the scientific revolution in America did not begin *after* WWII, as many conspiracy theorists would like to believe, but rather before — in the 1920s, to be exact.

Something happened to stimulate this activity. In the 1920s, during which time the Rockefellers and other monied groups made enormous contributions to "education in America," and during which the preparations for the Hitler drama were underway, another event that may or may not be connected took place. In Mark Hedsel's book *The Zelator*, we discover the "intimations" of this event. In Chapter Five of the referenced work, Hedsel informs us that:

> At the end of the last century *an astounding revelation was,* as a result of dissent among members of Secret Schools. Information, hitherto guarded jealously by the most enclosed of the Inner Orders, was made public. The secrets disclosed pertained to a far deeper level of knowledge than has hitherto been made exoteric by the Schools — even in this enlightened age.
>
> Our purpose here is not to document how so deep an esoteric idea was made public — or even to assess whether it was wise for this idea to be brought out into the open. All this has been dealt with in the literature.... (Hedsel and Ovason, 2000; this author's emphasis)

The speaker in Hedsel's book, his teacher, promises to provide titles via which this most curious item might be researched, but dies before doing so. He later mentions A.P. Sinnett, and Hedsel himself speculates that it is the Theosophical ideas of A.P. Sinnett and Helena Blavatsky. I don't agree, because of what the teacher says next:

> In a nutshell, *what was made public during this conflict* in the Schools was the truth that our Moon is a sort of counterweight to *another sphere, which remains invisible to ordinary vision.* This counterweighted sphere is called in esoteric circles the Eighth Sphere.
>
> The truth is that this Eighth Sphere does not pertain to anything we are familiar with on the physical plane, yet we must use words from our own vocabularies whenever we wish to denote its existence. Were we to use a word which fits most appropriately this Sphere, then we should really call it a vacuum. Certainly, Vacuum is a more appropriate term than sphere, for *the Eighth Sphere sucks things into its own shadowy existence.* [This author's emphases]

The reference to the Moon brings us back in a loop to the so-called Philadelphia Experiment which was promoted on the platform of the death of Morris K. Jessup. What the Philadelphia Experiment distracted attention away from was Jessup's ideas about the gravitational neutral zone of a three-body system which we have also identified as being connected to the Chandler wobble-cycle of 18.6 years, which then connects to the "dancing of the god" at Stonehenge every 19 years, which is called the Metonic Cycle. Of course, we are beginning to realize that all of this connects to the ideas of the Holy Grail, the Ark of the Covenant, and certain periods — cyclical in nature — in which *transfer of perimeters* is most easily accomplished! Remember what Jessup wrote:

> While I believe that these space islands probably use both earth and moon for their own convenience, I suggest that their most natural and permanent habitat is at the *gravitational neutral* of the earth-sun-moon three-body system which is well within the orbit of the moon.
>
> It has been postulated that gravitation need not be considered as acting with uniform continuity, from the center of the attracting body outward, even if subject to the inverse square law. Such a concept, today, would be especially horrendous to physics and astronomy. [...]
>
> Refinements of Bode's law indicate *nodes in the gravitational field,* at which planets, asteroids, and possibly comets and meteors tend to locate themselves. An extension of the theory to the satellite systems of the major planets indicates a similar system of nodes on smaller scales, where planets, rather than the Sun, are gravitational centers. [...] It might well be that these *gravitational nodes* are occupied to some degree by navigable construction. [...]
>
> Many researchers have extended the law so as to establish nodes *right down to the surface of the central bodies,* and in so doing the nodes become closer and closer together so that there may be many of them at short distances from the parent body. Thus, if the law or its derivatives

have significance, there could be a number of these orbital nodes between the moon and the surface of the earth. [...]

There may even be hints available to us regarding gravity. For instance, no final settlement has ever been made of the argument over the opposed wave and corpuscular theories of the propagation of light. *An assumption that the ether, a necessary adjunct to the wave theory, is identical with the gravitational field, whatever that may be, would reconcile the opposing theories and a quantum of light would then be merely a pulsation or fluctuation in the gravitational field.* Intense studies of the movements of space-navigable UFOs might furnish vital clues to such problems. (Jessup, 1955; this author's emphases)

Let us stop right here and consider the above remarks. Jessup has suggested that a "quantum of light would then be merely a pulsation or fluctuation in the gravitational field." This connects us directly to several remarks made by the Cassiopaeans, curiously in the context of questions about Sufism:

06-15-96

A: Now, learn, read, research all you can about unstable gravity waves. [...] We mean for you, Laura, to meditate about unstable gravity waves as part of research. [...] *Unstable gravity waves unlock as yet unknown secrets of quantum physics to make the picture crystal clear.* [...]

Q: (L) Gravity seems to be a property of matter. Is that correct? [...]

A: And antimatter!

Q: (L) Is the gravity that is a property of antimatter "antigravity"? Or is it just gravity on the other side, so to speak?

A: Binder. [...] Gravity binds all that is physical with all that is ethereal through unstable gravity waves!!! [...]

Q: (L) Is light the emanation of gravity?

A: No.

Q: (L) What is light?

A: Gravity. [...] If gravity is everything, what isn't it? Light is energy expression generated by gravity. [...]

Q: (L) According to what I understand, at the speed of light, there is no mass, no time, and no gravity. How can this be?

A: No mass, no time, but yes, gravity.

Q: (L) A photon has gravity?

A: Gravity supersedes light speed.

Returning now to Jessup's comments:

There is increasingly strong evidence that gravity is neither so continuous, so immaterial, nor so obscure as to be completely unnamable to use, manipulation and control. Witness not only the documented movements of UFOs in the form of lights, discs, nebulosities, etc., but the many instances of stones, paper, clothes baskets and many other things which have been seen to leave the ground without apparent cause. The lifting of the ancient megalithic structures, too, must surely have come through levitation. […]

It is my belief that something of the sort was done in the antediluvian past, through either research or through some fortuitous discovery of physical forces and laws which have not as yet been revealed to scientists of this second wave of civilization…

It is my belief that the possibility of gravity control, or at least gravity reactance, has been strongly indicated by the phenomena listed in this book.

There seems to be something of *periodicity in events of celestial and spatial origin.* This has been called to our attention by John Philip Bessor in the *Saturday Evening Post* as early as May 1949; but no one has thus far been able to catalogue and classify enough of this data to determine for certain whether such cycles exist, much less their time period or cause. It is not particularly astonishing that these phenomena should be cyclic, for practically everything astronomical is periodic. *If periodicity could be firmly established for these phenomena, that fact alone would be proof of their reality and integration with the organic world about us.* […]

It is no longer necessary to explain [aliens] as visitors from Mars, Venus, or Alpha Centauri. They are a part of our own immediate family — *a part of the earth-moon, binary-planet system.* They didn't have to come all of those millions of miles from anywhere. They have been here for thousands of years. Whether we belong to them by possession, like cattle, or whether we belong to each other by common origin and association is an interesting problem, and one which may soon be settled if we keep our heads.

In final summary, the UFOs have been around us for a long time and probably are a connecting link with the first wave of terrestrial civilization. (Jessup, 1955; this author's emphases)

Now, the question is: Who was it who revealed this "great secret of secrets" at the end of the last century, as Hedsel's teacher has remarked? We have been directed to think about Sinnett and Blavatsky, but we are already wise to such tactics. It is important to know *who* the "revealer" was in order to know what was revealed. This will then lead us to understand *what was done* with that knowledge and how it affects us today.

We have also been directed to look at a certain time period that would correspond to the activities of such sources, and so we might wish to look at the vast array of literature for internal clues. In reviewing all of this literature, in casting our net far and wide, there is only one source that "fits" the description: Gurdjieff's metaphor of "Food for the Moon." There is, in fact, a singular remark made by Gurdjieff in conversation with P.D. Ouspensky, recorded by the latter in his book, *In Search of the Miraculous*, which confirms that the information revealed by Gurdjieff was, in fact, related to the cyclic catastrophes and their relations to hyperdimensional realities. It also confirms that he was the one who really "knew" something:

> "The intelligence of the Sun is divine," said Gurdjieff. "But the earth can become the same; only, of course, it is not guaranteed and the earth may die having attained nothing."
>
> Gurdjieff's answer was very vague. "There is a definite period," he said, "for a certain thing to be done. If, by a certain time, what ought to be done has not been done, the earth may perish without having attained what it could have attained."
>
> "Is this period known?" I asked.
>
> "It is known," said Gurdjieff. (Ouspensky, 1949)

As we note from the remarks of Hedsel's teacher, the idea that there is "ancient wisdom" guarded by hidden custodians or Masters is often attributed to Helena Blavatsky. In fact, her disciple Annie Besant wrote a book about it. Thirty years later *The Mahatma Letters* — supposedly written by two of these "Masters" to A.P. Sinnett — was published. A careful investigation of the matters surrounding Blavatsky and Sinnett will bring the thoughtful person to the conclusion that they were either taken in by "wishful thinking," or were themselves part of some sort of disinformation campaign. Unfortunately, most so-called "occultists" have either knowingly or unknowingly relied upon the Blavatsky/Sinnett interpretations as foundations for their own ideas.

Joscelyn Godwin, in his books *Arktos* and *The Hermetic Brotherhood of Luxor*, attempted to trace the threads of these ideas to the original sources. Geoffrey Ashe, in his book *The Ancient Wisdom*, made a similar attempt with much better results. Ernest Scott, in *The People of the Secret*, states the problem as a legend that the ebb and flow of history are subject to purposive direction from a higher level of understanding, the process being manipulated by a hierarchy of intelligences, the lowest level of which makes physical contact with humanity.

We come now to Gurdjieff's comments about "planetary evolution," "secret schools," esoteric and exoteric teachings, in relation to our present subjects of "Alternative 3" and the "Secret Games of the Gods." In these remarks we will find, I believe, the Terrible Secret, the *astounding revelation*, the information hitherto so jealously guarded by the most enclosed of the Inner Orders, the secrets pertaining "to a far deeper level of knowledge than has hitherto been made exoteric by the Schools, even in this enlightened age." Read it slowly; ponder it; reread it; and consider the implications of how the denizens of the Eighth

Sphere might react to this revelation, how their "agents" among humanity might be manipulated to deal with it in order to conceal it, to cover it up, to distort it, to nullify it, and most of all, to make it come to naught.

Gurdjieff noted that standard scientific teachings tell us that life is "accidental." Such ideas fail to take into account the idea that there is nothing accidental or unnecessary in nature, that everything has a definite function and serves a definite purpose of Cosmic Consciousness. Gurdjieff then says:

"It has been said before that organic life transmits planetary influences of various kinds to the earth and that it serves to feed the moon and to enable it to grow and strengthen. But the earth also is growing; not in the sense of size but in the sense of greater consciousness, greater receptivity. The planetary influences which were sufficient for her at one period of her existence become insufficient, she needs the reception of finer influences. To receive finer influences a finer, more sensitive receptive apparatus is necessary. Organic life, therefore, has to evolve, to adapt itself to the needs of the planets and the earth. Likewise also the moon can be satisfied at one period with the food which is given to her by organic life of a certain quality, but afterwards the time comes when she ceases to be satisfied with this food, cannot grow on it, and begins to get hungry. [...] This means that in order to answer its purpose organic life must evolve and stand on the level of the needs of the planets, the earth, and the moon.

"We must remember that the ray of creation, as we have taken it, from the Absolute to the moon, is like a branch of a tree — a growing branch. The end of this branch, the end out of which come new shoots, is the moon. If the moon does not grow, if it neither gives nor promises to give new shoots, it means that either the growth of the whole ray of creation will stop or that it must find another path for its growth, live out some kind of lateral branch. [...]

"If organic life on earth disappears or dies, the whole branch will immediately wither. The same thing must happen, only more slowly, if organic life is arrested in its development, in its evolution, and fails to respond to the demands made upon it. The branch may wither. [...]

"General growth is possible only on the condition that the 'end of the branch' grows. Or, speaking more precisely, there are in organic life tissues which are evolving, and there are tissues which serve as food and medium for those which are evolving. Then there are evolving cells within the evolving tissues, and cells which serve as food and medium for those which are evolving. In each separate evolving cell there are evolving parts and there are parts which serve as food for those which are evolving. But always and in everything it must be remembered that evolution is never guaranteed, it is possible only and it can stop at any moment and in any place.

"*The evolving part of organic life is humanity.* Humanity also has its evolving part. [...] If humanity does not evolve it means that the evolution of organic life will stop and this in its turn will cause the growth of the ray of creation to stop. At the same time if humanity ceased to evolve, it becomes useless from the point of view of the aims for which it was created, and as such *it may be destroyed.* In this way *the cessation of evolution may mean the destruction of humanity.*[...]

"[E]xamining the life of humanity as we know it historically, we are bound to acknowledge that *humanity is moving in a circle.* In one century it destroys everything it creates in another and the progress in mechanical things of the past hundred years has proceeded at the cost of losing many other things which perhaps were much more important for it. Speaking in general there is every reason to think and to assert that *humanity is at a standstill and from a standstill there is a straight path to downfall and degeneration.* [...]

"[W]e see that a balanced process proceeding in a certain way cannot be changed at any moment it is desired. It can be changed and set on a new path *only a certain 'crossroads.'* In between the 'crossroads' nothing can be done. At the same time, if a process passes by a 'crossroad' and nothing happens, nothing is done, then nothing can be done afterwards and the process will continue and develop according to mechanical laws; and even if people taking part in this process foresee the inevitable destruction of everything, they will be unable to do anything. I repeat that *something can be done only at certain moments which I have just called 'crossroads.'*[...]

"Of course there are very many people who consider that the life of humanity is not proceeding in the way in which according to their views it ought to go. And they invent various theories which in their opinion ought to change the whole life of humanity. [...] All these theories are certainly quite fantastic, chiefly because they do not take into account the most important thing, namely, *the subordinate part which humanity and organic life play in the world process.*

"Intellectual theories put man in the center of everything; everything exists for him. [...] And all the time new theories appear evoking in their turn opposing theories; and all these theories and the struggle between them undoubtedly *constitute one of the forces which keep humanity in the state in which it is at present.* [...]

"Everything in nature has its aim and its purpose, both the inequality of man and his suffering. To destroy inequality would mean destroying the possibility of evolution. To destroy suffering would mean, first, destroying a whole series of perceptions for which man exists... and thus it is with all intellectual theories.

"The process of evolution... which is possible for humanity as a whole, is completely analogous to the process of evolution possible for the individual man. And it begins with the same thing, namely, *a certain group of cells gradually becomes conscious*; then it attracts to itself other cells, subordinates others, and gradually makes the whole organism serve its aims and not merely eat, drink, and sleep. This is evolution and there can be no other kind of evolution. In humanity as in individual man everything begins with the formation of a conscious nucleus. *All the mechanical forces of life fight against the formation of this conscious nucleus in humanity,* in just the same way as all mechanical habits, tastes and weaknesses fight against conscious self-remembering in man."

"Can it be that there is a conscious force which fights against the evolution of humanity?" [Ouspensky] asked.

"From a certain point of view it can be said," said G[urdjieff].

"There are two processes which are sometimes called 'involutionary' and 'evolutionary.' The difference between them is the following: An involutionary process begins consciously in the Absolute but at the next step it already becomes mechanical — and it becomes more and more mechanical as it develops; an evolutionary process begins half-consciously and conscious opposition to the evolutionary process can also appear at certain moments in the involutionary process. From where does this consciousness come? From the evolutionary process of course.

"The evolutionary process must proceed without interruption. Any stop causes a separation from the fundamental process. Such separate fragments of consciousnesses which have been stopped in their development can also unite and at any rate for a certain time can live by struggling against the evolutionary process. After all, it makes the evolutionary process more interesting.

"Instead of struggling against the mechanical forces there may, at certain moments, be *a struggle against the intentional opposition of fairly powerful forces* though they are not of course comparable with those which direct the evolutionary process. *These opposing forces may sometimes even conquer.* The reason for this consists in the fact that the forces guiding evolution have a more limited choice of means; in other words, they can only make use of certain means and certain methods. The opposing forces are not limited in their choice of means and they are able to make use of every means, even those which only give rise to a temporary success, and in the final result *they destroy both evolution and involution at the point in question.* [...]

"Are we able to say for instance that life is governed by a group of conscious people? Where are they? Who are they? We see exactly the opposite: that *life is governed by those who are the least conscious, by those who are most asleep.*

"Are we able to say that we observe in life a preponderance of the best, the strongest, and the most courageous elements? Nothing of the sort. On the contrary we see a preponderance of vulgarity and stupidity of all kinds.

"Are we able to say that aspirations towards unity, towards unification, can be observed in life? Nothing of the kind of course. We only see new divisions, new hostility, new misunderstandings.

"So that in the actual situation of humanity *there is nothing that points to evolution proceeding.* On the contrary when we compare humanity with a man, we quite clearly see a growth of personality at the cost of essence, that is, a growth of the artificial, the unreal, and what is foreign, at the cost of the natural, the real, and what is one's own.

"Together with this, we see a growth of automatism.

"Contemporary cultures require automatons. And people are undoubtedly losing their acquired habits of independence and turning into automatons, into parts of machines. It is impossible to say where is the end of all this and where the way out — or whether there is an end and a way out. One thing alone is certain, that man's slavery grows and increases. Man is becoming a willing slave. He no longer needs chains. He begins to grow fond of his slavery, to be proud of it. And this is the most terrible thing that can happen to a man.

"[A]s I pointed out before, the evolution of humanity can proceed only through the evolution of a certain group, which, in its turn, will influence and lead the rest of humanity.

"Are we able to say that such a group exists? Perhaps we can on the basis of certain signs, but in any event we have to acknowledge that it is a very small group, quite insufficient, at any rate, to subjugate the rest of humanity. Or looking at it from another point of view, we can say that humanity is in such a state that it is unable to accept the guidance of a conscious group."

"How many people could there be in this conscious group?" someone asked.

"Only they themselves know this," said G[urdjieff].

"Does it mean that they all know each other?" asked the same person again.

"How could it be otherwise?" asked G. "Imagine that there are two or three people who are awake in the midst of a multitude of sleeping people. They will certainly know each other. But those who are asleep cannot know them. How many are they? We do not know and we cannot know until we become like them. It has been clearly said before that *each man can only see on the level of his own being.* But two hundred conscious people, if they existed and if they found it necessary and legitimate, could change the whole of life on the earth. But either there are not enough of them, or they do not want to, or perhaps the time has not yet come, or perhaps other people are sleeping too soundly." (Ouspensky, 1949; this author's emphases)

Now, interestingly, this idea has repeatedly surfaced in UFO research and lore: the idea of mankind being "lunch" for the aliens. It is not quite that simple, but the Cassiopaean source indicates that there is an energy that is released when the soul separates from the body and this is ostensibly the reason that higher-density beings are often noted at times of great disaster and during wars; they are feeding on this awareness. Now, note, they are *not* feeding on the soul, but *the energy of awareness*! And, we have to look at the ideas of cyclic catastrophes in this light.

The Book of Revelation says:

> Then I saw a single angel stationed in the Sun's light, and with a mighty voice he shouted to all the birds that fly across the sky, Come, gather yourselves together for the great supper of God, that you may feast on the flesh of rulers, the flesh of generals and captains, the flesh of powerful and mighty men, the flesh of horses and their riders, and the flesh of all humanity, both free and slave, both small and great! [...] And all the birds fed ravenously and glutted themselves with their flesh. (19:17, 18 and 21, *Amplified*, Zondervan)

We now realize, of course, that we have found the answer to the questions we posed at the beginning of this chapter:

> The question is, Why? What, in the name of all things reasonable, would prompt anyone to wish to hide these matters? What kind of sick mind would divert the attention of humanity away from what is evident all over the planet to those with open eyes, and promote so assiduously ideas that mislead, misguide, and generally placate the populace into an assurance that either nothing is going to happen, or if it does, it will be preceded by a long period of approach by a body that is well organized and clearly seen?
>
> Well, the clue is right there: "Placate the populace." Control.
>
> But, heavenly days! What kind of lunatics would want to keep everything under control in that sense if they have some idea that they, themselves, might be destroyed in the very processes they are concealing?
>
> Obviously, they don't think so. Obviously, they think they have a plan. And that suggests that, obviously, they know a lot more about what's going on, what the possibilities and probabilities are, than the rest of us.

We begin to have a glimmer of understanding of how economics and Game Theory fit into the picture. It is by means of Game Theory strategies that control of economics is obtained. Since in our reality, Money is Power, those who control the economy control the world.

We begin to understand why members of the "elite" on our planet, having been apprised of this fact, immediately went to work to discover the ways and means for their own escape. We understand why they funded Princeton and other institutions of higher learning, and why they imported all the brains on the planet, to put them to work to devise a method that could be activated at a certain point in time to *transfer perimeters*. We also begin to understand why they have made so concerted an effort to keep the masses of humanity deaf, dumb and blind: they are the sheep who will be the "Food for the Moon" while the "Masters of the Game" escape.

Chapter 58
Alien Reaction Machines

We come now to a subject of immense importance: Psychopaths as "alien reaction machines," which directly relates to what we have all been learning from the activities of Frank Scott, Vincent Bridges and others we will meet as the series progresses. As noted in the previous chapter: intolerance and cruelty are *needed* to guarantee the "cover-up." A *certain kind of "human being"* acts on behalf of this cover-up; a certain kind of human who is a playing piece in the Secret Games of the Gods.

Allow me to bring your attention back to certain remarks of Gurdjieff that point out to us the danger, both from the activities of these "automatons," as well as our own reactions to them:

> "So that in the actual situation of humanity *there is nothing that points to evolution proceeding.* On the contrary when we compare humanity with a man, we quite clearly see a growth of personality at the cost of essence, that is, a growth of the artificial, the unreal, and what is foreign, at the cost of the natural, the real, and what is one's own.
>
> "Together with this, we see a growth of automatism. *Contemporary cultures require automatons.* [...] One thing alone is certain, that man's slavery grows and increases. Man is becoming a willing slave. He no longer needs chains. He begins to grow fond of his slavery, to be proud of it. And this is the most terrible thing that can happen to a man." (Ouspensky, 1949; this author's emphases)

In the process of coming to some understanding of how such individuals operate, based on the latest research, as well as our own experiences, I hope to clarify some of the "rules" they play by so that the reader will be better equipped to spot them and deal with them. It is clear from the correspondence we have been receiving that this sort of encounter is a lot more frequent than any of us would like to think, and it is only going to get worse in the coming years, as the reader will soon see.

But before we begin to analyze the playing pieces themselves, let me give a little outline of what seems to have been the emergence of "games" in the sense that we are coming to understand them: That they are part of the process of preventing humanity from having the knowledge needed to effectively deal with our reality, and that the moves of this game have been made over millennia in cyclic time-loops very likely via time-travel.

Shamanism, as we have noted in *Secret History*, seems to be the closest we can get to the clues about hypothesized archaic technology. We may be certain that it is corrupted by millennia of changes, and it is important to remember that they are only clues; we cannot take any of the activities at face value.

Let me try to give an example: In the myths of Hermes, we find a "god" who has sandals that enable him to fly. He also has a helmet of invisibility. The question we ought to ask is: Why would ancient peoples have suggested that a god needed to put something on his head to be invisible or something on his feet to fly?

Remember, other gods didn't need these things — they were "gods," after all, and could do as they liked. So, where the heck did these objects come from? That is not to say that we ought to think that sandals were what did it, nor a helmet. The point is, the myth tells us that some sort of object conferred the ability. It was technology.

I thought about that for a long time. The objects, as they were, didn't make any sense; but what did make sense was that the concept of some sort of device was vaguely remembered and was being conveyed in these objects.

In reference to our Grail Hallows, we think of the sword/lance and the cup/platter. A sword is a death dealing instrument, right? Well, so is a gun. So is a "death ray." A gun makes a big noise, so maybe guns or similar items that made noise and flashes of light were converted by ancient peoples who had lost their understanding of technology into lightning bolts or something noisy. But a sword, in the terms of the Grail Hallows, is obviously something different. In fact, a sword, a lance, and a shepherd's crook, all somewhat resemble antennae. Just imagine Grandpa describing to the grandchild antennae that pick up signals and transmit them. "Well, Junior, it was a long narrow thing... it had a base... it was made of metal... sorta like that knife over there, only longer... well, like a sword."

What about the dish? How about a satellite dish? Something that collects energy? A cup or cauldron? The oldest representations are "wells" and the "cauldron of regeneration." How about a chamber into which the energy is directed which heals or rejuvenates or even enables one to travel in time?

How much plainer do the myths have to make it?

In the most ancient of tribal societies, postdiluvian we must assume, the shaman was an extremely important figure. He combined the function of diviner, medicine-man, and mediator between the worlds of humans and transcendental powers. What is important is the fact that the shaman held this position long before the creation of established priesthoods and colleges of "magical technicians," such as astrologers, geomants, and augurs.

Another important thing to remember was that the shaman's role was *hereditary*. There are more recent local variations, but the most ancient remnants of shamanism demonstrate that, in archaic times, it was wholly hereditary. There are examples of Siberian shamanism found throughout the world, including many examples of Native American shamanism being of the same type. The features include a separation from normal society and powers that enable the shaman to see beings and events beyond the boundaries of normal space and time.

In shamanic variations where inheritance is no longer of main importance, there is an element that gives us a small clue to the processes of corruption. In these types of shamanism, the shaman is "initiated," instead of born. In most such instances, the prospective shaman is led up a local holy mountain or goes to some desolate place. At the appropriate place and time, he is given his clothing of office, drum, and stick, and then swears an oath to the elder shaman responsible for his instruction. The elder reveals to him the secrets of his shamanic calling, and teaches him rituals, and at the end, he is sprinkled with blood from a sacrificial animal, and often dons the skin of the flayed beast.

In the oldest and purest forms, there was no initiation. Shamanism was handed from father to son or mother to daughter, with instruction being given from birth. Initiation was of almost no importance, and bloodline was supreme. The Altai shamans received their "initiations" spontaneously from higher sources, without ceremony.

We also note the connection between this hereditary shamanism and the *Vatis* or "shaman-diviner" of Britain. Here, too, there was no initiation. Bards and Druids *were* initiated, the Bard being the recipient of transmitted discipleship, and the Druid being the recipient of transmitted priesthood. The *Vatis*, however, was outside this initiatory activity. The *Vatis* was "born." He was not a traditionalist or a fundamentalist. He was a person who was open to everything new that might be worthwhile, and thereby improved and increased his art in response to the changing environment.

The shaman was also known to journey out of the body. These travels were said to be both physical as well as astral. In this way, he was able to become personally familiar with the many landscapes of the physical and ethereal worlds. While on these travels, the shaman was able to converse with the denizens of the non-material worlds and to gain knowledge and accumulate energies that were valuable to themselves and to the tribe. Because of his ability to travel "into the air," so to say, the shaman was often represented as a bird. The bird symbolized the "out of body" experience or "flight." We should note that among the most ancient depictions of the Mother Goddess are statuettes that portray her with the head of a bird.

The *direct communication with other realms* of existence is the oldest form of divination. It was later corrupted to formalized systems in an attempt to give the same "powers" to people who were not of the shamanic bloodline. Divination stands in an uneasy relationship with formalized religion as we know it. The reason for this is the fact that religions of our experience are based on the concept of Divine Providence and stability of our world. Divination cannot be tolerated by such a system because of its elements of uncertainty and the possibility of a result that contradicts the wishes of the priesthood.

As formal religion was established, divination was transferred from the bloodline seers of other realms to that of the Divine, accessible only by the approved priesthood. It was at this point that the grid, or checker board, became the symbol of the diviner. This grid is found in association with ancient representations of many gods and saints. The stag-gods of Mesopotamia, central Asia, and Europe, all represent the conversion from the hereditary shamanism to the priestly shaman as intermediary between the people and the god who, by this symbol of the grid, had dominion over space and time. The Urim and Thummim of ancient Judaism is a case in point.

Whether hereditary or priestly, divination assumes that hyperdimensional powers express themselves, or control our reality. This viewpoint was criticized as early as the first century AD, by the Roman statesman, Seneca, who wrote: "The difference between us and the Etruscans is the following: Whereas we believe lightning to be released as the result of the collision of clouds, they believe that clouds collide to release lightning; for as they attribute all to the deity, they are led to believe not that things have meaning in so far as they occur, but rather that they occur because they must have a meaning."

The Cassiopaeans have given some fascinating clues about these matters, which I will insert at this point. I expect the reader will be able to make all kinds of connections.

06-21-97

Q: Change of subject: I am tracking the clues through the various languages and alphabets. I would like to know which of these alphabets — Runic, Greek, or Etruscan — preceded the others, and from which the others are derived?

A: Etruscan.

Q: Well, who were the Etruscans?

A: Templar carriers.

Q: What does that mean?

A: Seek and ye shall find.

Q: Well, how am I supposed to do that? I can't find anything else on the Etruscans!

A: No.

Q: What do you mean "No"? You mean there is more out there on the Etruscans?

A: Yes.

Q: Okay. What are "Templar carriers"?

A: Penitent Avian Lords.

Q: What does that mean?

A: For your search. All is drawn from some more ancient form.

08-16-97

Q: We have the phoenix, cranes, herons, doves, ravens, and all are related somehow to speech or writing. Why are all these birds related this way?

A: Pass the test.

Q: What do you mean "Pass the test"?

A: Discover.

Q: Well, writing is related to the words for cutting and inscribing and even shearing and sharks. You called the Etruscans "Penitent Avian Lords," who were also "Templar Carriers." Is this related to these bird images? Then related to speech, writing, and shearing?

A: Pass the test.

Q: So, if you are writing, and you pass the test, then you can be a phoenix, dove, or whatever?

A: Discover.

Q: Ark suggested that the Etruscans may have gotten their alphabet as a mirror image. Could it be that they lived on the "other side" of the mirror?

A: Latter is closer.

We see in the above the amazing juxtaposition of "Avian" or "birdlike" with "Templar." The Templars were designated in certain of the Grail legends as "Guardians of the Grail." So again we find a connection between the Holy Grail and a thread that travels back in time to Central Asia and Stonehenge.

The flight of a bird was often seen as a shaman in an out-of-body state or as a guardian spirit in bird form. The myths of farseeing messenger birds have been preserved in many traditions, including the birds released by Noah and Odin's ravens. The imagery of birds was later transferred to arrows, and from there to sticks or staves, and then to runes.

But coming back to Seneca's negative view of hyperdimensional realities: He was critical of the idea that transcendental powers expressed themselves through bolts of lightning, storms, earthquakes, strange celestial phenomena, the flight of birds, or through human beings themselves. However, Seneca and many others were quite willing to accept the determination of meaning in life based on that which was pronounced as true by a formalized system of augury, such as a priesthood.

In the earliest shamanic practices, as we have seen, it seems that the people conceived of the gods as benevolent and communicative; interested and participatory in all aspects of their lives. The "adorable Maruts" as shamans "danced," and the heavens delivered blessings. The god danced with the people, and there was peace and plenty.

Somewhere along the way, this changed drastically and the gods became fearsome and vengeful, and potentially very dangerous. At this point, the idea that a correct relationship between human beings and the hyperdimensional beings was important took an interesting turn. In the days of the old shamans, if there were dangerous gods, the shaman was empowered to fight them, to defeat them, and to protect the people from their depredations. However, at this point, the idea that the shaman could battle dark forces was replaced with the concept of *propitiation via sacrifice*. This coincided with the creation of shamans by external initiation, which then led to formalized priesthoods. With the coming of the priesthood the only propitiation was that effected by the priest according to well-defined rules and regulations. In this respect, the correct relationship was achieved when the prescribed rituals and taboos were observed at the appropriate places and times. We see the earliest example of this idea in the "star clocks" of the ancient Egyptians, which were observed so that the proper rituals could be performed at the right hour of the night.

The priest, or "chief of the observers," was concerned with the observation and interpretation of signs in the heavens. These signs were observed from a location called a *templum*, which was an outdoor viewing mound. The sky was *divided into squares* viewed through a *lituus*, a ceremonial staff which, when held at arm's length, divided the horizon into sections. By the use of this staff in relation to known direction markers on the horizon, the chief of the observers could determine in which section of the sky the observed phenomenon manifested itself. This was also related to the time of day or night, and the day of the lunar calendar. All of these provided the material by which the omen was to be interpreted, and we might guess that it very often suggested more sacrifices or gifts to the priesthood in order to propitiate the gods. The squares were later transferred to the ground, and divinatory methods were devised to take the place, or to augment, the observing of signs in the skies.

And so, with this brief review we come to the idea that the emergence of "games" or game boards, is a "sigil" of the dark, mechanical forces. Most of what we call "games" were originally developed as means of formalized divination by priests, as opposed to natural shamans. The chessboard originated as a "locator" in space time. The gods were known as "those who measure," and we see this symbolism of the Secret Games of the Gods in the checkerboard floor of Masonic lodges, and in other "occult" lore.

What is less apparent is the identity of the players: The shamanic bloodline of the benevolent goddess vs. the ritual priesthood of the vengeful god. This brings us back to a comment I made in the previous chapter: What seems to have happened is that, through repeated cataclysms, man has been brought low, relegated to darkness regarding his history, and at the very point when he began to study and analyze his environment objectively, *religion stepped in and put a period to such ideas.*

Over and over again we come up against that little problem — religion and belief systems that have to be defended against objective evidence or the beliefs of others. "Why did man, through thousands of years, wherever he built scientific, philosophic, or religious systems, go astray with such persistence and *with such catastrophic consequences*? [...] The answer lies somewhere in that area of our existence which has been *so heavily obscured by organized religion* and put out of our reach. Hence, it probably lies in the relation of the human being to the cosmic energy that governs him." (Reich, 1949).

Please note Reich's use of the term "catastrophic consequences." Anyone familiar with the history of religion, and looking at the matter with objectivity, will affirm that the introduction of, the spreading of, the enforcing of, religion is the cause of nearly all the evils on our planet. It's that simple. Jesus said: "By their fruits you shall know them." That's a pretty bitter fruit. Carlos Castaneda brings our attention to the very same matter in a far more direct way:

> Think for a moment, and tell me how you would explain the contradiction between the intelligence of man the engineer and the stupidity of his systems of beliefs, or the stupidity of his contradictory behavior. Sorcerers believe that the predators have given us our systems of beliefs, our ideas of good and evil, our social mores. They are the ones who set up our hopes and expectations and dreams of success or failure. They have given us covetousness, greed, and cowardice. It is the predators who make us complacent, routinary, and egomaniacal.
>
> In order to keep us obedient and meek and weak, the predators engaged themselves in a stupendous maneuver — *stupendous, of course, from the point of view of a fighting strategist.* A horrendous maneuver from the point of view of those who suffer it. *They gave us their mind!* Do you hear me? The predators give us their mind, which becomes our mind. [...] Through the mind, which, after all, is their mind, *the predators inject into the lives of human beings whatever is convenient for them.* (Castaneda, 1998; this author's emphases)

In Book One of *The Wave*, I quoted the purported remarks of an entity that presented itself as a "demon," but which gave strong indications of being similar to what we are calling "aliens." This "creature" said something that was rather astonishing to me at the time I read it; but later, after I had learned so much more, I realized that it might in fact be true. I was so fascinated by this case that I contacted the author of the book to see if I could determine if any part of it was confabulated or sensationalized. What I learned was that, in fact, much of the more disturbing parts of the book had been cut by the editor. And, in fact, the author had suffered some serious psychic backlash when she insisted on leaving in the part I found so interesting. The creature was known as the "Lady," and the individual who was interacting with her (in the same way that many abductees interact with cute little Gray aliens) was Ann Haywood. In an interview with a member of the press, Ann was trying to explain how the Lady transported her in time to distant places:

> "She puts the robe around me and then my mind separates from my body. I can look back and see it lying there. Then we go up through the ceiling, pop out the roof, and fly into space. One night the Lady took me back in time. We were in a foreign country and the people wore old-fashioned clothes. The Lady took on the appearance of a beautiful woman in a blue robe. She performed miracles for them."

Suddenly Ann's face turned ashen and she asked to be excused. Her scream of pain was heard from the bathroom where she had taken refuge. When Ann came out, she was sniffling and holding her abdomen. The Lady had savagely attacked her for revealing that down through history, creatures like the Lady have taken the form of saints. They then use the gullibility of humankind to misguide and misinform people so that they believe they are seeing miracles performed. Ann begged the newsman to delete that portion of the interview. (Osborn, 1982)

The fact is, when we study religions, religious visions, the appearance of new religions, they nearly always occur in a context that is not much different from so-called "alien" interactions. This is what led Carl Jung to propose his ideas of UFOs as being representations of an archetype, and the clues to the creation or revivification of a grand myth of sorts.

However, since we also have the idea that the alien phenomenon is a hyperdimensional one, and that hyperdimensional capabilities include mastery of space and time (perhaps within certain limits, but we don't know for sure), then it seems only logical to consider the possibility that any religion could be "created" by the appearance of such beings masquerading as benevolent performers of miracles.

And, in fact, that seems to be something that is suggested by Gurdjieff's remarks about the so-called "Secret Masters" and whether or not they can or cannot help us.

"[T]he life of humanity to which we belong is governed by forces proceeding from two different sources: First, planetary influences which act entirely mechanically and *are received by the human masses as well as by individual people quite involuntarily and unconsciously*; and then, influences proceeding from inner circles of humanity whose existence and significance the vast majority of people do not suspect any more than they suspect planetary influences. [...]

"Can it be that there is a conscious force which fights against the evolution of humanity?" [Ouspensky] asked.

"From a certain point of view it can be said," said G[urdjieff].

"Instead of struggling against the mechanical forces there may, at certain moments, *be a struggle against the intentional opposition of fairly powerful forces* though they are not of course comparable with those which direct the evolutionary process. *These opposing forces may sometimes even conquer.* The reason for this consists in the fact that the forces guiding evolution have a more limited choice of means; in other words, they can only make use of certain means and certain methods. The opposing forces are not limited in their choice of means and they are able to make use of every means, even those which only give rise to a temporary success, and in the final result *they destroy both evolution and involution at the point in question.*" (Ouspensky, 1949; this author's emphases)

This reminds us of the all-important point of free will. Those forces that guide evolution cannot violate free will. This seems to be the "limitation" that Gurdjieff intends in the above paragraph. However, we notice that the opposing forces, the forces of STS (Service to Self) or "darkness," are *not* limited in their choice of means. Lying and tricks and miracles, and every kind of imitation of what is "positive" is not only allowed; it is considered to be a better "game" if they cheat and stack the cards in their own favor. When one considers the ideas of Machiavelli in the deepest sense, it is only logical — actually mandatory — that such forces would be behind the creation of religions.

"The humanity to which we belong, namely, the whole of historic and prehistoric humanity known to science and civilization, in reality constitutes only the outer circle of humanity, within which there are several other circles... consisting so to speak of several concentric circles.

"The inner circle is called the 'esoteric'; this circle consists of people who have attained the highest development possible for man, each one of whom possesses individuality in the fullest degree, that is to say, an indivisible 'I,' all forms of consciousness possible for man, full control over these states of consciousness, the whole of knowledge possible for man, and a free and independent will. *They cannot perform actions opposed to their understanding or have an understanding which is not expressed by actions. At the same time there can be no discords among them*, no differences of understanding. Therefore their activity is entirely coordinated and leads to one common aim without any kind of compulsion because it is based upon a common and identical understanding." (Ouspensky, 1949; this author's emphasis)

In the paragraph above, we see that Gurdjieff was describing exactly what the Cassiopaeans talk about in terms of a "network" and "colinearity" of understanding. It also reminds us of their remarks about Stonehenge:

12-08-96

Q: (L) Well, we talked about Stonehenge before, that it was an energy transducer, so to speak. So, was Stonehenge put there because of the location, or did Stonehenge create...... (T) Why don't you just ask what it is about Stonehenge? (L) Okay, what is it about Stonehenge?

A: Location attracted those spirit types on the proper frequency, who in turn, placed stones in proper location to receive the coded communications in code telepathically, in order not to have to chase around the countryside reading encoded pictographs.

Q: (L) What was the technique used within the circle to receive the information telepathically? [Planchette spiraled in, and spiraled out.]

A: Transcendent focused thought-wave separation. [...] The spiral serves to translate message by slowing down the wave and focusing thought-wave transference energy. Utilizes/transduces electromagnetic waves, the conduit, by breaking down signal from universal language of intent into language of phonetic profile. This is for multiple-user necessity.

Q: (L) Multiple-user necessity implies that a number of people must do the spiral. Is that correct?

A: No. Must hear and feel and understand precisely the same thing. The molecular structure of the rock, when properly sculpted, sing to you.

Please note that most important remark: "Multiple user" means a number of people must "hear and feel and understand precisely the same thing." That is exactly what Gurdjieff is talking about in his discussion of the inner circle of ascended masters.

"The next circle is called the 'mesoteric,' that is to say, the middle. People who belong to this circle possess all the qualities possessed by the members of the esoteric circle with the sole difference that their knowledge is of a more theoretical character. This refers, of course, to knowledge of a cosmic character. They know and understand many things which have not yet found expression in their actions. They know more than they do. But their understanding is precisely as exact as, and therefore precisely identical with, the understanding of the people of the esoteric circle. *Between them there can be no discord, there can be no misunderstanding.* One understands in the way they all understand, and all understand in the way one understands. But as was said before, this understanding compared with the understanding of the esoteric circle is somewhat more theoretical.

"The third circle is called the 'exoteric,' that is, the outer, because it is *the outer circle of the inner part of humanity.* The people who belong to this circle possess much of that which belongs to people of the esoteric and mesoteric circles but their cosmic knowledge is of a more philosophical character, that is to say, it is more abstract than the knowledge of the mesoteric circle. A member of the mesoteric circle calculates, a member of the exoteric circle contemplates. Their understanding may not be expressed in actions. But there cannot be differences in understanding between them. What one understands all the others understand." (Ouspensky, 1949; this author's emphases)

We see in the above paragraph what seems to be a description of what we might think are "ascended masters," such as great yogis, saints, healers, and so on. Obviously, our understanding of such things is very limited.

"In literature which acknowledges the existence of esotericism humanity is usually divided into two circles only and the 'exoteric circle' as opposed to the 'esoteric,' is called ordinary life. In reality, as we see, the 'exoteric circle' is something very far from us and very high. For ordinary man this is already 'esotericism.'

"The outer circle is the circle of mechanical humanity to which we belong and which alone we know. *The first sign of this circle is that among people who belong to it there is not and there cannot be a common understanding.* Everybody understands in his own way and all differently. This circle is sometimes called the circle of the 'confusion of tongues,' where no one understands another and takes no trouble to be understood. In this circle *mutual understanding between people is impossible excepting in rare exceptional moments or in matters having no great significance,* and which are confined to the limits of the given being. *If people belonging to this circle become conscious of this general lack of understanding and acquire a desire to understand and to be understood, then it means they have an unconscious tendency towards the inner circle* because mutual understanding begins only in the exoteric circle and is possible only there. But the consciousness of the lack of understanding usually comes to people in an altogether different form.

"So that the possibility for people to understand depends on the possibility of penetrating into the exoteric circle where understanding begins." (Ouspensky, 1949; this author's emphases)

In the following remarks, what Gurdjieff seems to be saying is that, in addition to the "three ways" whereby an individual can penetrate to the inner circles of humanity — from the outside in — there is something called the Fourth Way which, by implication, is the result of direct communication from the inner circle of masters to one or more individuals in the outer circle of general humanity — from the inside out, producing the necessary "work" that results in the finding of the gate.

"If we imagine humanity in the form of four concentric circles we can imagine four gates on the circumference of the third inner circle, that is, the exoteric circle, through which people of the mechanical circle can penetrate.

"These four gates correspond to the four ways.

"The first way is the way of the fakir, the way of people number one, the people of the physical body, instinctive-moving-sensory people without much mind and without much heart.

"The second way is the way of the monk, the religious way, the way of people number two, that is, of emotional people. The mind and the body should not be too strong.

"The third way is the way of the yogi. This is the way of the mind, the way of people number three. The heart and the body must not be particularly strong, otherwise they may be a hindrance on this way.

"Besides these three ways yet a fourth way exists by which can go those who cannot go by any of the first three ways.

"The fundamental difference between the first three ways, that is, the way of the fakir, the way of the monk, and the way of the yogi, and the fourth way consists in the fact that they are tied to permanent forms which have existed throughout long periods of history almost without change. At the basis of these institutions is religion. Where schools of yogis exist they differ little outwardly from religious schools. And in different periods of history various societies or orders of fakirs have existed in different countries and they still exist. These three traditional ways are permanent ways within the limits of our historical period.

"Two or three thousand years ago there were yet other ways which no longer exist and the ways now in existence were not so divided, they stood much closer to one another.

"The fourth way differs from the old and the new ways by the fact that it is never a permanent way. It has no definite forms and there are no institutions connected with it. It appears and disappears governed by some particular laws of its own." (Ouspensky, 1949)

In this last remark, we find a reflection of the definition of the work of the *Vatis*, "having his degree under the privilege of genius; discipleship shall not be required in respect to him." Strabo (IV, 4) quoting Poseidonius: "Among all the tribes, generally speaking, there are three classes of men held in special honor: the bards, the *vates* and the druids."

"The fourth way is never without some work of a definite significance, is never without some undertaking around which and in connection with which it can alone exist. When this work is finished, that is to say, when the aim set before it has been accomplished, the fourth way disappears, that is, it disappears from the given place, disappears in its given form, continuing perhaps in another place in another form. Schools of the fourth way exist for the needs of the work which is being carried out in connection with the proposed undertaking. They never exist by themselves as schools for the purpose of education and instruction." (Ouspensky, 1949)

As I read the above paragraph, I thought of the work of the now unknown man who lived in the Middle East, who did a great work for humanity, and whose true life has been obliterated by the myths of a man named Jesus, after the Egyptian religion. Right there I pause and consider the Secret Games of the Gods. When one side makes a move, the other moves in with full capabilities of cheating and deceiving. Never has that been more evident than in the myths compiled into that document we call the Bible.

"The work itself of schools of the fourth way can have very many forms and many meanings. In the midst of the ordinary conditions of life the only chance a man has of finding a 'way' is in the possibility of meeting with the beginning of work of this kind [...]

But no matter what the fundamental aim of the work is, the schools continue to exist *only while this work is going on*. When the work is done the schools close. The people who began the work leave the stage. Those who have learned from them what was possible to learn and have reached the possibility of continuing on the way independently begin in one form or another *their own personal work*.

But it happens sometimes that when the school closes a number of people are left who were round about the work, who saw the outward aspect of it, and saw the whole of the work in this outward aspect. Having no doubts whatever of themselves or in the correctness of their conclusions and understanding they decide to continue the work. To continue this work they form new schools, teach people what they have themselves learned, and give them the same promises that they themselves received. All this naturally can only be outward imitation." (Ouspensky, 1949; this author's emphases)

Again I am reminded of the work of the original "Jesus":

"When we look back over history it is almost impossible for us to distinguish where the real ends and where the imitation begins. Strictly speaking almost everything we know about various kinds of occult, Masonic, and alchemical schools refers to such imitation. We know practically nothing about real schools excepting the results of their work and even that *only if we are able to distinguish the results of real work from counterfeits and imitations*. [...]

The idea of initiation, which reaches us through pseudo-esoteric systems, is also transmitted to us in a completely wrong from. The legends concerning the outward rites of initiation have been created out of the *scraps* of information we possess in regard to the ancient Mysteries. The Mysteries represented a special kind of way in which, side by side with a difficult and prolonged period of study, theatrical representations of a special kind were given which depicted in allegorical forms the whole path of the evolution of man and the world.

Transitions from one level of being to another were marked by ceremonies of presentation of a special kind, that is, initiation. But *a change of being cannot be brought about by any rites. Rites can only mark an accomplished transition*. And it is only in pseudo-esoteric systems in which *there is nothing else except these rites*, that they begin to attribute to the rites and independent meaning.

It is supposed that a rite, in being transformed into a sacrament, transmits or communicates certain forces to the initiate. This again relates to the psychology of an *imitation* way. There is not, nor can there be, any outward initiation. In reality only self-initiation, self-presentation exist. Systems and schools can indicate methods and ways, but no system or school whatever can do for a man the work that he must do himself. Inner growth, a change of being, depend entirely upon the work which a man must do on himself." (Ouspensky, 1949; this author's emphases)

If we consider certain of the remarks of the Cassiopaeans that I have included in this series, in the context of what Gurdjieff has said above, we come to the idea that the Cassiopaean material is a work of the True Fourth Way. And, as such, it is naturally constrained by certain principles and understandings. However, we also realize that the forces of the "other side" are not so constrained. It is becoming more and more evident why I was kidnapped as a child in an attempt to install self-destruct programming and locks on knowledge, by an individual who was identified as being connected to an "economic legion." We realize now why Frank Scott was sent to destroy and/or derail me, and why when he failed, an economist was sent to "scope me out"; and why, immediately thereafter, Vincent Bridges was sent in along with a supporting cast of players, to make the next move in the Secret Games of the Gods: Attempt to either destroy the Cassiopaean work, or us.

Now, let's talk about Game Theory and what kinds of minds develop and operate with these "rules." To me, that has been the most interesting question. What kind of person would think up something like that?

John von Neumann first wrote about Game Theory in the late 1920s. Since this issue of economics and its relation to Game Theory came up, I have been reading many of the original sources, and I was quite taken aback to read what Von Neumann actually proposed in his famous seminal work:

The purpose of this book is to present a discussion of some fundamental questions of economic theory which require a treatment different from that which they have found thus far in literature. ... They have their origin in the attempts to find an exact description of the endeavor of the individual to obtain a maximum of utility, or ... a maximum of profit. [...]

We believe that it is necessary to know as much as possible *about the behavior of the individual* and about the simplest forms of exchange. [...] It does not seem to us that these notions are qualitatively inferior to certain well established and indispensable notions in physics, like *force, mass, charge,* etc. That is, while they are in their immediate form merely definitions, *they become subject to empirical control* through the theories which are built upon them — and in no other way. [...]

The individual who attempts to obtain these respective maxima is also said to act "rationally." But it may safely be stated that there exists, at present, no satisfactory treatment of the question of rational behavior. There may, for example, exist several ways by which to reach the optimum position; they may depend upon the knowledge and understanding which the individual has and upon the paths of action open to him. [...]

We hope, however, to obtain a real understanding of the problem of exchange by studying it from an altogether different angle; this is, from the perspective of a "game of strategy." [...] Let us look at the type of economy which is represented by the "Robinson Crusoe" model, that is an economy of an isolated single person or otherwise organized under a single will. [...] The problem is to obtain a maximum satisfaction. [...] Crusoe is given certain physical data (wants and commodities) and his task is to combine and apply them in such a fashion as to obtain a maximum resulting satisfaction. [...] Thus Crusoe faces an ordinary maximum problem, the difficulties of which are of a purely technical — and not conceptual — nature. [...]

Consider now a participant in a social exchange economy. His problem has, of course, many elements in common with a maximum problem. But it also contains some, very essential, elements of an entirely different nature. He too tries to obtain an optimum result. But in order to achieve this, he must enter into relations of exchange with others. If two or more persons exchange goods with each other, then the result for each one will depend in general not merely upon his own actions but on those of the others as well. Thus each participant attempts to maximize a function of which he does not control all variables. This is certainly no maximum problem, but a peculiar and disconcerting mixture of several conflicting maximum problems. Every participant is guided by another principle and neither determines all variables which affect his interest. [...]

A particularly striking expression of the popular misunderstanding about this pseudo-maximum problem is the famous statement according to which the purpose of social effort is the "greatest possible good for the greatest possible number." A guiding principle cannot be formulated by the requirement of maximizing two or more functions at once.

Such a principle, taken literally, is self-contradictory. [...] The general theory must cover all these possibilities, all intermediary stages, and all their combinations. [...] Every participant is allotted a set of variables, "his" variables, which together completely describe his actions, i.e. *express precisely the manifestations of his will.* (Von Neumann and Morgenstern, 1953; this author's emphases)

In short, Game Theory as applied to economics is all about gaining control of the free will of others. It is, of course, "disguised" as "economics," but as we all know, Money rules our world. As I continued to read, I couldn't help but recall: "Also he compels all, both small and great, both the rich and the poor, both free and slave, to be marked with an inscription on their right hands or on their foreheads, so that no one will have power to buy or sell unless he bears the stamp of the beast or the number of his name."

But isn't that from the Bible? Isn't that a prophecy of God? Isn't that a warning from the "good guys"? The reader who is familiar with our book, *Secret History*, may already be formulating certain ideas. Others may have decided that even if we can analyze the Bible and determine how it was assembled and when, it doesn't detract from their beliefs in God as presented in the Bible. After all, God is more than the Bible, right? And faith in God is not dependent on the Bible, right?

What if that Bible, that system of understanding, was *planted* via time travel as a strategic move in the Secret Games of the Gods? When we asked the question in the previous chapter about what a powerful elite under the control of mechanical forces would do if they discovered what Gurdjieff revealed to be the *truth*. Of course, covering it up comes immediately to mind. But, more than that, they would want to ensure that the "event" goes down in their favor. That means that they have to line up about six-billion other people and get them to behave in a certain way at a certain point in time.

But manipulating vast masses of people is not as simple as organizing a croquet game. What is more, it's a one shot deal, and with that kind of money they can buy all the brains they need to figure out how to make it stick. And so, we come to time travel. You have to manipulate the past to control the future. That reminds me of something very significant that Gurdjieff said:

> Shortly before his death, one of his students asked him: "Mr. Gurdjieff ... the 'I' which I am trying to develop ... is this the soul that survives after death?"
>
> He waited a long time before he asked "How long have you been with me?"
>
> "Almost two years."
>
> "Too short the time. You are not able yet to understand. Use the present to repair the past and prepare the future. Go on well; remember all I say." (Patterson, 1998)

What would be the objective of time travel into the past in order to control the future? Well, the simplest way to do it would be to create a religion, do some miracles to make sure it "took," and maybe help your new believers out in destroying the opposition. But there is a more insidious reason for creating religions, as we will see in the following description of certain mind control procedures. When I first read the following segment, I had a huge "aha!" moment in realizing that it was a model of how people are "driven" into the "religious fold" via hyperdimensional manipulation. Forget the use of religion for control by top secret organizations — that's just the disinformation designed to distract our attention away from the hyperdimensional realities and blame everything on human agencies — and just think about the life pressures, the emotional manipulations from hyperdimensional beings, that can be brought to bear on a person to convince them that being "born again" is the answer and "faith" is the key. Read my own story, *Amazing Grace*, to see how these pressures were brought to bear on me, and how, at a certain point, I "followed the program"; only in my case, the inner nature of questioning brought me out of it on the other side, wiser for the experience. But how many people do not have the courage to escape such controls?

So, keeping in mind that it is very likely a disinformation model of hyperdimensional control agendas, have a look at how religion is used as a mind control, social programming tool:

1. The NSA's behavioral modification process starts with identification and qualification of the subject. The NSA used to choose subjects based on the subject's net present value to the agency in public visibility, financial resources, political clout, or other intelligence and counter-intelligence reasons. Additional considerations are given to minimizing security risks of exposure, the subject's post-hypnotic suggestibility index, the subject's intelligence and reasoning ability, moral and superstitious beliefs, and the subject's social status and the weakness of the subject's primary support groups (family). Now a recent report referenced in the March 26th Business section of the Orange County Register from the National Sleep Foundation reports that 40% of Americans are experiencing sleeping problems. This news could indicate that the NSA is broadening its influence to the greater public. As explained below in this document, the NSA always starts its behavioral modification process with REM Deprivation.

2. After selection, the subject is subjected to long periods of REM Sleep Deprivation and reinforced torturing post-hypnotic suggestions that will breakdown the subject's will, confidence, self-reliance, and moral values. Meanwhile, the subject is increasingly isolated from their familiar and trusted peer groups causing the subject to experience depression, apathy, and ultimately social and financial failure.

3. Typical posthypnotic induced delusions reported by subjects are tingling in various areas of the body, which are thought to be resulting from microwave beams. Hearing ticks, thumps, or cracks from walls, ceilings, clocks, lights, etc. Beliefs that the subject's neighbors are conspiring against them, or that the subject is being followed. Sometimes subjects believe that the various perceptions, feelings and experiences are the result of "Implants" in their body. It is important for the subjects to understand that the NSA controls this technology from nuclear hardened underground shelters and the neighbors next door have nothing to do with the subject's experiences. Nobody has the time or inclination to

follow a subject around with a microwave gun to tickle various parts of the body.

We are saturated with microwaves all the time from television stations, communication satellites, etc. and yet we do not have any symptoms because microwaves do not have the ability to trigger localized synaptic responses in our brains. Furthermore, when the subject is in a room surrounded by several people, and the subject is the only one experiencing the "thoughts," tingling feelings, etc., then obviously a delivery method is being employed that affects only the subject; high-speed acoustic delivered hypnosis.

4. After a while, the subject has an emotional breakdown and a new support group is built around the subject. The new support group is typically a church with doctrines centered in the Bible but the NSA also uses cults and other social groups. *The NSA prefers Christian churches because the doctrines allow "God or Jesus to speak directly to the subject"* and the negative reinforcement can be attributed with Satan and the positive rewards can be considered to be blessings from God thereby masking the NSA's technology and processes. When the NSA uses other relationships within which the subject experiences a religious awakening and "Gives their Life to Christ" and the NSA achieves total control of the subject.

5. The subject is slowly released from the damaging uncomfortable hypnosis and it is replaced with positive rewarding hypnosis as "God and Jesus works in their life." Soon, the subject has complete loyalty to Jesus (AKA: NSA) and will do anything on command from Jesus (NSA).

6. The subject is required to give daily status reports in the form of prayers in the privacy of their home, office, or car where the NSA's electronic surveillance system captures and sorts the prayers by "Keywords." The NSA then delivers additional hypnosis in the form of punishments or rewards or directs the subject accordingly to "God's will." If the subject resists the NSA's instructions, additional punishments are inflicted on the subject.

7. The subject is institutionalized in this system where any nonconformances committed by the subject are watched, critiqued, and reported on through prayer by other "Christians" to the NSA. Thus, the new church peer group acts as a behavioral reinforcing mechanism that will bring any of the subject's problems to the NSA as they have been trained themselves (this is similar to the Nazi Gestapo of World War 2 and other fascist approaches).

8. A subject that has successfully completed the NSA's behavioral modification program lives out the rest of their mediocre life in service to Jesus (NSA) and never causes any waves in the church or news media for fear of reprisal from the NSA. The subject's lives are relatively unproductive because their focus is on their "Life after Death" and not what they accomplish while they are alive. They avoid "worldly activities," and usually are confused and disjointed in rational thoughts and concepts. For instance, they don't believe in anything that is not in

the Bible, i.e. dinosaurs, evolution, space travel, even though they ride on airplanes and watch television both of which are not referenced in the Bible. (Bamford, 1982; this author's emphasis)

If such information has been "allowed" to be leaked suggesting that religion is used by such groups, again using the rules of disinformation, we might think that the issue that is being concealed is that religions are *created* by hyperdimensional being via time travel. We cannot exclude the human factor as well.

At this point, I want to present a short and selective list of books and papers on Game Theory so that the reader will have some idea of how this theory is being used, by reading the titles of the articles. Keep in mind, this is only what is allowed to be published:

1928, John von Neumann: "Zur Theorie der Gesellschaftsspiele."

1930, F. Zeuthen: "Problems of Monopoly and Economic Warfare."

1944, John von Neumann and Oskar Morgenstern: *Theory of Games and Economic Behavior.*

1950–53, John Nash wrote four papers: "Equilibrium Points in N-Person Games" (1950); "Non-cooperative Games" (1951); "The Bargaining Problem" (1950); and "Two-Person Cooperative Games" (1953).

1953, H. W. Kuhn: *Extensive Games and the Problem of Information*; includes the formulation of extensive form-games which allow the modeller to specify the exact order in which players have to make their decisions and formulate the assumptions about the information possessed by the players in all stages of the game.

1954, L. S. Shapley and M. Shubik: "A Method for Evaluating the Distribution of Power in a Committee System." They use the Shapley value to determine the power of the members of the UN Security Council.

1954–55, "Problems of forming and solving military pursuit games." Rand Corporation research memoranda, by Isaacs, RM-1391 (November 30, 1954), RM-1399 (November 30, 1954), RM-1411 (December 21, 1954), and RM-1486 (March 25, 1955); all entitled, in part, "Differential Games."

1955, R. B. Braithwaite: "Theory of Games as a Tool for the Moral Philosopher."

1959, Martin Shubik: "Strategy and Market Structure: Competition, Oligopoly, and the Theory of Games."

1960, Thomas C. Schelling: *The Strategy of Conflict.*

1961, R. C. Lewontin: "Evolution and the Theory of Games."

1962, D. Gale and L. Shapley: "College Admissions and the Stability of Marriage."

1965, Rufus Isaacs: "Differential Games: A Mathematical Theory with Applications to Warfare and Pursuit, Control and Optimization."

1966, John Harsanyi: *A General Theory of Rational Behavior in Game Situations.* A game is cooperative if commitments — agreements, promises, threats — are fully binding and enforceable. It is non-cooperative if commitments are not enforceable.

1967–68, John Harsanyi: *Games with Incomplete Information Played by 'Bayesian' Players, Parts I, II and III.* This laid the theoretical groundwork for information economics that has become one of the major themes of economics and Game Theory.

1972, John Maynard Smith: "Game Theory and The Evolution of Fighting."

1973, John Maynard Smith and G. Price: *The Logic of Animal Conflict.*

1973, Gibbard: "Manipulation of Voting Schemes: A General Result."

1974, R. J. Aumann and L. S. Shapley: *Values of Non-Atomic Games.* It deals with values for large games in which all the players are individually insignificant (non-atomic games).

1976, Robert Aumann: *Agreeing to Disagree.* An event is common knowledge among a set of agents if all know it and all know that they all know it and so on *ad infinitum*.

1981, R. J. Aumann: *Survey of Repeated Games.* This survey firstly proposed the idea of applying the notion of an automaton to describe a player in a repeated game.

1985, J.-F. Mertens and S. Zamir: *Formulation of Bayesian Analysis for Games with Incomplete Information.* Shows that it is not possible to construct a situation for which there are no sets of types large enough to contain all the private information that players are supposed to have.

1988, Tan and Werlang: *The Bayesian Foundations of Solution Concepts of Games.* Formally discusses the assumptions about a player's knowledge that lie behind the concepts of Nash equilibria and rationalizability.

As it happens, the economist who visited us, and who left with the Cassiopaean transcripts in hand, was an expert in Bayesian Logic.

Strategic behavior arises when two or more individuals interact, and each individual's decision turns on what that individual expects the others to do. *Incomplete information is the central problem in Game Theory.* Control of what information is available and whether it is or is not utilized amounts to "stacking the cards" in favor of the one in control of the information. People make choices based on "pay-offs." They are "rational" in the sense that they consistently prefer outcomes with higher pay-offs to those with lower pay-offs. People make decisions based on their beliefs about what others will do. When we come to the problem of beliefs, we begin to enter the arena that is controlled by what is known as Bayesian Logic. This solution builds on the idea of positing the knowledge and/or beliefs of the various players.

In Game Theory, the best way to know what knowledge or beliefs the players have or believe is most easily controlled by *creating* the beliefs that assist in the covering up of information that would assist the player in formulating a winning strategy.

It is as likely that those who are creating the New Age religion and/or running the New Age COINTELPRO operation are part of this Game Strategy of making sure that humanity is limited in terms of information, and are able to be manipulated in terms of belief. In such a way, the Control System can play their "Dominant Strategy" with assurance that the other players will respond in a very precise way.

Religion is the Devil's greatest achievement. In the guise of religion he has pulled off his most audacious coup. He has flagrantly masqueraded as God. He has had us bow down and worship him. He has had us commit every type of evil in the name of holiness. He has passed off his bigotry as God's opinions. He has had us segregate humanity into the "ins" and the "outs," believers and non-believers, the saved and the damned. He has convinced us that God likes us but not them. And convinced them that God likes them but not us. And then, in a stroke of dark brilliance, he warns his faithful flock of sheep: "Be sure you do not pay heed to anyone but me, for the Devil is a wily wolf and he will surely trick you." (Freke and Gandy, 2001)

World War II was known as the "scientists' war." After the Rockefeller and Bamberger money had set things up to make Princeton the center of the mathematical universe here on Earth, Princeton mathematicians were involved in ciphers and code breaking, the science of ballistics, statistical analysis of enemy positions, and so on. The scientists from Princeton made breakthroughs in radar, infrared detection, bomb-delivering airplanes, long-range rocketry, torpedoes and other instruments of destruction. Mathematicians were needed to evaluate assessment of the effectiveness of weapons, efficient use of weapons, how many tons of explosives were needed to kill the most people, and so on.

Then, of course, there was the A Bomb.

At the end of WW II, there was no longer any doubt in the minds of those running the American government that mathematics was the king of sciences. New theories and sophisticated math gave them the needed edge to win the war, and mathematicians — mainly those at Princeton — partook of the prosperity that followed. What's more, mathematics was no longer the activity of gentlemen of leisure — it was a wide-open field for anyone with talent. Being Jewish, foreign, or from the streets of Brooklyn, no longer mattered. If you had talent, Princeton wanted you and would foot the bill. At this point, there was nothing they couldn't get money for: topology, algebra, number theory, computer theory, operations research, and, of course, Game Theory.

Naturally, they wanted Nash. Again we ask, "Why?"

I read Sylvia Nasar's well-researched biography of John Nash with enormous interest. What I found most fascinating was the section about his childhood. As far as I can tell from the reference notes, Ms. Nasar went to original sources — including Nash's sister, teachers, childhood acquaintances, school records, local newspapers, and so on — for her information. She did an extremely thorough job of gathering the information, but made little attempt to interpret it. She was simply recounting what she was told and what she discovered. It was an excellent job of pure reporting. And, as I noted in the previous chapter, I realized that I was reading an almost verbatim description of the childhood of Frank Scott, as described to me by Frank himself.

Because of all of the recent research we had done on psychopathy (a.k.a "antisocial personality disorder"), as I read I recognized the psychopathic personality being described in detail in Nasar's account, right down to Nash's admission to another mathematician at MIT that, as a child, he "enjoyed torturing animals."

By the time I had finished the book, I had the idea that psychopathy and "mind control" activity was a better explanation for Nash's so-called "paranoid schizophrenia," which was somewhat mysterious in terms of onset, symptoms and later remission. As I was reading, I kept thinking about another article I had read describing psychopaths as "alien reaction machines" (Horne, 2000). Several other connections were made, as the reader will see, and whether it is a conscious conspiracy or not, I will leave to the reader's judgment. I just know that the picture that is being revealed is frightening.

Of course, I am not a psychiatrist nor a psychologist. I am a very good "technician" in terms of hypnotherapy, but I have always advocated following the therapeutic models of professional clinicians, which I studied, implemented and observed in order to come to some ideas about what did or did not work. And so, again, rather than just make a bold statement that "I think this is what the problem is," since I am neither trained nor qualified to diagnose mental conditions, I will just present the information I have gathered on the subject and leave it to the reader to judge.

Based on the research I have read, the importance of psychopathy in the present day cannot be overstated. Simply put, it is a growing phenomenon and it is going to impact every single one of us individually and collectively in the not-too-distant future. It is also extremely important to understand psychopathy in order to be able to fully understand Game Theory and how it is the underlying dynamic being used at the present time to move all the pieces into place for the Secret Games of the Gods. There is so much literature on the subject, that what I am going to include here will only skim the surface. I will, however, urge every single reader to do their own research on this subject at the soonest possible opportunity. As the Cassiopaeans say, "Knowledge protects" — and the knowledge of the functional modes of the psychopath could save your life.

After reading through a slew of books and papers, reviewing in my mind the many experiences I have had with other people, searching for clues and assessing interactions, my personal preference is for the work of Dr. Robert Hare. Because of my own life experiences, it is my opinion that his professional opinion and presentation is the most concise and to the point, and also most realistic with practical hands-on understanding conveyed in simple language. One of the most important things about his work is the fact that he makes it quite clear that we cannot excuse psychopathy based on environment. The "Nature vs. Nurture" argument is overwhelmingly answered as "nature." The implications this has for all of us are deep and profound. Something, or somebody, seems to have set things up so that these kinds of genes will propagate widely and produce many "offspring" *at this point in our history.* When we have a look at Nash, something of an "experiment" in their program, we will have a much better idea of what we are facing.

> Psychopathy is a personality disorder defined by a distinctive cluster of behaviors and inferred personality traits, most of which society views as pejorative. It is therefore no light matter to diagnose an individual as a psychopath. Like any psychiatric disorder, diagnosis is based on the accumulation of evidence that an individual satisfies at least the minimal criteria for the disorder. (Hare, 1999)

Notice that Hare says that the only thing that can really be said about the psychology of psychopaths is inferred from a "cluster of behaviors." This is one of the big problems of the subject. Psychopaths just simply do not ever consider that there is anything wrong with them, and as a result, they do not ever willingly seek psychiatric help — unless it is part of a plan to deceive or con someone, or if they are incarcerated and obliged one way or another to submit to psychiatric examination. Even then, because of the nature of the psychopath, it is highly questionable that what they reveal in such interviews is an accurate representation of what really goes on inside them. And so, having a real live psychopath turn himself in and tell the truth about what goes on in his head just isn't part of the reality. However, there have been a number of recent studies of psychopathic prisoners, using advanced brain study technology, and this has helped to sort out the "real" from the "fake" to a great extent.

Many people think that all psychopaths are dangerous criminals like Ted Bundy or John Wayne Gacy, and this is due to the fact that the only psychopaths we know about are criminals. The important point to make here is that not all criminals are psychopaths, and it seems to be so that most psychopaths are not adjudicated criminals. The facts seem to be that the only ones who ever get studied are the ones who are "less successful," shall we say, and who therefore get caught and incarcerated. The really *good* ones aren't in prison. That ought to scare us to death! It is very difficult to estimate the numbers of true psychopaths in the population, but one thing seems certain: the numbers are increasing rapidly.

> In spite of more than a century of clinical study and speculation and several decades of scientific research, the mystery of the psychopath still remains. Some recent developments have provided us with new insights into the nature of this disturbing disorder, and its borders are becoming more defined. But the fact is, compared with other major clinical disorders, little systematic research has been devoted to psychopathy, even though it is responsible for far more social distress and disruption than all other psychiatric disorders combined. (Hare, 1999)

That is a startling statement — that psychopaths have a more detrimental effect on our society than all other psychiatric disorders *combined*. So few people are even aware of this fact. They may know all about schizophrenia, or bipolar disorders, or ADHD, because all of those things can be medicated and controlled to one extent or another. Also, they are disabling to the individual. Conversely, the chief thing about psychopathy is that it is not disabling to the individual unless certain other factors are present. In general, psychopaths always manage to do very well for themselves. People ask: "Isn't psychopathy maladaptive?" The terrifying answer is: *It may be maladaptive for society, but it is adaptive for the psychopaths themselves.*

The hallmark of the psychopath is a stunning lack of conscience and their objective of self-gratification at the expense of others. Psychopaths are possessed of a cold, calculating rationality combined with an inability to even conceive of others as thinking, feeling beings. To witness such incomprehensible behavior produces a feeling of bewilderment and helplessness. What is important, however, is that *most of us never "witness" such an "inside" view*, unless we have been burned a sufficient number of times to develop an acute awareness that all is not as it seems on the surface. To actually identify one of these people, we must become very, very aware, and make certain "tests" of the behavior — "systematic harassment," as Don Juan calls it — of those we suspect may not have our best interests at heart because *their acts don't match their words.* Through such tests, psychopaths can be "flushed out" into the open, fully displaying their true nature. However, even when in full display, most people simply cannot believe what they are seeing. Only the victims know the truth, and their insights are generally discounted, as we will see.

[G]ood people are rarely suspicious: they cannot imagine others doing the things they themselves are incapable of doing; usually they accept the undramatic solution as the correct one, and let matters rest there. Then too, the normal are inclined to visualize the [psychopath] as one who's as monstrous in appearance as he is in mind, which is about as far from the truth as one could well get. [...] These monsters of real life usually looked and behaved in a more normal manner than their actually normal brothers and sisters; they presented a more convincing picture of virtue than virtue presented of itself — just as the wax rosebud or the plastic peach seemed more perfect to the eye, more what the mind thought a rosebud or a peach should be, than the imperfect original from which it had been modeled. (William March, *The Bad Seed,* quoted by Hare, 1999)

Only by reading the literature on the subject can the average person truly begin to grasp the nightmare of living with or dealing with a true psychopath. Lying, deceiving, and manipulation are their natural talents. They have vivid imaginations that are focused entirely on themselves and getting what they want, and they are unbelievably unconcerned with the possibility — or, in some cases, the certainty — of being found out. When caught in a lie, or challenged with the truth, they are almost never embarrassed at all! They simply shift the attention of the questioner, change the story, or rework the facts to be more consistent with the original lie. The end result is that the listener is confused, and they are then vulnerable to being convinced that the confusion is their own fault!

Psychopaths also tell lies that are so liberally sprinkled with emotional trigger-words that the listener is completely taken in. Even sophisticated psychologists and psychiatrists are very often hornswoggled by psychopaths! One case cited by Hare is really funny in a horrible sort of way because the psychologist wrote such things about the subject as: "very impressive; sincere and forthright," "possesses good interpersonal skills," "intelligent and articulate," and so on. He was later humiliated to discover that virtually none of what was told to him by the psychopath was true. He had fallen for every word — hook, line and sinker!

Studying the words used by psycyopaths to convey emotion is revealing. As Sylvia Nasar wrote about John Nash:

> But, as in so many other relationships in his life, Nash's intentions weren't always matched by the emotional means to carry them out satisfactorily. Even as he tried to draw his son closer, he said and did things that could only be called insensitive and alienating. [...] [At present] the self-deprecating humor suggests greater self-awareness. The straight-from-the-heart talk with friends about sadness, pleasure and attachment suggests a wider range of emotional experiences. The daily effort to give others their due, and to recognize their right to ask this of him, bespeaks a very different man from the often cold and arrogant youth. (Nasar, 1998)

What we surmise from the above is that now that he is old and finds that he must look to others to survive, Nash has again "adapted" like the true psychopath. Psychopaths very often talk at great length about their "feelings" and they claim to experience strong emotions. However, a careful listener or interrogator will discover, if they are clever in their questioning, that the psychopath is unable to describe the subtleties of various emotional states. They will equate love with sexual arousal, or sadness with frustration. The conclusion of researchers is that the emotions of the psychopath are so shallow as to be little more than "proto-emotions," or primitive responses to self-centered needs.

For "normal people," it is our awareness of emotional consequences — fear of being hurt or of hurting someone else — that guides our choices of actions in life. The "inner voice" that tells us "how things are done" *when one is involved with other beings who have rights and feelings* is developed via a complex system of socialization. We can call this our "conscience." It acts as a sort of "inner policeman" to regulate our behavior, even in the absence of external controls. It is a sort of inner self that presents *a series of perceptions about what others expect of us, as well as what we expect of ourselves.* Psychopaths do not seem to have this "inner guidance" system. They may calculate coldly what could or could not happen, but they act primarily based on achieving immediate satisfaction, pleasure, or relief of some sort.

Please note: The psychopath does not act based on a consideration of the rights and feelings of others. They do *not* have a conscience, and it is *our* conscience that makes us prey! It is *our* perception of *them* as human beings with feelings, that restrains our actions, that makes us "consider" them in the way *we* would like to be considered; but all the while, *they* are not so constrained! In terms of Game Theory, they have a Dominant Strategy that takes into account this very weakness of conscience that they *know* will prevent their victims from responding in kind.

Psychopaths consider the rules and expectations of society to be inconvenient and unreasonable impediments to what they want or need. But, as noted, they don't always break the law so as to land in jail. They are generally too smart for that. Instead, they do things that are unethical, immoral and harmful to others, but in ways that are not illegal. The problem with behavior of this sort is that it is cruelly destructive to all around them, but almost impossible to document or explain to outsiders.

Obligations and commitments mean nothing to the psychopath. They don't honor formal or implied commitments to people and organizations, or even principles. They are also irresponsible parents. They may insist that they love their children, but typically, they will leave them for extended periods of time either alone, or with unreliable or inappropriate people. The story of Nash's behavior toward his mistress (with whom he had a child that he suggested she ought to "give up for adoption"), as well as his son by his wife (who went for a year without a name, and who was shuttled back and forth between his mother and his grandparents for most of his life), demonstrate clearly that Nash most definitely was not merely an "irresponsible parent" — he was cruelly neglectful. Of course, the excuse is made that he was "psychotic." But it is most curious that his psychosis began just shortly before the birth of his second son and within a marriage, where societal expectations of care and responsibility would be most likely brought to bear on him. It was almost as if the very idea of actually being expected to give something of himself to another human being was sufficient to drive him to self-destruction.

Nash suggested to Eleanor that she give John David up for adoption. [...] It was a cold-blooded suggestion, and it all but killed any remaining love Eleanor felt for Nash. One only hopes that among Nash's considerations in putting it forward — apart from eliminating any financial responsibility he might face for his child, which prompted Eleanor to say that Nash "wanted everything for nothing" — might have been a genuine belief that John David's chances in life would be greater with some middle-class couple than with his single, working mother. [...] He doubtless behaved selfishly, even callously. His son and others later attributed his acknowledgment of paternity and desire to maintain a bond, even while failing to protect his child from poverty and periodic separation from his mother, to a pure narcissism. (Nasar, 1998)

We note above that Nash obviously was saying the words, but that those words were contradicted by his actions — one of the signs of the psychopath.

Psychopaths are extraordinarily successful in talking their way out of trouble. They will say over and over again, "I've learned my lesson," or "You have my word that it won't happen again," or "It was all just a big misunderstanding, can't we forget it and go forward?" Usually this works, and one wonders how many times it had worked for Nash before he found an instance in which it did *not* work and, in fact, ended in the withdrawal of his top-secret security clearance and loss of a lucrative consulting contract.

In 1950, Nash was hired as a consultant for RAND Corporation, a secretive civilian think-tank funded by the Air Force. This was where the big brains worked out problems of nuclear war and Game Theory. The RAND ideal was a militarized worship of the rational life, geopolitical obsession, paranoia and megalomania. Its mission was to apply rational analysis and the latest quantitative methods to the problem of how to use nuclear weapons most effectively — as instruments of destruction or deterrence. Nasar suggests that RAND may have been the model for Isaac Asimov's *Foundation* series about hyper-rational social scientists or psycho-historians, who think it is their job to save the galaxy from chaos.

Nash was initiated into the secret world of RAND along with a host of other mathematicians. After World War II, many of the mathematicians and scientists recruited for the war continued to be employed by military research organizations. And, of all the ideas that had come along during the war, Game Theory was seen as the most sophisticated tool. RAND was privy to the military's most highly guarded secrets.

Nash stood out as the oddest of the bunch at RAND. It seems that, on a number of occasions, the maintenance crew reported that Nash was observed to be "tiptoeing exaggeratedly along the avenue, stalking flocks of pigeons, and then suddenly rushing forward to try to kick them."

Sounds like a real swell guy. Early cruelty to animals is almost a dead giveaway to the psychopathic personality. When they become adults, they may even tell others about their childhood cruelty as an "ordinary event" of growing up. They will describe it as having been "enjoyable," and possibly assume that it is a common feature of maturation. The fact that Nash continued to regard living creatures as something to stalk and hurt is seriously disturbing.

In 1954, Nash was arrested in Palisades Park. He was charged with "indecent exposure." Richard Best, RAND's security manager, was informed of the arrest, and was reported to have said that Nash went into a public restroom and came on to another man by taking out his penis and masturbating; the only problem was, the man was a cop.

Nash had a top-secret security clearance. The security guidelines forbade anyone suspected of homosexual activity to hold a security clearance because, at the time, vulnerability to blackmail was an issue. Aside from that, the reckless nature of Nash's act indicated poor judgment. When Best confronted Nash with the news that his security clearance had been canceled, that he would have to go right then — that very minute — *Nash was neither shaken nor embarrassed.* Another sign of the psychopath. They don't have feelings, so they can never be embarrassed!

Nash didn't take it all that hard. He denied that he had been trying to pick up the cop and tended to scoff at the notion that he could be a homosexual. "I'm not a homosexual," Best quotes Nash as saying. "I like women." He then did something that puzzled Best and shocked him a little: "He pulled a picture out of his wallet and showed us a picture of a woman and a little boy. 'Here's the woman I'm going to marry and our son'...."

Best ignored the obvious psychopathic ploy and asked Nash for his version of the "event." Nash kept repeating that he was "merely observing behavioral characteristics." Yeah, right. That and a buck will get you a cup of coffee!

Sylvia Nasar writes about the incident, asking questions about Nash's possible internal reaction to this event. She asks:

What was going through Nash's mind in that interval? Was he angry? Depressed? Frightened? [...] Did he try to have RAND's decision reversed? Generally, of course, people did not. Fearful of scandal and aware of the contempt with which any hint of homosexuality was viewed, people in Nash's shoes were usually only too happy to slink away without murmur or protest.

In the end, Nash did what he had learned to do in less extreme circumstances. He acted, weirdly, *as if nothing had happened.* He played the observer of his own drama, as if it were all a game or some intriguing experiment in human behavior, focusing neither on the emotions of people around him nor on his own, but on moves and countermoves. [...] At some point he told his parents he'd had trouble with his RAND security clearance, blaming it on the fact that his mentor at MIT, Norman Levinson, was a former communist who had been hauled before HUAC that year. (Nasar, 1998; this author's emphasis)

A more typical description of the behavior and actions of a psychopath could hardly be imagined. But Nasar, like the rest of us, sought answers to Nash's behavior by assuming that he was like other people. She wondered about him being afraid of scandal or contempt. She just didn't get it that Nash did not have a conscience. Those things simply were not part of his make-up.

Conscience seems to depend on the ability to imagine consequences. But most "consequences" relate to pain in some way, and psychopaths really *don't understand pain in the emotional sense*. They understand frustration of not getting what they want, and *to them, that is pain*. But the fact seems to be that they act based solely on a sort of Game Theory evaluation of a situation: *What will they get out of it, and what will it cost?* These "costs" have nothing to do with being humiliated, causing pain, sabotaging the future, or any of the other possibilities that normal people consider when making a choice. In short, *it is almost impossible for normal people to even imagine the inner life of the psychopath*.

This leads us to what psychopaths *do* have that is truly outstanding: An ability to give their undivided attention to something that interests them intensely. Some clinicians have compared this to the concentration with which a predator stalks his prey. This is useful if one is in an environment with few variables, but most real life situations require us to pay attention to a number of things at once. *Psychopaths often pay so much attention to getting what they want that they fail to notice danger signals.*

> For example, some psychopaths earned reputations for being fearless fighter pilots during World War II, staying on their targets like terriers on an ankle. Yet, these pilots often failed to keep track of such unexciting details as fuel supply, altitude, location, and the position of other planes. Sometimes they became heroes, but more often, they were killed or became known as opportunists, loners, or hotshots who couldn't be relied on — except to take care of themselves. (Hare, 1999)

Nash demonstrated this quality to an extreme degree in the field of mathematics. However, Nash wasn't interested in mathematics for the sake of mathematics itself — the problem had to be important in the opinion of others, and thereby *likely to garner attention and glory to himself*. Nash wouldn't work on a problem unless he was assured that it was sufficiently important to "deserve" his attention. But, once he had decided, his "attention" was prodigious.

> His tolerance for solitude, great confidence in his own intuition, indifference to criticism — all detectable at a young age but now prominent and impermeable features of his personality — served him

well. [...] The most eloquent description of Nash's single-minded attack on the problem comes from Moser:

The difficulty [that Levinson had pointed out], to anyone in his right mind, would have stopped them cold and caused them to abandon the problem. But Nash was different. If he had a hunch, conventional criticisms didn't stop him. He had no background knowledge. It was totally uncanny. Nobody could understand how somebody like that could do it. He was the only person I ever saw with that kind of power, just brute mental power. (Nasar, 1998)

Again, it should be emphasized that psychopaths are interesting as all get out — even exciting! They exude a captivating energy that keeps their listeners on the edge of their seats. Even if some part of the normal person is shocked or repelled by what the psychopath says, they are like the mouse hypnotized by the torturing cat. Even if they have the chance to run away, they don't. Many psychopaths "make their living" by using charm, deceit and manipulation to gain the confidence of their victims. Many of them can be found in white-collar professions, where they are aided in their evil by the fact that most people expect certain classes of people to be trustworthy because of their social or professional credentials. Lawyers, doctors, teachers, politicians, psychiatrists and psychologists, generally do not have to earn our trust because they have it by virtue of their positions. But the fact is, psychopaths are found in such lofty spheres also!

At the same time, psychopaths are good impostors. They have absolutely no hesitation about forging and brazenly using impressive credentials to adopt professional roles that bring prestige and power. They pick professions in which the requisite skills are easy to fake, the jargon is easy to learn, and the credentials are unlikely to be thoroughly checked. Psychopaths find it extremely easy to pose as financial consultants, ministers, psychological counselors and psychologists. That's a scary thought.

Psychopaths make their way by conning people into doing things for them; obtaining money for them, prestige, power, *or even standing up for them when others try to expose them.* But that is their claim to fame. That's what they do, and they do it very well. What's more, the job is very easy because *most people are gullible with an unshakable belief in the inherent goodness of man.*

Manipulation is the key to the psychopath's conquests. Initially, the psychopath will feign false emotions to create empathy, and many of them study the tricks that can be employed in the "empathy" technique. Psychopaths are often able to incite pity from people because they seem like "lost souls," as Guggenbuhl-Craig writes. So the pity factor is one reason why victims often fall for these "poor" people.

Hare cites a famous case where a psychopath was "Man of the Year" and president of the Chamber of Commerce in his small town. (Remember that John Wayne Gacy was running for Jaycee President at the very time of his first murder conviction!) The man in question had claimed to have a Ph.D. from Berkeley. He ran for a position on the school board, which he then planned to parlay into a position on the county commission, which paid more.

At some point, a local reporter suddenly had the idea to check up on the guy — to see if his credentials were real. What the reporter found out was that *the only thing that was true about this up-and-coming politician's "faked bio" was the place and date of birth.* Everything else was fictitious. Not only was the man a complete impostor, he had a long history of antisocial behavior, fraud, impersonation and imprisonment. His only contact with a university was a series of extension courses by mail that he took while in Leavenworth Federal Penitentiary. What is even more amazing is the fact that before he was a con-man, he was a "con-boy." For two decades he had dodged his way across America one step ahead of those he had hoodwinked. Along the way he had married three women and had four children, and he didn't even know what had happened to them. Now, he was on a roll! But darn that pesky reporter!

When he was exposed, he was completely unconcerned. "These trusting people will stand behind me. A good liar is a good judge of people," he said. Amazingly, he was right. Far from being outraged at the fact that they had all been completely deceived and lied to from top to bottom, the local community he had conned so completely — in order to accrue benefits and honors to himself that he had not earned — rushed to his support!

I kid you not! It wasn't just "token support." The local Republican Party chairman wrote about him: "I assess his genuineness, integrity and devotion to duty to rank right alongside of President Abraham Lincoln." As Hare dryly notes, this dimwit was easily swayed by words, but was blind to deeds.

We understand this phenomenon from direct personal experience. The above case is almost an item-by-item mirror of our interaction with Vincent Bridges. After questions were raised about his credentials — side by side with our observation of his many activities, including vociferously blaming the victims (us) for refusing to be further victimized — we became acutely aware of his capacity for lying. It was, in fact, his publicly posted lies, as well the lies of both him and others in his gang, witnessed by ourselves and many others, that clued us in to his true nature. Had he behaved otherwise, he would be well on his way to more and better con-jobs with our blessings, given out of ignorance. However, our observation of the deceitful nature of his written discourses — the endless lies stacked on lies — naturally led to the idea that maybe everything he said was a lie, including his credentials. This idea turned out to be correct, but it didn't seem to matter. Surprisingly (to us, at least), there was no lack of people who were willing to compare Bridges to Abraham Lincoln because of his "genuineness, integrity and devotion to duty." That factor, of course, is what contributes to the success of the psychopath.

We observed this for some months, shaking our head in wonder at how many people seem to *want* to be duped, to be made fools of, and that is partly why we undertook to study the phenomenon more deeply. We wanted to know what kind of psychological weaknesses drive people to prefer lies over truth.

This may have something to do with what is called "cognitive dissonance." Leon Festinger developed the theory of cognitive dissonance in the 1950s when he apparently stumbled onto a UFO cult in the midwest. They were prophesying a coming world cataclysm and "alien rapture." When no one was raptured and no cataclysm occurred, he studied the believers' response and detailed it in his book, *When Prophecy Fails*. Festinger observed:

> A man with a conviction is a hard man to change. Tell him you disagree and he turns away. Show him facts or figures and he questions your sources. Appeal to logic and he fails to see your point.
>
> We have all experienced the futility of trying to change a strong conviction, especially if the convinced person has some investment in his belief. We are familiar with the variety of ingenious defenses with which people protect their convictions, managing to keep them unscathed through the most devastating attacks.
>
> But man's resourcefulness goes beyond simply protecting a belief. Suppose an individual believes something with his whole heart; suppose further that he has a commitment to this belief, that he has taken irrevocable actions because of it; finally, suppose that he is presented with evidence, unequivocal and undeniable evidence, that his belief is wrong: what will happen? The individual will frequently emerge, not only unshaken, but even more convinced of the truth of his beliefs than ever before. Indeed, he may even show a new fervor about convincing and converting other people to his view. (Festinger, 1964)

It seems that part of the problem has to do with ego and the need to be "right." People with a high need to be "right" or "perfect" seem to be unable to acknowledge that they have been conned. "There is no crime in the cynical American calendar more humiliating than to be a sucker." People will go along with and support a psychopath, in the face of evidence that they *are* being conned, because their own ego structure depends on being right, and to admit an error of judgment would destroy their carefully constructed image of themselves.

Even more amazing is the fact that when psychopaths do get exposed by someone who is not afraid to admit that they have been conned, the psychopath is a master at painting their victims as the "real culprits." We have experienced this first hand also with both Frank Scott and Vincent Bridges, as well as others, as will be seen further along. We were, indeed, interested to discover that we weren't the only ones. Hare cites a case of the third wife of a forty-year-old high-school teacher:

> "For five years he cheated on me, kept me living in fear, and forged checks on my personal bank account. But everyone, including my doctor and lawyer and my friends, blamed *me* for the problem. He had them so convinced that he was a great guy and that I was going mad, I began to believe it myself. Even when he cleaned out my bank account and ran off with a seventeen-year-old student, a lot of people couldn't believe it, and some wanted to know what I had done to make him act so strangely!"
>
> Psychopaths just have what it takes to defraud and bilk others: they can be fast talkers, they can be charming, they can be self-assured and at ease in social situations; they are cool under pressure, unfazed by the possibility of being found out, and totally ruthless. And even when they are exposed, they can carry on as if nothing has happened, often making their accusers the targets of accusations of being victimized by *them*.
>
> I was once dumbfounded by the logic of an inmate who described his murder victim as having benefited from the crime by learning "a hard lesson about life." (Hare, 1999)

The victims keep asking: "How could I have been so stupid? How could I have fallen for that incredible line of baloney?" And, of course, if they don't ask it of themselves, you can be sure that their friends and associates will ask: "How on earth could you have been taken in to that extent?" The usual answer — "You had to be there" — simply does not convey the whole thing. Hare writes:

> What makes psychopaths different from all others is the remarkable ease with which they lie, the pervasiveness of their deception, and the callousness with which they carry it out.
>
> But there is something else about the speech of psychopaths that is equally puzzling: their frequent use of contradictory and logically inconsistent statements that usually escape detection. Recent research on the language of psychopaths provides us with some important clues to

this puzzle, as well as to the uncanny ability psychopaths have to move words — and people — around so easily. (Hare, 1999)

Here are some examples:

When asked if he had ever committed a violent offense, a man serving time for theft answered, "No, but I once had to kill someone."

A woman with a staggering record of fraud, deceit, lies, and broken promises concluded a letter to the parole board with, "I've let a lot of people down.... One is only as good as her reputation and name. My word is as good as gold."

A man serving a term for armed robbery replied to the testimony of an eyewitness, "He's lying. I wasn't there. I should have blown his fucking head off." (Hare, 1999)

From an interview with serial killer Elmer Wayne Henley:

Interviewer: "You make it out that you're the victim of a serial killer, but if you look at the record you're a serial killer."

Henley: "I'm not."

I: "You're not a serial killer?"

H: "I'm not a serial killer."

I: You're saying you're not a serial killer now, but you've serially killed."

H: "Well, yeah, that's semantics."

And so on. The point that the researchers noted was that psychopaths seem to have trouble monitoring their own speech. What is more, they often put things together in strange ways, such as this series of remarks from serial-killer Clifford Olson: "And then I had *annual* sex with her." "Once a year?" "No. Annual. From behind." "Oh. But she was dead!" "No, no. She was just *unconscientious*." About his many experiences, Olson said, "I've got enough *antidotes* to fill five or six books — enough for a *trilogy*." He was determined not to be an "*escape goat*" no matter what the "*migrating facts*." (Hare, 1999)

Those of us who have had experiences with psychopaths know that the language of the psychopath is only two-dimensional. They are, as someone once said, as "deep as a thimble." An analogy is given of the psychopath as a color-blind person who has learned how to function in the world of color by special strategies. They may tell you that they "stopped at a red light," but what it really means to them is that they knew that the light at the top means "Stop," and they stopped. They call it the "red" light like everyone else, but *they have no experience of what "red" really is.*

A person who is color-blind and has developed such coping mechanisms, is virtually undetectable from people who see colors. I was shocked when my brother told me, *when we were in our thirties*, that he had been refused certain flight-related training in the Navy because he was color-blind. All I could think of was the many model cars he assembled when we were kids, and how he carefully selected the colors to paint them; all the while he was saying to me, "Isn't that a pretty red?" he had *no* idea what "red" really was. It was only in the Navy, when the tests were administered, that even *he* learned that he was color-blind. He still doesn't know what "red" is, though we have discussed endlessly his perceptions of color.

Psychopaths use words about emotions the same way people who are color-blind use words about colors that they cannot perceive. Psychopaths not only learn to use the words more or less appropriately, they learn to *pantomime* the emotion. But they never *have* the emotion.

This quality of the mind of the psychopath has been extensively tested with word-association tests while the subjects are hooked up to an EEG. In normal individuals, words that have emotional content evoke larger brain responses than do neutral words, which is apparently a reflection of the large amount of information that can be packed into a word. For most of us, the word *cancer* can instantly bring to mind not only the description of the disease, but also fear, pain, concern, or whatever, depending upon our experiences with cancer, whether we or someone we love has had it, or if it had some impact on our lives, and so on. The same is true with many words in our collective and individual vocabularies. And, unless we had a traumatic experience with it, a word such as box or paper will be neutral.

Psychopaths *respond to all emotional words as if they were neutral.* It is as if they are permanently condemned to operate with a Juvenile Dictionary. Hare writes:

> Earlier I discussed the role of "inner speech" in the development and operation of conscience. It is the emotionally charged thoughts, images, and internal dialogue that give the "bite" to conscience, account for its powerful control over behavior, and generate guilt and remorse for transgressions. This is something that psychopaths cannot understand. For them, conscience is little more than an intellectual awareness of rules others make up — empty words. *The feelings needed to give clout to these rules are missing.* (Hare, 1999; this author's emphasis)

What is more, just as the color-blind individual may never know he is color-blind unless he is given a test to determine it, the psychopath is *unable to even be aware of his own emotional poverty.* They assume that their own perceptions are the same as everyone else's. They assume that their own lack of feeling is the same for everyone else. Make no mistake about it: You can *not* hurt their feelings — because they don't have any! They will pretend to have feelings if it suits their purposes or gets them what they want. They will verbalize remorse, but their actions will contradict their words. They know that "remorse" is important, and "apologies" are useful, and they will give them freely, though generally in words that amount to blaming the victim for needing to be apologized to.

This is why they are so good at using Game Theory, and unless we learn the rules of how they think, they will continue to use it on us with devastating results. Normal people *hurt* when treated cruelly and insensitively. Psychopaths can only feign being "hurt" in the way that most people experience it — because they can only perceive "hurt" as not getting what they wanted, and tried to get by manipulation!

In the book *Violent Attachments*, women and men have noted the particular stare of the psychopath: It is an intense, relentless gaze that seems to precede his destruction of his victim or target. Women, in particular, have reported this stare, which is related to the "predatorial" (reptilian) gaze; it is as if the psychopath is directing all of his intensity toward you through his eyes, a sensation that one woman reported as a feeling of "being eaten." They tend to invade peoples' space either by their sudden intrusions or intimidating look-overs (which some women confuse for sexuality).

Another extremely interesting study had to do with the way psychopaths move their hands when they speak. Hand movement can tell researchers a lot about what are called "thought units." The studies indicate that psychopaths' thoughts and ideas are organized into small mental packages. This is handy for lying, but makes dealing with an overall, coherent and integrated complex of deep thoughts virtually impossible.

Most people are able to combine ideas that have consistent thought themes, but psychopaths have great difficulty doing this. Again, this suggests a genetic restriction to what we have called the Juvenile Dictionary. Not only are they using extremely restricted definitions, they cannot, by virtue of the way their brains work, do otherwise. Virtually all of the research on psychopaths reveals *an inner world that is banal, sophomoric, and devoid of the color and detail that generally exists in the inner world of normal people*. This goes a long way to explain the inconsistencies and contradictions in their speech.

"The situation is analogous to a movie in which one scene is shot under cloudy conditions and the next scene — which supposedly takes place a few minutes later — is shot in brilliant sunshine. […] Some moviegoers — the victims of psychopaths — might not notice the discrepancy, particularly if they are engrossed in the action." (Hare, 1999)

Psychopaths are notorious for not answering the questions asked them. They will answer something else, or in such a way that the direct question is never addressed. They also phrase things so that some parts of their narratives are difficult to understand. This is not careless speech, of which everyone is guilty at times, but *an ongoing indication of the underlying condition, in which the organization of mental activity suggests something is wrong.* It's not what they say, but how they say it, that gives insight into their true nature.

But this again raises the question: If their speech is so odd, why do smart people get taken in by them? Why do we fail to pick up the inconsistencies?

Part of the answer is that the oddities are so subtle that our general listening mode will not normally pick up on them. But my own experience is that some of the "skipped," oddly arranged words, or misused words are automatically reinterpreted by *our* brains in the same way we automatically "fill in the blank" space on a neon sign when one of the letters has gone out. We can be driving down the road at night and ahead see "M_tel," and then mentally put the "o" in place and read "Motel." Something like this happens between the psychopath and their victim. We fill in the "missing humanness" by filling in the blanks with our own assumptions, based on what *we* think, feel and mean. And, in this way, *because there are these "blank" spots*, we fill them in with what is inside us, and thus we are easily convinced that the psychopath is a great guy — because *he is just like us*! We have been conditioned to operate on trust, and we always try to give the "benefit of the doubt." So, when there are blanks, we "give the benefit of the doubt," and we are thereby hoisted on our own petard.

"Psychopaths view any social exchange as a 'feeding opportunity,' a contest or a test of wills in which there can be only one winner. Their motives are to manipulate and take, ruthlessly and without remorse" (Hare, 1999).

One psychopath interviewed by Hare's team said quite frankly: "The first thing I do is I size you up. I look for an angle, an edge, *figure out what you need and give it to you.* Then it's pay-back time, with interest. I tighten the screws." Another psychopath admitted that he never targeted attractive women; he was only interested in those who were insecure and lonely. He claimed he could smell a needy person "the way a pig smells truffles."

The callous use of the old, the lonely, the vulnerable, the disenfranchised, the marginalized, is a trademark of the psychopath.

When any of these victims wake up to what is happening, they are generally too embarrassed to complain.

One of the chief ways psychopaths prey on others is to make use of the normal person's need to find meaning or purpose in life. They will pose as grief counselors, or "experts" of various sorts that attract followings of people who are looking for answers. They are masters of recognizing the kind of "hang-ups" and self-doubts that most people experience, and they will brazenly pander to them to gain a "follower" to use later. Hare tells of a staff psychologist in a mental hospital whose life was destroyed by a psychopathic patient. He cleaned out her bank account, maxed out her credit cards, and then disappeared. How did he get to her? She said that her life had been "empty" and she had just simply succumbed to his sweet words and verbal caresses. As we already know, such words are cheap legal-tender to the psychopath. They can say "I'll pray for you," or "I love you," just to create an impression. It really, *really* doesn't mean a thing. But some people are so lonely and so desperate that even imitations are better than nothing.

Then, of course, there are people who are just simply so psychologically damaged themselves that the psychopath is the obvious choice for a partner. They may have a need to be treated badly, excited by danger, or to "rescue" or "fix" somebody whose soul is in obvious peril.

"In a book about Richard Ramirez, the Satan-worshipping 'Night Stalker,' the author described a young coed who sat through the pretrial hearings and sent love letters and photographs of herself to Ramirez. '*I feel such compassion for him. When I look at him, I see a real handsome guy who just messed up his life because he never had anyone to guide him,*' she is reported to have said" (Hare, 1999).

Sadly, as we see, psychopaths have no lack of victims, because so many people are ready and willing to play the role. In many, many cases, the victim simply refuses to believe the evidence that they are being victimized. Psychological denial screens out knowledge that is painful, and persons with large investments in their fantasies are often unable to acknowledge that they are being deceived, because it is too painful. Most often, these are women who rigidly adhere to the traditional role of the female, with a strong sense of duty to be a "good wife." She will believe that if she tries harder or simply waits it out, her husband will reform. When he ignores her, abuses her, cheats on her, or uses her, she can simply just decide to "try harder, put more energy into the relationship, and take better care of him." She believes that if she does this, eventually he will notice and will see how valuable she is, and then he will fall on his knees in gratitude and treat her like a queen.

117

Dream on.

The fact is, such a woman, with her fierce commitment to such a man, her dedication to being a proper wife, has allowed such fairy tales to distort her sense of reality. The reality is that she is doomed to a lifetime of abuse and disappointment until "death do us part."

One of the basic assumptions of psychotherapy is that the patient needs and wants help for distressing or painful psychological and emotional problems. *The psychopath does not think that they have any psychological or emotional problems, and they see no reason to change their behavior to conform to standards with which they do not agree.* They are well-satisfied with themselves and their inner landscape. They see nothing wrong with they way they think or act, and *they never look back with regret or forward with concern.* They perceive themselves as superior beings in a hostile world in which others are competitors for power and resources. They feel it is the optimum thing to do when they manipulate and deceive others in order to obtain what they want.

Most therapy programs only provide them with new excuses for their behavior, as well as new insights into the vulnerabilities of others. Through psychotherapy, they learn new and better ways of manipulating. What they do *not* do is make any effort to change their own views and attitudes.

One particular psychopath studied by Hare and his team of researchers was in a group-therapy program in a prison. The prison psychiatrist had written in his record: "He has made good progress.... He appears more concerned about others and to have lost much of his criminal thinking."

Two years later, Hare's staff member interviewed the man. At this point it ought to be made clear that, in order to make the research more accurate, the terms were that nothing said by the subjects to Hare or his staff could or would be repeated to the prison authorities; and they kept to their agreement in order to insure that the subjects felt free to talk to them. Psychopaths, if they know that they won't be penalized for what they express, are very happy to boast about their prowess in deceiving others. The man, assessed above by his prison psychiatrist as having made such remarkable improvement, was described by Hare's staffer as "the most terrifying" offender she had ever met and that he openly boasted about how he had conned the prison staff into thinking that he was well on the road to rehabilitation. "I can't believe those guys," he said. "Who gave them a license to practice? I wouldn't let them psychoanalyze my dog! He'd shit all over them just like I did."

Psychopaths are not "fragile" individuals, as Robert Hare reminds us after years of research. What they think and do is produced from a "rock-solid personality structure that is extremely resistant to outside influences." Many of them are protected for years from the consequences of their behavior by well-meaning family and friends. As long as their behavior remains unchecked or unpunished, they continue to go through life without too much inconvenience.

Some researchers think that psychopathy is the result of some attachment or bonding difficulty as an infant. Dr. Hare has turned that idea around, after all his years digging into the background of psychopaths: "In some children the very failure to bond is a symptom of psychopathy. It is likely that these children lack the capacity to bond readily, and that their lack of attachment is largely the result, not the cause, of psychopathy" (Hare, 1999). In other words: They are born that way and you can't fix them.

To many people, the idea of a child psychopath is almost unthinkable. But the fact is, *true* psychopaths are born, not made. Oh, indeed, there are psychopaths that are "made," but they are generally different from the "born psychopath" in a number of ways.

The fact is, clinical research clearly demonstrates that psychopathy does not spring into existence in adulthood, unannounced. The symptoms reveal themselves in early life. It seems to be true that parents of psychopaths *know* something is dreadfully wrong, even before the child starts school. Such children are stubbornly immune to socializing pressures. They are "different" from other children in inexplicable ways. They are more "difficult," "willful," or "aggressive," or "hard to relate to." They are difficult to get close to, cold and distant, and self-sufficient.

One mother said: "We were never able to get close to her even as an infant. She was always trying to have her own way, whether by being sweet, or by having a tantrum. She can put on a sweet and contrite act...."

The fact is, childhood psychopathy is a stark reality, and failing to recognize it can lead to years of vain attempts to discover what is wrong with a child, and the parent blaming themselves. Hare writes:

> As the signs of social breakdown grow more insistent, we no longer have the luxury of ignoring the presence of psychopathy in certain children. Half a century ago Hervey Cleckley and Robert Lindner warned us that our failure to acknowledge the psychopaths among us had already triggered a social crisis. Today our social institutions — our schools, courts, mental health clinics — confront the crisis every day in a thousand ways, and the blindfold against the reality of psychopathy is still in place. [...]
>
> The last decade has seen the emergence of an inescapable and terrifying reality: a dramatic surge of juvenile crime that threatens to

overwhelm our social institutions. [...] Children under the age of ten who are capable of the sort of mindless violence that once was reserved for hardened adult criminals. [...] At this writing, a small town in a western state is frantically searching for ways to deal with a nine-year-old who allegedly rapes and molests other children at knife point. He is too young to be charged and cannot be taken into care because "such action may only be taken when the child is in danger, not his victims," according to a child protection official. (Hare, 1999)

Why does it seem that we have a veritable epidemic of psychopaths? Sociobiologists are suggesting that increasing psychopathy is an expression of a particular genetically-based reproductive strategy. Simply put, most people have a couple of children and devote a lot of time and effort to their care. Psychopaths systematically mate with and abandon large numbers of women. They waste little of their energy raising children, and in this way, psychopathic genes are being propagated like wildfire. The sociobiologists aren't saying that the sexual behavior of people is consciously directed, only that "nature" has made them a certain way so that it will happen effectively.

The behavior of female psychopaths reflects the same strategy. "I can always have another," one female psychopath coldly replied, when questioned about an incident in which her two-year-old daughter was beaten to death by one of her many lovers. When asked why she would want to have another child (two had been taken into protective custody), she said "I love children." Again we see that the expressed emotion is in contradiction to the behavior.

Cheating skills seem to have an adaptive value in our society. The fact is, psychopaths often end up on the top of the heap — John Forbes Nash, for example.

At the present time, there is something very scary going on in the metaphysical community: Talk about the so-called "Indigo Children." One of the chief promoters of this idea, Wendy Chapman, writes:

Indigo Children are the current generation being born today and most of those who are eight years old or younger. They are different. They have very unique characteristics that set them apart from previous generations of children. [...] These are the children who are often rebellious to authority, nonconformist, extremely emotionally and sometimes physically sensitive or fragile, highly talented or academically gifted and often metaphysically gifted as well, usually intuitive, very often labeled ADD, either very empathic and compassionate *or* very cold and callous, and are wise beyond their years. Does this sound like yourself or your child?

Indigos have come into this world with difficult challenges to overcome. Their extreme levels of sensitivity are hard to understand and appreciate by parents who don't share this trait. Their giftedness is unusual in such high numbers. Their nonconformity to systems and to discipline will make it difficult to get through their childhood years and perhaps even their adult years. It is also what will help them accomplish big goals such as changing the educational system, for instance. Being an Indigo won't be easy for any of them, but it foretells a mission. The Indigo Children are the ones who have come to raise the vibration of our planet! These are the primary ones who will bring us the enlightenment to ascend.[27]

Sounds like a severe case of denial and wishful thinking, in my opinion. But, as we already understand, the psychological reality is merely a tool for the "theological reality." I suspect that the reader already has jumped ahead of me here and realizes what a big snow-job this "Indigo Children" deal is. Ms. Chapman has kindly provided a check-list to determine an "Indigo Child." After learning what we have about psychopaths, let's have a look at her list:

Have strong self esteem, connection to source.

Know they belong here until they are told otherwise.

Have an obvious sense of self.

Have difficulty with discipline and authority.

Refuse to follow orders or directions.

Find it torture to wait in lines, lack patience.

Get frustrated by ritual-oriented systems that require little creativity.

Often see better ways of doing things at home and at school.

Are mostly nonconformists.

Do not respond to guilt trips, want good reasons.

Get bored rather easily with assigned tasks.

Are rather creative.

Are easily distractible, can do many things at once.

Display strong intuition.

Have strong empathy for others or *no* empathy.

Develop abstract thinking very young.

Are gifted and/or talented, highly intelligent.

Are often identified or suspected of having ADD or ADHD, but can focus when they want to.

Are talented daydreamers and visionaries.

Have very old, deep, wise-looking eyes.

Have spiritual intelligence and/or psychic skills.

Often express anger outwardly rather than inwardly and may have trouble with rage.

Need our support to discover themselves.

---

[27] http://www.metagifted.org/ topics/metagifted/indigo/

Are here to change the world, to help us live in greater harmony and peace with one another and to raise the vibration of the planet.

What we see above is a list that includes certain definitive psychopathic behaviors along with behaviors of gifted children. We have to wonder at the attempt to weave the two together.

Where did this idea of "Indigo Children" come from? The phrase "Indigo Child" was coined by Nancy Ann Tappe in her book, *Understanding Your Life Through Color* (1982), and refers to the color in these children's auras. Ms. Tappe was interviewed by Jan Tober for her book *The Indigo Children* (1999), and said: "These young children — every one of them I've seen thus far who kill their schoolmates or parents — have been Indigos."

That didn't stop Tober from writing her book and declaring that these children are "Spiritual Masters, beings full of wisdom, here to teach us a new way of being." The way the followers of her ideas justify the fact that "not all Indigo children are filled with unconditional love, tolerance and non-judgment," is by declaring that they require "special" treatment, handling with kid gloves because they are so special and delicate and sensitive.

In a pig's eye. They are psychopaths and they are here for an altogether different reason. Somehow, some force is trying to make sure that its offspring are well cared for, and that a lot of psychopaths grow up without being identified for what they are.

Nevertheless, there is no explaining the extremes that "true believers" will go to in order to find excuses for inexcusable things. Elizabeth Kirby, a businesswoman in southern California, who has "studied and practiced metaphysics for the last 21 years," writes:

In hearing about the school shootings, I knew Indigo children were pulling the triggers. The Columbine High School shooting was so horrific it caught everyone's attention. At the time my eldest daughter said to me, "Because they (Eric Harris and Dylan Klebold) were Indigos they wanted to do it, so they just did it. No remorse, no guilt, they just went ahead and shot all those people because they wanted to and felt they needed to." Indigo children don't have guilt to keep them in check and because they balk at authority they don't believe they have to follow the rules.

Writers in mainstream America like Jonathan Kellerman are lumping the Indigo school shooters with the psychopaths; the dark entities who are bullies, con-men, stalkers, victimizers, serial killers and those who kill for thrills. I don't believe these Indigo children who have taken weapons to school to harm other children are psychopaths. They have been bullied and teased and have an avenger attitude seeking justice for injuries inflicted on them. They aren't killing just for the thrill of killing.

These kids know changes have to be made within the school system and they chose violence to make their statement, to give us a wake up call. Some of these metaphysical Indigo children are not hesitant about using violence to bring about change, and to bring us to enlightenment.

Indigo violence is here and it will continue, at least with this present generation of Indigo children. We are seeing with the current Indigo violence how the school system needs to be changed and how imperative it is to address the issues of bullying and intimidation in school. As the Indigo children grow to adulthood, their agendas will move out of the school system into our other systems, our social, political and judicial systems. Timothy McVeigh, the Oklahoma City bomber, is an Indigo.[28]

Amazing, huh? Did you catch the remark, "Some of these metaphysical Indigo children are not hesitant about using violence to bring about change, and to bring us to enlightenment"? Don't we find that just a tiny bit contradictory? Aren't we stretching a bit? How about diving straight into denial?

Oy.

At the present moment in history, the appeal of the psychopath has never been greater. Movies about psychopaths are all the rage. Hare asks: "Why? What accounts for the terrific power that the personality without conscience has over our collective imagination?" One theorist proposes that people who admire, believe, or identify with psychopaths, are partly psychopathic themselves. By interacting with a psychopath, even peripherally, they are able to voyeuristically enjoy an inner state not dominated by the constraints of morality. Such people are enabled to enjoy aggressive and sexual pleasures at no cost.

For normal people, such movies may serve to remind them of the danger and destructiveness of the psychopath. They will shiver with the sense of something cold and dark having breathed on their neck. For others, people with poorly-developed inner selves, such movies and glorification of psychopathic behavior only serves as a role model for serious acts of violence and predation against others.

That brings us back to Nash. Remember Nash? Sylvia Nasar proposes some small rationalizations for Nash's psychopathic behavior, suggesting that he was just an "inward looking child" who reacted to "intrusive adults by withdrawing further into his own private world." Funny, I thought the same thing about Frank Scott. Nasar does admit that, based on the fact that there is no evidence of any trauma, any abuse, and only the most loving and educational environment, that Nash's temperament must have been "one that he was born with."

---

[28] "Understanding Indigo Violence,"
http://www.planetlightworker.com/articles/elizabethkirby/article1.htm

Well, no surprise there! However, the only difference that family background seems to make is how the psychopath expresses himself. A psychopath who grows up in a stable family and has access to positive social and educational resources might become a white-collar criminal, or perhaps a somewhat shady entrepreneur, politician, lawyer, judge or other professional. Another individual with the same traits and a deprived background, might become a common con-artist, a drifter, mercenary or violent criminal.

The point is, social factors and parenting practices only shape the expression of the disorder, but *have no effect on the individual's inability to feel empathy or to develop a conscience.*

In Nash's case, he became a mathematician. The story of how this happened is extremely interesting. I would like to suggest the exercise of reading Robert Hare's book, *Without Conscience,* followed immediately by Sylvia Nasar's book about John Nash. It will soon become apparent how a psychopath could make such a contribution to science and win a Nobel Prize in *economics.*

Robert Hare once submitted a paper to a scientific journal. The paper included EEGs of several groups of adult men performing a language task. The editor of the journal returned the paper saying, "Those EEGs couldn't have come from real people." But they did. They were the EEG's of psychopaths.

Some people have compared psychopathy to schizophrenia. However, there is a crucial distinction, as we will see:

Schizophrenia and psychopathy are both characterized by impulsive, poorly planned behavior. This behavior may originate from a weak or poorly coordinated response inhibition system. We tested the hypothesis that schizophrenia and psychopathy are associated with abnormal neural processing during the suppression of inappropriate responses.

The participants were schizophrenic patients, non-psychotic psychopaths, and non-psychotic non-psychopathic control subjects (defined by the Hare "Psychopathy Checklist, Revised"), all incarcerated in a maximum security psychiatric facility. We recorded behavioral responses and event-related potentials (ERPs) during a Go/No-Go task.

Results: Schizophrenic patients made more errors of commission than did the non-psychopathic offenders. As expected, the non-psychopathic non-psychotic participants showed greater frontal ERP negativity (N275) to the No-Go stimuli than to the Go stimuli. This effect was small in the schizophrenic patients and *absent in the psychopaths.* For the non-psychopaths, the P375 ERP component was larger on Go than on No-Go trials, a difference that was absent in schizophrenic patients and *in the opposite direction in psychopaths.*

Conclusions: These findings support the hypothesis that the neural processes involved in response inhibition are abnormal in both schizophrenia and psychopathy; however, *the nature of these processes appears to be different in the two disorders.* (K. A. Kiehl, A. M. Smith, R. D. Hare, P. F. Liddle, "An event-related potential investigation of response inhibition in schizophrenia and psychopathy," *Biological Psychiatry,* 2000, vol. 48, n° 3, pp. 173-183 (2 p.), pp. 210-221. This author's emphases)

"More and more data are leading to the conclusion that psychopathy has a biological basis and has many features of a disease," says Sabine Herpertz, a psychiatrist at the RWTH-Aachen University in Germany (*Nature*, 410, March 15, 2001).

The brain imaging techniques of positron emission tomography (PET) and magnetic resonance imaging (MRI) provide the opportunity to investigate psychopathy further. They might allow researchers to discover whether psychopaths' physiological and emotional deficits can be pinned down to specific differences in the anatomy or activation of the brain.

Among researchers who are starting to explore this area, there are two main theories of psychopathy. One, championed by Adrian Raine of the University of Southern California in Los Angeles and supported by the work of Antonio Damasio of the University of Iowa, gives a starring role to a brain region called the orbitofrontal cortex. This is part of an area of the brain, known as the prefrontal cortex, involved in conscious decision-making.

The other theory, promoted by James Blair of University College London, holds that the fundamental dysfunction lies within the amygdala, a small almond-shaped structure that plays a critical role in processing emotion and mediating fear. Recently, using PET scanning, Blair has shown that activation of the amygdala in normal volunteers is involved in responding to the sadness and anger of others, and he hypothesizes that amygdala dysfunction could explain the lack of fear and empathy in psychopaths.

Blair points out that the two theories may not be mutually exclusive, as the orbitofrontal cortex, which does the "thinking", and the amygdala, which does the "feeling," are highly interconnected.

Following widespread concern that the criminal justice and mental health systems are failing to deal effectively with dangerous psychopaths, there is a movement in several countries to instigate fundamental legal reform. The most controversial suggestion is to make it possible for individuals who have severe personality disorders to be detained in secure mental institutions, even if they have been accused of no crime. Although these particular provisions have alarmed civil-liberties campaigners, the raft of measures also includes a major initiative within the prison service to improve the handling of those with Antisocial Personality Disorder (APD) — including psychopaths.

According to one individual who suffered at the hands of a psychopath: "The World has only one problem, Psychopaths. There are two basic types of Psychopaths, Social and Anti-Social. The essential feature of Psychopaths is a Pervasive, Obsessive-Compulsive desire *to force their delusions on others*. Psychopaths completely disregard and violate the Rights of others, particularly the Freedom of Association which includes the right not to associate and the Right to Love."

We have come full circle. Over and over again we come up against that little problem: Religion and belief systems that have to be defended against objective evidence or the beliefs of others. We have to ask ourselves where these belief systems came from that so evidentially are catastrophic. Then, we have to think about the fact that now, in the present day, when many of these systems are breaking down and being replaced by others that similarly divert our attention away from what *is*, it becomes necessary to "enforce" a certain mode of thinking. That is what psychopaths do best.

Psychopaths dominate and set the standard for behavior in our society. We live in a world based on a psychopathic, energy-stealing food chain, because that's just the way things are. Most people are so damaged they no longer have the capacity to even imagine a different system, based on a symbiotic network.

> They are not only damaged by others, but also by the thousand little evils they have done to others to survive. For them to see the system for what it is would require them to see the part they have played in perpetuating it. That is a lot to ask of a fragile ego. Also, those who are not psychopaths, still want to make human connections but are afraid to, for fear of being taken advantage of and stolen from energetically speaking. [Thank you S.M. for such a clear explication!]

With the brief historical review we have examined, we are acutely aware that this is *not* a phenomenon confined to our present "time." It is a trans-millennial program that, step by step, has brought us to our present position. What emerges in the present day is just Machiavellian diversion that focuses the attention of those who are easily deceived. This is reinforced by the "clappers" in the audience, and there seems to be an entire army of psychopaths among us whose job it is act as vectors of attention and direction. We hope that the readers of these pages will give themselves permission to imagine, research and implement a different way of being, and to stand up for themselves while doing it.

12-14-96

Q: (L) Along the lines of some of the things that I have been working on recently, I'd like to ask if there's any more information you can give to us about the hypnotic-opener strobe-effect, and what it is preventing us from seeing. Is this one of the things that keeps us from expanding into the next density, in terms of awareness?

A: Not related to that. You see, the souls that are affected by all these "cloaking" techniques are vibrating on a low level anyway. The point is to block those who are blockable.

Q: (T) We're not blockable? (L) Is there anything we can do to avoid this blocking? (T) We're not being blocked....

A: You are not blockable.

Q: (T) We are not being blocked. We're beyond the blocking.

A: If you were, would you be doing this?

Q: (T) That type of blocking technique doesn't work on us. There may be other blocking techniques, but that particular stuff doesn't work. We either see or don't see stuff, because we are either meant or not meant to see it. We don't see UFOs anymore, because we don't need to.

A: Not necessarily true.

Q: (L) OK, what is not necessarily true? Why don't we see them any more?

A: "Don't" does not equal "won't." If a Buick does not go by, you don't see that, either! And if you are inside doing the laundry when Mr. Jones decides to take the old "Electra" for a spin, you do not see him, or his precious car, do you?

Q: (L) I don't care, I've seen enough!

A: Oh, yes, you do care!!!

Q: (L) OK, yes, I care, but I've seen enough. I believe, I believe!!!

A: It is not up to you whether you want to see them or not. If they want you to see them, you will!

Q: (T) So, if they want us to see them, we'll see them!

A: Yes, and they will, and you will!

Q: (T) They will and we will... yes, but, there's a blocking technique being used on people to lower the vibrational frequency to prevent them from seeing them, right?

A: The blocking technique is for many things.

Q: (T) So that people do not understand what's going on around them.

A: Yes. [...] That is it, in a nutshell. See and know and think or... see, know and think that which is desired.

05-03-97

Q: The way I understood it is that a person can be an EM [electromagnetic] vector. Is that possible?

A: "Vector" means focuser of direction.

Q: Could that mean that EM waves can be vectored by a human being simply by their presence? I also noticed that several of us have been involved with persons and relationships that seem designed to confuse, defuse and otherwise distort our learning, as well as drain our energy. Basically, keeping us so stressed that we cannot fulfill our potential. Is there some significance to this observation?

A: That is elementary, my dear Knight! [...]

Q: One of the things I have learned is that these individuals seem to attach via some sort of psychic hook that enters through our reactions of pity. Can you comment on the nature of pity?

A: Pity those who pity.

Q: But, the ones who are being pitied, who generate sensations of pity, do not really pity anybody but themselves.

A: Yes...?

Q: Then, is it true, as my son said, when you give pity, when you send love and light to those in darkness, or those who complain and want to be "saved" without effort on their own part, when you are kind in the face of abuse and manipulation, that you essentially are giving power to their further disintegration, or contraction into selfishness? That you are powering their descent into STS?

A: You know the answer!

Q: Yes. I have seen it over and over again. Were the individuals in our lives selected for the extremely subtle nature of their abilities to evoke pity, or were we programmed to respond to pity, so that we were blind to something that was obvious to other people?

A: Neither. You were selected to interact with those who would trigger a hypnotic response that would ultimately lead to a drain of energy.

Q: (T) Well, it is a fact, because my energy sure is drained. (L) What is the purpose of this draining of energy?

A: What do you think?

Q: (T) So you can't concentrate or do anything. You can't get anywhere with anything.

A: Or, at least not the important things. [...]

Q: (L) Why is it that one of the primary things about us that seems to prevents us from acting against such situations, is our fear of hurting another person? [...] Why are we so afraid of hurting someone's feelings if they are hurting us?

A: Not correct concept. You do not need to "act against them," you need to act in favor of your destiny.

Q: But, when you do that, these persons make you so completely miserable that there seems to be no other choice but a parting of the ways.

A: Yes, but that is not "acting against." Quite the contrary. In fact, remember, it takes two to tango, and if you are both tangoing when the dance hall bursts into flames, you both get burned!!!

Q: Why is it that when one tries to extricate from such a "tango"... why is there is such violent resistance to letting you go when it is obvious, clearly obvious, that they do not have any feeling for you as a human being?

A: It is not "they." We are talking about conduits of attack. [...]

Q: Is it true that being in the presence of such people, that one is under the influence of an energy, an emanation from them physically, that befuddles the mind and makes it almost impossible to think one's way out of the situation?

A: It is the draining of energy that befuddles the mind.

Q: Where does this energy drain to?

A: 4th density STS.

Q: They drain our energy from us, and 4th density STS harvests it from them?

A: "They" do nothing!!!! 4th density STS does it all through them!

Q: (T) Well, I would like to know what it is in us that makes us attracted to such people.

A: It was the idea of 4th density STS.

Q: That means that they can control your thoughts and emotions, put ideas into your head, and you think it is a good idea to "save" someone. You don't know. It is taught in our religions and culture to give until it hurts, and, in fact, to give *because* it hurts. The whole situation is designed and controlled from another level. Any further comment on this subject?

A: Once you have truly learned the program, just plug it in.

Q: I guess once you have truly learned what is being said here, just plug it in....

A: No. We mean that all you have to do is learn the patterns of behavior, the subtle signs, and you will always have the ability of avoiding it. Your own as well as others. [...] Lesson number 1: Always expect attack. Lesson number 2: Know the modes of same. Lesson number 3: Know how to counteract same. [...] When you are under attack, expect the unexpected, if it is going to cause problems. [...] But, if you expect it, you learn how to "head it off," thus neutralizing it. This is called vigilance, which is rooted in knowledge. And, what does knowledge do?

*

"Why did man, through thousands of years, wherever he built scientific, philosophic, or religious systems, go astray with such persistence and *with such catastrophic consequences*? [...] The answer lies somewhere in that area of our existence which has been *so heavily obscured by organized religion* and put out of our reach. Hence, it probably lies in the relation of the human being to the cosmic energy that governs him." (Reich, 1949)

*

"I want to appeal to your analytical mind.... Think for a moment, and tell me how you would explain the contradiction between the intelligence of man the engineer and the stupidity of his systems of beliefs, or the stupidity of his contradictory behavior. Sorcerers believe that the predators have given us our systems of beliefs, our ideas of good and evil, our social mores. They are the ones who set up our hopes and expectations and dreams of success or failure. They have given us covetousness, greed and cowardice. It is the predators who make us complacent, routinary, and egomaniacal.

"In order to keep us obedient and meek and weak, the predators engaged themselves in a stupendous maneuver — *stupendous, of course, from the point of view of a fighting strategist.* A horrendous maneuver from the point of view of those who suffer it. *They gave us their mind!* Do you hear me? The predators give us their mind, which becomes our mind. [...] Through the mind, which, after all, is their mind, *the predators inject into the lives of human beings whatever is convenient for them.*" (Castaneda, 1998)

*

"So that in the actual situation of humanity there is nothing that points to evolution proceeding. On the contrary, when we compare humanity with a man, we quite clearly see a growth of personality at the cost of essence, that is, a growth of the artificial, the unreal, and what is foreign, at the cost of the natural, the real, and what is one's own.

"Together with this, we see a growth of automatism. Contemporary culture requires automatons. [...] One thing alone is certain, that man's slavery grows and increases. Man is becoming a willing slave. He no longer needs chains. He begins to grow fond of his slavery, to be proud of it. And this is the most terrible thing that can happen to a man." (Ouspensky, 1949)

*

Intolerance and cruelty are *needed* to guarantee the "cover-up." As we've already repeated: A certain kind of "human being" acts on behalf of this cover-up. In this sense, psychopaths, as Alien Reaction Machines, are the playing pieces in the Secret Games of the Gods.

*

Note: Since writing the chapters on John Nash, we have discovered the work of Polish psychologist Andzrej Lobaczewski, author of *Political Ponerology.* According to Lobaczewski, Polish psychologists had a highly developed understanding of psychopathy that was suppressed and destroyed by the communist regime. A reading of his book shows that their understanding of psychopathy was way ahead of its time. They had already determined that psychopathy was an inherited (not behavioral, and thus not "curable") personality type that shows itself in a spectrum. The obvious psychopaths (like those described by Cleckley and Hare) play a lesser role in mass social problems, while the "socially-compensated" psychopaths are much harder to identify. Their "masks of sanity" are more consistent. Just recently, Robert Hare and Paul Babiak wrote a book on this type, called *Snakes in Suits*.

This science also included an understanding of various different types of inherited "psychopathies." These include various "personality disorders," known to Westerners as "borderline," "schizoid," "obsessive-compulsive," "histrionic," "dependent," etc. However, Lobaczewski points out that the Western "types" often overlap and are imprecise. As such, various disorders (like frontal brain damage) are often confused with psychopathy, which Lobaczewski calls "essential psychopathy."

One such psychopathy is "schizoidal psychopathy," which seems to better describe John Nash. Lobaczewski describes schizoidal psychopathy in the following terms (this author's emphasis in *italics*):

From the beginning, [schizoidia, or schizoidal psychopathy] was treated as a lighter form of the same hereditary taint which is the cause of susceptibility to schizophrenia. However, *this latter connection could neither be confirmed nor denied* with the help of statistical analysis, and no biological test was then found which would have been able to solve this dilemma. [...] Carriers of this anomaly are *hypersensitive and distrustful*, while, at the same time, *pay little attention to the feelings of others*. They tend to *assume extreme positions*, and are *eager to retaliate* for minor offenses. Sometimes *they are eccentric and odd*. Their poor sense of psychological situation and reality leads them to superimpose erroneous, pejorative interpretations upon other people's intentions.

They easily become involved in activities which are ostensibly moral, but which actually inflict damage upon themselves and others. *Their impoverished psychological worldview makes them typically pessimistic regarding human nature.* We frequently find expressions of their characteristic attitudes in their statements and writings: "Human nature is so bad that order in human society can only be maintained by a strong power created by highly qualified individuals in the name of some higher idea." Let us call this typical expression the "schizoid declaration."

Human nature does in fact tend to be naughty, especially when the schizoids embitter other people's lives. When they become wrapped up in situations of serious stress, however, the schizoid's failings cause them to collapse easily. *The capacity for thought is thereupon characteristically stifled, and frequently the schizoids fall into reactive psychotic states so similar in appearance to schizophrenia that they lead to misdiagnoses.*

The common factor in the varieties of this anomaly is *a dull pallor of emotion* and lack of feeling for the psychological realities, an essential factor in basic intelligence. This can be attributed to some incomplete quality of the instinctive substratum, which works as though founded on shifting sand. *Low emotional pressure enables them to develop proper speculative reasoning, which is useful in non-humanistic spheres of activity, but because of their one-sidedness, they tend to consider themselves intellectually superior to "ordinary" people.* [...]

A schizoid's ponerological activity should be evaluated in two aspects. On the small scale, such people cause their families trouble, easily turn into tools of intrigue in the hands of clever and unscrupulous individuals, and generally *do a poor job of raising children.* Their tendency to see human reality in the doctrinaire and simplistic manner they consider "proper" — i.e. "black or white" — transforms their frequently good intentions into bad results. However, *their ponerogenic role can have macrosocial implications if their attitude toward human reality and their tendency to invent great doctrines are put to paper and duplicated in large editions.*

Chapter 59
An Encounter with the Unicorn

The reader now knows a whole lot more than I did in the spring of 1999 when Vincent Bridges first wrote to me, after Ray Flowers forwarded to him my post to the Ancient Wisdom discussion list. And, as the reader who has read the emails will note, I was sending him a *lot* more information than I was receiving. Until this period of recapitulation, I never connected the interaction with Bridges to the things that followed immediately after, but looking back, it all seems very strangely synchronous, in the same way as the black cat walking by twice in the movie *The Matrix*.

In early June of 1999, a correspondent wrote and asked if I would like to join a discussion group on the subject of UFOs and aliens. For the most part, I steered clear of such things because I didn't see too much fruitful research being done, and I had also been "burned" by my interaction with the Gray-hugging crowd at Mike Lindemann's ISCNI ("Institute for the Study of Contact with Non-human Intelligence"). Nevertheless, because I was asked, I went over and had a look at the discussion board, and decided to make one "trial post." Since I was crusading on behalf of Karla Turner, I thought that I would include a quote from her writings about abductions in an effort to make more people aware of some of her conclusions.

Looking back, of course, it's easy to see that Karla's death on January 9, 1996, is a perfect example of how "forces" might deal with someone who was getting entirely too close to the truth. She wasn't "martyred," in the obvious way of Jessup and Schneider and others, she was just quietly "disposed of." In 1995, Karla contracted a very dangerous form of breast cancer immediately following an abduction experience. She died at the age of 48. Karla's husband, Elton, had published the following remarks, which are more timely today than ever (the first portion of this article is quoted in book one of *The Wave*):

We, as a race, have never been free to discover our own true identity. Every social advance we attempt is thwarted by some maniac who springs up with almost divine grace to lead us into madness. Saint Paul, for instance, seems to have taken the real message of Jesus and his earliest followers and distorted it into something that we kill, lie and cheat for. And, in spite of all that, we still aspire for redemption of our souls. The followers of that doctrine — Christians, they call themselves — are not the only ones who behave in such a manner. Every major religion has managed to find an excuse in its teachings to destroy non-believing fellow human beings. A part of me shudders every time I hear of yet another killing based on 2,000-year-old hatreds.

What law allows us to continue with such atrocities? What influence keeps such hatred and fears alive? Why are we abductees so afraid to ask for real help from our own society?

We have been INVADED — but I do not yet believe that the battle is over. Invasion with sticks, knives or guns is a human reality, not necessarily a universal one. There are very sophisticated mechanisms being used in the invasion of our world. Why should our invaders use pointed sticks against us when they can get us to sharpen sticks and use them against each other? We provide them with everything they want from us, and they take none of the blame for our misery. They just zip around in their wonderful flying machines, dazzling us with their magical abilities and filling us with awe at their insight.

Can there be a more successful military campaign than one in which no shot is (apparently) fired and in which the conquered populace gladly and openly welcomes their enslavers? We are being programmed mentally and socially to accept our invaders as saviors, not a conquering force. I truly believe we are being deceived by smoke, mirrors and sleight-of-three-fingered-hand movements. Are we going to sell our birthright to some sneaky beings who appear on our shores in marvelous ships and offer us a few glitzy baubles?

The researcher asked if I personally knew of harm that has come to anyone at the hands of, or because of, the aliens. Yes, harm has come. My early youth was damaged severely by the unconscious knowledge that I was being used by some non-human agency. It took me 40 years to recognize that the fears which guided me daily were not of my own making and that the rebellion I constantly felt was engendered by my contempt for the powerful invisible agents that forced me to do things that I knew were wrong, even as I was doing them.

For example, I did not want to marry the person who became my first wife, yet I had no control over the decision. Before we were married, we were jointly abducted and subjected to severe programming. The results brought no happiness to either of us. We both starved for love and companionship, even though we tried with all our might to find them. My son (now 25) was also one of their subjects, and is miserable and lost. He is an artistic person with so many unknowable fears that he is paralyzed. I know of abductees MURDERED by mutilations (reports of which are suppressed immediately and completely), by cancers that no physician has ever seen before, and by madness that has led to suicide. In my opinion, these acts were not caused by "brothers" of any sort.

I do not believe that all is lost, however. I have felt a guiding hand that helped me to discover happiness and inner peace amid all this chaos and misery. What I have come to understand is that that hand is only there when I take responsibility for my own happiness and do something about whatever is bothering me.

Reality left in the hands of the invaders is neither what we need nor what we want. It is time that we think hard about ourselves and what we have on this gem of the universe, our home — our planet. There are laws governing the actions of our invaders, rules guiding their actions and patterns of behavior we can discover if we will make a concerted effort to discern them. We humans have something valuable that is desirable to, and usable by the alien forces acting on us. I feel it is time we take back that which is ours, that we use all our resources to discover the laws that govern reality and become the beings that we intrinsically know we are. (Turner, 1994)

The post I made to the UFO discussion board brought about an amazing flood of correspondence from abductees desperate for more rational information. I was overwhelmed with the realization that there were so many people looking for answers, and so much nonsense being propagated.

Apparently, one of the members of that discussion forwarded my post to a woman named Eve Lorgen, who had been working on a book titled *The Love Bite*, about the very types of situations described above by Elton Turner, wherein people are "programmed" to marry the wrong people, and then, because of the religious programming we receive from infancy, remained in marriages that were quite simply designed to diminish their capacity to grow as human beings. This is where we encounter a problem.

Over the years, as I had read about many instances of "personal growth" and "self-realization," accompanied by a complete breakdown and restructuring of the life of the individual, most often including divorce, the idea nagged at the back of my mind that "if this is a good thing, how come so many people are getting hurt by it?" Until I had faced my own realization that "love" (as human beings think of it and try to manifest it) really has nothing to do with real spiritual love at all, I was stuck in the idea that a manifestation of enlightenment ought to confer on a person the almost magical ability to "fix" anything or anybody! If they were married to a monster, their "enlightenment" ought to enable them to "soothe the savage breast." If they were in a state of some kind of suffering for financial or health reasons, their enlightenment ought to show them the way out and back to physical and fiscal health. And, of course, it ought to all be accomplished without anybody getting hurt! There should be no divorce, no breaking of promises and commitments, and most of all, they ought to all be just dancing around the Maypole and gazing into one another's eyes in mutual admiration and delight.

By this time, I already knew that didn't work. In fact, continuing to support those on the STS pathway only fueled their descent into deeper STS, drained the energy of the "codependent STO" person so that they had nothing to share with those who *were* like them, which prevented the establishment and growth of an STO dynamic, and effectively only continued to feed and perpetuate STS energy in our lives and our world. But Eve was now talking about "alien-orchestrated human-bonding dramas." In short, she was saying that a lot of what we call "falling in love" is merely being chemically manipulated for nefarious purposes.

So, Eve wrote to me and wanted to talk shop. We did, extensively. She described her work, and her forthcoming book (www.alienlovebite.com), and I was very interested to discover that there were some few — very few — people who had a handle on some of these things, and were putting some of the pieces of the puzzle together in interesting ways.

A few days later, seemingly out of the blue, I received the following:

From: REngelm@....

Date sent: Tue, 15 Jun 1999 22:46:02 EDT

Subject: Being a guest on my radio program

My name is Ron Engelman.

I do a four-hour radio program on the Talk Radio Network from 2-6AM PDT, Monday through Friday, in 29 markets across the country. I have read small portions of your web site and I would be interested in having you as a guest on my show. I would be more than happy to allow you to promote any publications you may have, your web site, or any books you have in publication.

I look forward to hearing from you.

Ron

I wasn't too sure I wanted to do such a thing. I had the idea that just putting the material on the website, letting those who were actively looking to just "find it," and leave it at that. I wasn't interested in any kind of promotion or hoopla (still am not), and even if I had been told that I had a talent for public speaking, it wasn't among my favorite things to do. Yes, radio isn't exactly "public," but I had done it and all that had resulted in was Frank pouting and criticizing for weeks afterward, apparently because *he* hadn't been asked to be a guest on a radio show. I didn't see it as part of our "mission," whatever *that* was, and I didn't want to offend him.

Yes, I know, I was being manipulated by such emotional pressures, but it just seemed easier to "keep the peace" than to explain to Frank that nobody invited him to do anything because he didn't do any work, had nothing interesting to say, and when he did start talking, rambled so much that everyone in the room was soon fidgeting and wishing to be somewhere else. When you feel sorry for somebody who is constantly complaining that everyone has always been mean to them, and no one has ever listened to them, it's kind of hard to tell them that maybe there is a reason for it. Of course, you always think that if you just tolerate all the glitches, help them build up their self-esteem, that eventually they *will* become that dynamic and interesting person they say that they have always wanted to be.

But, rather than just say "No," in the event that the Universe was in favor of this radio interview, I decided to leave it open, but still register my reservations. Maybe if I acted hesitant, the guy would say "Forget it," and the decision would be out of my hands anyway? I wasn't even sure how he thought I could do this. Did I have to travel somewhere? If so, never mind. Funny how we play these mind games with ourselves. I wrote back to Mr. Engelman:

To Ron Engelman:

Date sent: Wed, 16 Jun 1999 19:51:28 -0400

Hi!

Thank you for your interest. I would like to know what the format is and the general "trend" of the subjects you cover. I would also like to know exactly HOW such an "appearance" would be engineered?

I don't really have anything to "promote" except my website, which is a total mess today after I was making changes.... And, I don't really think there is much of a market for my husband's latest book: Conference Proceedings of the Xth Max Born Symposium, "Quantum Future." It's not exactly gonna be a best seller!

And, I ought to tell you that I DID a talk show once in Tampa... WFLA...everyone agreed that it was a tremendous success... lots of callers and all that... but the Deejay was "terminated" very soon after and has never worked in radio again. He is now selling real estate. Of course, it COULD be unrelated... but he had a good "following."

And then, there was the television appearance... the newscaster who decided that I would be an interesting person to interview for a "human interest" segment on the evening news ALSO is no longer employed in the media. The only one who has survived interviewing me is a St. Pete Times journalist who has not finished the writing project, so nothing is yet in print. And may never be....

I would like to think about it after I have some more information, if that is okay.

Best,

Laura Knight-Jadczyk

I feel guilty to this day for Johnny Dollar getting canned from WFLA in Tampa back in 1995. The way it all happened was that I had a hypnosis client who came for the standard "relaxation" therapy. He turned out to be an executive at the corporate offices of a radio station conglomerate that owned, at the time, most of the radio stations in the area. It now owns *all* of them, or so I understand. In any event, this client, in addition to having a lot of fun with hypnosis, was very interested in all the other things we were doing and asked many questions. Over time, he came to know the full range of the material and the Cassiopaeans' perspective, and he just decided one day I ought to talk on the radio. I told him I didn't think I could do it, and he said he was sure I could, and he would set me up with a real pro who would manage everything so smoothly that I wouldn't have to do a thing!

Since Tom French was still hanging out with us, and I had been pretty much sworn to secrecy about his project, I thought I had better ask him what he thought about it, since I didn't want to have anybody "scoop" him, so to say. I had committed to his project, and if that meant I had to turn down other things, that was fine with me.

Well, Tom, of course, didn't want me to mention his research, but he was excited about the idea of me doing a radio show. He wanted to be there.

Frank, as usual, started a major program of instructing me on what to say and not say; it was pretty clear from everything he was saying that he must have been convinced I was an idiot. I think he still thinks so. But, at the time, I figured that he was just feeling threatened, and being in the "save Frank" mode, I was more inclined to try to find ways to appease him than anything else. I said that I would only do the radio show if the whole gang could go with me. Frank beamed. He was already dreaming of radio stardom. I decided, if that is what makes him happy, I'm going to try to finagle it so he can talk on the radio. I was told that the space in the room that the broadcast was done was limited, so with Frank, another group member, Tom French and Cherie Diez, the Times photographer, it was going to be pretty tight; and, as my friend in the business confided to me: Better keep Frank off the air or he would make all of us look stupid. I was a little bit offended on Frank's behalf, but I knew exactly what he meant. Frank had to be protected even from himself. But, I didn't have to tell him that. He could still go and be there for the show and have a little fun. Terry and Jan agreed to listen and tape the show from home.

When we arrived at the station to do the show, I was surprised to discover that Johnny Dollar had made a sort of "event" out of it. He had made arrangements for Mike Lindemann and some other woman to be participants via telephone, and it somehow ended up being a debate over whether or not aliens were here to help us or not. A good segment of the show was devoted to the subject of the alien autopsy film, and at the end of the debate, the consensus was that I had made a very respectable presentation. I was just glad that I hadn't gotten all tongue-tied and hadn't made a fool of myself. The main thrust of my comments were that the alien reality was a hyperdimensional one, and it seemed, based on the evidence so far, that humanity was *not* at the top of the food chain. Even Johnny Dollar was convinced.

And the next day, without warning, he was fired.

My friend, the executive at the station offices, still stays in touch and visits occasionally. He is still convinced that Johnny was fired because of that show, and both of them still talk about it on the occasions when they have lunch together. The only way my friend managed to keep his own job was because he arranged for Johnny to invite me himself, and he was never "connected" to the event.

But back to the story. Summer of 1999: Eve Lorgen had suggested that I might want to get in touch with the fellow who was publishing her book, a Mr. Hank Harrison, to inquire about publishing the Cassiopaeans' material. I wrote to him, and he wrote back giving me a website address. I checked his site and noticed that he had published a book or two on the subject of the Holy Grail. I was intrigued. After a few more exchanges, I ordered his book. He then suggested we talk via telephone and sent his number. We played phone tag for a few days and finally connected for a discussion.

Meanwhile, Ark and I were watching the fascinating parade of sometimes-crazy antics of guys-with-brains on the "Sarfatti" list. There was a guy on the list named Dick Farley, who I had been introduced to some years ago by my friend Blue. Blue had told me that Dick Farley (a.k.a. "Cloudrider") was as close to "Deep Throat" as anybody in the UFO community could get, and that his "take" on subjects connected to same, as well as covert intelligence, was very insightful. I had my doubts about anybody who would talk about such things openly really being an "insider," but I usually did find Cloudrider's commentary to be interesting, if not downright entertaining on occasion. At this point in time, Dick made some remarks that qualified as both:

Date sent: Mon, 21 Jun 1999 10:28:25 -0700

Subject: Re: Tabloid disinformation, but a question about context...anybody?

Dr. S,

The supermarket tabloid "Weekly World News" (WWN) has long played an easily demonstrable role in disinformation, often government-related, and with excellent timing disclosing or obfuscating pending releases of previously secure information, for deflections. "UFOs," parapsychology, psi-spying and exotic physics are routine.

The current issue has an uncharacteristically interesting "disclosure" related to the topics frequently discussed on your list. I'm just tossing it out for informal assessing by anybody interested. It suggests "time travel," with the usual WWN chaff & fluff.

The story claims the FBI is holding a "time traveler" from our future at a remote farm, apparently in New England. The story, ostensibly reported by "Australian journalist Theodore Dumane," claims that a male, age 47, "appeared" on 24 Jan 1999 next to a dumpster at a supermarket parking lot in Woburn, MA. His "time machine" had had a breakdown, and his appearance in our time was "an accident." He was stuck here.

The guy reportedly showed up at "a Harvard research facility," where security people believed he could be a spy and called in the Feds. The FBI is said to have located the guy in Boston, where he was taken into custody and remains today, WWN says.

Of course, the WWN story suggests the FBI is grilling the guy and learning all sorts of cool stuff about the next 423 years, since the fellow reportedly came back from Year 2422. Although not up there with the "Bat Boy" and "Hillary has alien baby," the piece is fascinating in terms of content analysis in context with previous WWN stuff which later related accurately to eventual "mainstream" disclosures. (Lockheed's "stealth ship" is one example; Dr. Leo Sprinkle, UFO/abduction therapist in Laramie, Wyoming being "outed" on WWN's cover, in 1988, is another, as was CIA "psi/spy" story breaking in WWN disinformation context to deflect mainstream "retail" news.)

Some of you folks on this list are apparently ready to "do the math" that establishes the theoretical basis for what WWN is suggesting and confabulating, so if past is prelude to the future, somebody might be getting "too close," or stumbling into areas where "mad scientists" from contraband programs are institutionalized and, in this case, one may have "gotten loose." Hey, it's happened before. (And then there were the Marconi "suicides" in the U.K., and last week's "suicide" of Australia's topmost intelligence official in this country...but those are other stories, for other times.)

If you have any insights into this, or contextual information, it would be appreciated. Thanks, Dick Farley.

P. S. — The issue of WWN of which I speak is easy to find. The cover story this week is that somebody "dug up" Elvis, and the body in the grave at Graceland wasn't his!!!

P. P. S. — Personally, I wondered whether the "time traveler" might not have been a Dan Smith clone, sent back from the future to locate the "real" Dan Smith now-time. But seeing that someone at least using Dan's identity is still transmitting messages, that theory will have to fall to one perhaps somebody on your list will put forward. DF

The initial reaction of the physicists on the list was that the story was probably false, but that the idea ought to be investigated. Jack Sarfatti wrote:

This is an interesting story. Probably fiction. If true FBI and or CIA should have me and my people interview the guy ASAP. But consider the source. It's probably not true. It may be part of a disinformation effort to debunk what I am doing in which case we should expect such things to really happen with increasing frequency. This is part of what Dan Smith calls the "Eschaton." Jacques Vallee has described it more scientifically in his books.

Dick Farley wrote back in response to Jack's comments:

Dr. S.,

I agree that it's most likely false, but your sensitivity to the possibility that it may be indirectly targeting the spheres in which you're exploring is an excellent reflex. Back in 1988, when some friends (and benefactors) and I were approached by Linda Howe about funding a video on the "Mars Face," we had an encounter with the WWN.

Linda arranged for us (three, including myself) to accompany her to The Analytical Sciences Corporation (TASC), near Boston, where Dr. Marc Carlotto was working on image analysis, and where the "Mars Face" photo-interp had been done. Our team had interest in something scientifically balanced, at least raising the questions but at the same time including Marc's "caveats" about the image density and uncertainties.

Even more interesting, to us and to Marc at that juncture (August 1988), was what he was exploring about fractal analyses when applied to the larger Mars field of view. He had just identified several other possible anomalous formations. That in itself was scientifically intriguing for any movie...but Linda and her cadre of handlers didn't see it that way, apparently. Even though "our team" had agreed already to fund half of the $95,000 Linda had said she'd need to do the film, the deal "fell through." Seemed to me that she and her handlers preferred the "sensationalized pseudo-science" of Dick Hoagland and other disinformationists, feeding into the more bizarre and mystical of theories as we've seen promulgated in the intervening eleven years.

Jump to WWN!

A few short weeks after our visit to TASC, when it appeared that the "Mars Face" in context would get a viewing in the video WE were interested in helping to get made, the WWN launched a series of cover stories about the "Mars Face Talking to Earth!" Imagine the impact of that on mainstream interest? My sense has been that Linda is part of the cadre which spins such yarns, acting either as an asset for "intell" or she is a dupe. She was too good of a reporter to fall for the whole thing, at least initially. But she is now well hooked, if her behaviors at Rockefeller's JY Ranch symposium (which I'd had a hand in organizing when I was with the Human Potential Foundation, 9/13/93) are any indication. Since then, she's worked for Bigelow and with Art Bell?

So, the WWN has an uncanny sense of timing when it comes to highlighting REAL scientific breakthroughs or potentials therefore, in areas which touch on the sensitive.

Heads up!

Regards,

Dick Farley

My head was certainly up and I was definitely paying attention to that thread! Time travel? Linda Howe being co-opted? And, of course, we see some of the early hints about the Stargate Conspiracy. The next couple of posts were somewhat confusing, however. Obviously, a conversation was taking place over my head:

Date: Tue, 22 Jun 1999 16:14:04 EDT

From Ira Einhorn

Dick wrote:

As for Jacques and Colin, both of whom I respect immensely and in Jacques' case, whom I've had the pleasure of meeting and exchanging information with for a time, it must be remembered that both of them play for the same team.

Ira wrote:

What team? I can't speak for Jacques [Vallee] any more, BUT we were real close back then and he played for no team to my knowledge.

Colin [Wilson] and I are still close and TEAM? Please define the parameters better.

Thanks for the info. on Linda whom I know not at all and not a line of her work. I know she got screwed; that is all. She has spent time with my closest friend who felt she was an alien.

ira

Who was this Ira guy? What did he know, having been "close" to Jacques Vallee and still close to Colin Wilson, a writer whose work I admire? The response from Dick Farley to Ira's post was puzzling in the extreme:

From: Dick Farley

Date: Tue, 22 Jun 1999 16:14:04 EDT

Ira,

So, what's an "alien" THESE days, eh? An honest American?

Regarding Linda being "an alien." Having read the book "about you" (The Unicorn's Secret, albeit with caveats to you that as a writer, I could see through the chaff and half-told stories, glosses, etc.) and some of your remnant stuff still in circulation, I'd suggest that while you're whiling away your time in France, you consider a new front of consideration. Kind of the "poetic" side of the quantum-entity thing, which is not a far cry from Vallee's and Wilson's thoughts about this.

From Jacques I learned about Phyllis Schlemmer and her earlier experiences with monks, who used her gifts to aid in their (monks) discernments about whether certain folks coming to the monastery for help ("personality" problems) were maybe "possessed" (i.e., resident entities).

Vallee's published stuff has led the UFO discussion in this direction for thirty years, and it is what attracted me to him and caused me to solicit his participation in the early aspects of Laurance Rockefeller's "UFO Disclosure Initiative" to the Clintonista White House, which I staffed from October '92 until I departed on my own accord in April of '94, information which I provide only for chronological context.

Jacques came east and consulted a couple of times, then met with Rockefeller (and John Mack), but then told us (me, personally, etc.) he was uncomfortable with the direction those other folks wanted to take it...i.e., they were taking it like "true believers." I've not shared that "online" as my CloudRider persona, but what we were and are dealing in, as you WELL know, is "something that's real," (although pushing the definitions of reality), and which powerful oligarchs and their court wizards are desperate to control or at least be the interlocutors of planetary and interdimensional contact. That would be spelled "P-R-I-E-S-T-H-O-O-D" if we were having a cultural anthropology chat, eh?

What the primary "trap" is for those who engage interdimensionality as an operant or contextual paradigm for scholarship and analysis, is that we can "handle it" with our three-dimensional egos and a dash of creativity and intellectual "wild west" bravado.

I suggest your own experiences, even as described by your friends interviewed and I acknowledge undoubtedly selectively quoted or misquoted by the author with a bias, describe someone essentially "out of control" and tumbling through hyper-space with no gyroscope to point the way to "consensus reality" where there WAS none to find. Add in the psycho-cybernetics and pharmacological enhancement experimentation, and to use a sailing phrase, you were sailing downwind in a full gale also in full sail.

When I got to the same place, albeit without the pharmacological component at that juncture, I "reefed" my sails, hove to and let the storm blow itself out.

And my family and I almost didn't make it, anyway. We took a lot of "green water over the bow," as the Dark Side did its damnedest to derail and destroy us. We've survived thus far, still right side up and pointy end going first...which is about the barest one needs to still be considered "sailing" and not shipwrecked.

Personally, I could have been you, not to paraphrase Linda Tripp, but in all seriousness. I identified considerably with your excesses and ego-overdrive as you confronted and became enmeshed in a "reality" that is excruciatingly enthralling and exciting when compared to "standard 3-D now."

And the sensual overloads, the powerful intuitive and "psycho-sexual" components (which seem to be an "occupational hazard" for certain male types exploring in these realms...I mean in all seriousness, which ought to be explored and is rather well so in Janet & Chris Morris's "The Stalk," which I've mentioned in past postings), all of these are consistent with my own experiences and not a few other men I've known.

Without the discipline of a philosophical and psycho-physical context, we Westerner types were tumbling in that interdimensional milieu like a cat in a clothes dryer. It's all warm and fuzzy, but terrifyingly so...and we can't make it stop when we're scared.

Others, like Leary, Murphy and now-aging "counter-cultural" explorers who now are "the Establishment" New Agers and marketers of "alt.lifestyle," got off the elevator at "ladies lingerie" or "sporting goods."

Guys like you kept pushing, climbing, seeking up and up, above the tree-line, as it were. Eventually, you were rock-climbing onto the highest Himilayan ledges and ridges, looking for God or whoever it is the Dalai Lama and others are communing with as they seek guidance and carry out orders of the "King of the World," (spelled "priest-kings") and following threads in a tapestry all but invisible to the rest of us uninitiated. Power untamed and undisciplined is deadly!

Which brings me to something that helped me develop a working vocabulary, albeit one which is heavily loaded with semantic and cultural baggage...but which once I'd worked it though, proved helpful. A dentist-turned-hypnotherapist-cum Ph.D., a man with great compassion and keen insight, plus a lot of inherent integrity (at least in the early stages of his work, and still so for the most part now), took a clean, fresh look at some of the stuff Roger Woolger and others were doing. His name is Dr. Bill Baldwin, and he developed his doctoral thesis into a "Spirit Releasement Therapy & Past Life Regression Manual," which is available. He and his wife Judith do the usual trainings and teaching sessions, but have avoided the "cult-like" self-adulatory styles of so many others who come upon useful operational paradigms. You may find Bill's book interesting, perhaps helpful, as there is a rather extensive, well-indexed series of clinical and anecdotal narratives, session transcripts and contextual information. It goes where Vallee and Wilson have pointed, is consistent with a lot of what "The Nine" and variants have described, but it remains VERY "human centered" and solid.

Bill's approach is, "Hey, it doesn't matter whether you 'believe' this or not, it's simply my clinical experience." And some of the results he's achieved were independent of the client's "belief system" or theological context, actually in most of his cases are.

The point? Linda Howe's perhaps being "an alien" is not a joke, in the sense that she may have...indeed, my sense after spending various amounts of time with her over the years, in very different contexts spanning now twelve years...opened herself to a kind of "attachment" of the sort Baldwin describes, especially in the "E.T. context."

Baldwin puts phraseologies into this context that transcend all of the old Catholic or "demonic" biases, although he's very clear about the right of humans NOT to have to be manipulated or have energy drained by attaching or embedded entities of various energies and intentionalities.

With your encyclopedic background, I recommend this. Baldwin's book is most readily available from Headline Books, Inc., which took over publishing of it after the Human Potential Foundation Press foundered at the demise of that foundation. Publisher Bob Teets also has a couple of other books you ought to have, if you don't already: "UFOs and Mental Health," (1997), to which I and Walt Bowart contributed in addition to Bob's breezy but focused narratives; and "West Virginia UFOs: Close Encounters in the Mountain State" (1995), an excellent, very readable journalistic style narrative of various stories...mostly firsthand, not the old rehashes of Gray Barker's and John Keel's stuff (although Keel liked the book and is complimentary of Bob's (et al.) work... and which has more than just UFO stories in it.

Bob's (and my) experiences working with Bill Baldwin color his approach to what he wrote in "West Virginia UFOs," and the mix is a fascinating exercise in what we called, "Applied Jacques Vallee." In short, Baldwin's approach is "good armament."

Give it a try. Give Bob a shout, or have somebody stateside do it:

Headline Books, Inc.,

P. O. Box 52, Terra Alta, WV, 26764.

Order 1-800-570-5951.

Tell 'em Dick sent you. Ask for some back issues of "The American UFO Newsletter," also. It's NOT your average "UFO rag." Bob's read a lot of your stuff also, and knows I was tracking "their" tracking of you since at least 1988. Why? Because of what I described to you in the preceding...I recognized some of what happened to you, then and subsequent.

There but for the grace of The Force, God or whoever, could have gone I. Maybe still. One must keep up one's discernment and screen for "dark ones," wherever it may be that "they" germinate...i.e., in the provinces of the mind...or "someplace else." Keep up your guard...Guard your mind!

And remember, humor is a sign of "sanity" yes? It all goes toward the "weird" elements reported variously as happening to Kit Green & Pat and those guys, at the fringes. And it's what the "debunkers and skeptics" are perhaps well-intentionedly working so hard to keep from being considered seriously.

"Who" killed Holly? Perhaps the answers reside somewhere "in there," as described. Baldwin's work with Vietnam-era veterans is "worth the price of admission," and I've heard his audio tapes he uses in his training sessions, (one of which I've completed).

Whatever is going on, and something IS going on, if you explore without considering what Bill's developed for protections and appreciations of the "dangers," it's like free-climbing a sheer face. It may be a macho thrill, but it's a VERRRRY long way down.

As noted, I was somewhat confused by all this. I didn't "get it" that Dick was suggesting to Ira that he was possessed. There were references being made that were obscure to me, but the one thing I was certain about was that I agreed with Dick's assessment of the UFO/alien reality as having something of the "flavor" of demonic possession, though I wanted to emphasize to them that it would be dangerous to think that this was all there was to it. I wrote back to Dick privately with these concerns, and then, since this fellow "Ira" seemed to be connected in some way, I copied to him also. I quoted some of the Cassiopaean material on the para-physical nature of the hyperdimensional reality, and thought that would be that. A day or so later, I received a private communication from Ira to me, via Ark.

From Ira Einhorn

Subject: Re: For laura

Date sent: Thu, 24 Jun 1999 14:23:52 -0400

I would to be happy to dialogue with you, after going through your comments more carefully, as I think we are essentially in agreement.

I lived through years of the most fantastic phenomena without ever buying the mythology attached to them. The phenomena were very real and of almost daily occurrence; the myth attached to them was pure projection, BUT the unknown frightens people and living in vanilla reality forces people to project, so when the paranormal becomes the daily reality the desire to have an explanation grows exponentially.

It began almost 30 years ago in a very different social climate, and the intell. groups were initially freaked. If that different context is not understood, the rest becomes very confusing.

There is so much noise about that the signal tends to get lost which is unfortunate for those who establish real connections with sources: a byproduct of our particular moment in time, alas.

ira

Meanwhile, I had received an email from Ron Engelman who assured me that, since he had stayed online with David Koresh during the Waco event — and that the government hadn't shut him down over that, though they tried — I could be certain that he was not easily intimidated, and if he wanted me on his show, I didn't have to worry about him losing his job. The show was scheduled for July 2. I wrote back to Ron that the info about Waco was interesting, maybe he would like to read the Cassiopaeans' comments on it. I enclosed them, *including the subsequent comments about O.J. Simpson*, just because they had proven to be correct and were of general interest.

10-05-94

Q: (L) Did the United States government deliberately murder the Branch Davidians at Waco?

A: Close. Led them to destroy themselves.

Q: (L) How?

A: Psychological warfare tactics.

Q: (L) Did the US government set their compound on fire?

A: No.

Q: (L) Who set the compound on fire?

A: Branch Davidians. Drove them crazy.

Q: (L) Were ELF or subliminals used?

A: Yes. As well as other means.

Q: (L) Did OJ Simpson kill his wife?

A: Yes.

Q: (L) Did he take the murder clothes and weapon to Chicago and leave it there in a bag as some people are saying?

A: No.

Q: (L) Where is it?

A: L.A. dump. One of them.

Q: (L) Is OJ Simpson going to be found guilty?

A: No.

Next, I finally connected with Hank Harrison to discuss publishing matters. During the course of the conversation, I brought up the subject of hyperdimensional realities, and he snorted something about "Ira Einhorn." I was taken aback, and asked him did he know the guy, and if so, what was the story?

Hank proceeded to enlighten me about Ira, and claimed to have spent time with him while Ira was in hiding in Ireland. Hank went on to talk about his close friendship with Jack Sarfatti, which surprised me; and Ira was, it seems, a convicted murderer on the lam and, as far as Hank was concerned, he was guilty and no amount of claiming "the CIA set me up" was sufficient to ameliorate that guilt.

I pointed out that Eve's ideas about "alien orchestrated dramas" might lend credibility to Ira's theory, and Hank went off about Eve's book in a very negative way, which surprised me since he was supposed to be publishing it. At the end of the discussion, I was sure of one thing and one thing only — Hank Harrison was *not* the person to handle the Cassiopaean material, and I wondered why Eve was letting him handle hers. But, more than that, having all of these strange things connecting in some kind of crazy circle was disconcerting, to say the least.

I went on a search for information about Ira Einhorn, and after discovering the details of his crime, I was pretty shocked. I was also really curious: He was declaring vociferously that he didn't do it, that he was set up. I could certainly admit to such a possibility even if the evidence in his case really made him look guilty. I decided to just leave that alone and wait and see what happened if he ever wrote back to me.

The very next thing that happened was that Ron Engelman called to tell me that for some strange reason the time he had slotted for a show had been canceled — the whole schedule had been switched around and he had no idea what was going on. I just said: "I told you so!" Ron said: "I'll tell you this, you *will* be on the air, and soon. I promise."

Meanwhile, Eve wrote about a whole series of disconcerting events in her life related to her book. Her computer crashed, she started getting the "bum's rush" from some former supporters, and just a general melee was going on in her life. I decided to ask the Cassiopaeans about this very strange series of oddly connected events. Why was it that the instant Eve and I began comparing notes, all of these strange little "connections" popped up? Why did it all circle around Ira Einhorn?

07-03-99

Q: I would like to find out what was the deal with this radio show? Ron Engelman invited me to be on his show, and the day before it was scheduled, he called telling me his schedule had been changed and I had been canceled. Why was it canceled?

A: Maybe others complained.

Q: Who were these others who might have complained

A: Others scheduled to appear.

Q: What did they complain about?

A: Your purported agenda.

Q: What agenda did they object to?

A: Us, my dear.

Q: Well, that does seem to be true. We do get more than our share of attack. I just want you to know that, for the most part, it is a thankless task.

A: Which means you are on the right track.

Q: [Laughter] Well, if the Gray Huggers don't bomb us, or the scientific types don't bomb us, then somebody else does! Okay, recently there were some serious questions being brought up on one of the mailing lists about Linda Howe being 'co-opted' for disinformation purposes. A number of seemingly reputable people have pointed out coincidences in clear disinformation and her work. I have always felt that, of all the people out there doing research, she was truly sincere and clear headed. Could you comment on that possibility?

A: Linda is not co-optable.

Q: Why is she hanging out with folks who *do* seem to definitely be co-opted and promulgating such clear disinformation?

A: The "modus" is not to be confused with the "operandi."

Q: There was a strange little series of events last week involving Ron Engelman, Eve Lorgen, Hank Harrison, Dick Farley, and Ira Einhorn who wrote me a couple of e-mails. I would like to know, what is the deal with Ira? Either this guy is completely crazy to have lived for 18 months with a dead body in his apartment, completely egotistical, and believes nobody else is as smart as he is, or the whole deal was planted to shut him up, or he was Greenbaumed. Now, did I miss any possibilities, is it one of these, or is it something else altogether?

A: Maybe he did what he did, and then convinced himself that he did not *a la* O.J. Simpson.

Q: In this whole situation, who or what was the portal connecting all of these things?

A: To find portal, retrace steps until you find the ray peeking through the blinds.

Q: (A) Was it coming through me?

A: No.

Q: Well… it all started when I began to communicate with Eve.

A: No. It was not Eve. She is under attack for her revelations.

Q: (L) Can you give me another clue? "The ray peeking through the blinds…"

A: Yes.

I was a bit startled at the mention of O.J. Simpson, since I had included the Simpson clip in the email to Ron. But, more than anything, I wanted to know what was the "connector." The Cassiopaeans said to "retrace steps until you find the ray peeking through the blinds."

I was completely baffled. The whole thing seemed like so improbable a series of odd connections that I *knew* there was something at play here. I just couldn't figure out what. And, in fact, until I began to analyze things in terms of the chronology, I just didn't get it. The *"ray"* peeking through the blinds?

Of course: Ray Flowers and Vincent Bridges. The "portal" connecting all of these things.

Now, if my brain had been firing on all cylinders at the time, I would have immediately known the "connecting principle."

Another odd connection that occurs to me as I write all of the above is the fact that I had my last communication with Karla Turner at about the same time as I did the radio show with Johnny Dollar. In the above series of events, it was a post about Karla Turner's work that led Eve Lorgen to write to me, followed almost immediately by the email from Ron Engelman about another radio show. Then, there was the Ira Einhorn thing right in my face, connected to time travel and co-opting and people being manipulated by other forces. Oy. The Universe speaks to us in oh, such mysterious ways!

Again, I want to make it clear that it is impossible for me to say whether or not human beings are "conscious" in these Theological Dramas. In many, if not most, cases, I think that everybody is just doing what they have to do according to how their internal circuitry is laid out. And, some people love light, while some people love the darkness.

So, in order to freely explore this strange connection that only today begins to make sense, in terms of the activation of the crypto-geographic being in the Secret Games of the Gods, let's have a look at the Unicorn. We will see that there *may* indeed be an archetypal connection between Ira Einhorn, Vincent Bridges and the entire subject of psychopathy that we have been exploring.

Chapter 60
The Unicorn's Closet

On September 11, 1977, Paul Herre heard a very loud thumping on the ceiling of his apartment — "like somebody being bounced against the floor." This was instantly followed by a "blood-curdling scream." Since he lived in a "college neighborhood," and such things were not terribly uncommon, and since the "uproar" did not continue, he did nothing and thought no more about it. After all, the upstairs neighbor, Guru Ira Einhorn, always had strange visitors and activity in his apartment. He was also always fighting with his girlfriend, Holly Maddux.[29]

Later, the same day, Ira went for a drive with Jill and Sharon, two young women, just out of high school, who had been given Ira's name as a contact in the world of the paranormal. Ordinarily, they came to his apartment for sessions in which Ira supposedly taught them "psychic arts" such as meditation or astral travel. But, on this particular day, instead of having a "session" in Ira's apartment at 3411 Race Street, Ira suggested a drive in Jill's car to the West River Drive along the Schuylkill River in Philadelphia. According to Jill, Ira "seemed like he had a sense of urgency. He said he wanted to ask us something."

"I'm in some really, really deep trouble," Ira told them. "I've got a steamer trunk that has some very, very valuable documents in it. They're documents that belong to the Russians, and I need to get rid of it."

He wanted them to put the trunk in Jill's car and take it and dump it in the river.

"He almost had us convinced to get this thing [and do it]. Sharon and I were looking at each other, like, this is really weird. I mean, why did he want to get rid of something like this?"

---

[29] All quotations are from Steven Levy's *The Unicorn's Secret*. For more details on the case and additional bibliographical references check *The Crime Library* website at http://www.crimelibrary.com.

As it turned out, the trunk wouldn't fit in Jill's car, so the project was abandoned.

The next day, September 12, Bea Einhorn, Ira's mother, received a call from Ira saying: "Mom, I'm upset. Holly didn't come home last night. She didn't take anything. She didn't have any money. Where is she?" Bea ran through the list of places he ought to have checked such as hospitals, friends and so on, and Ira assured her that he had made all such inquiries. According to his mother, Ira was panic stricken and as emotional as she had ever heard him.

As it happens, Ira had *not* called any of Holly's friends. He had not even called her place of employment — her "destination," according to him, when he last saw her. And while he was *not* calling her friends to express concern about her, some of her friends *were* calling him and Ira was telling them not to worry, there was nothing unusual about Holly's disappearance!

On September 14, one of Holly's friends, Saul Lapidus, who had expected her in New York, asked Andrija Puharich, a mutual friend of his and Ira's, to intercede and try to get information from Ira about Holly's possible whereabouts or plans. Puharich reported that "[Ira] didn't want to talk about Holly, and he shut me off. He was very terse. He wasn't trying to be communicative at all. He just said she left. So I told Saul, and Saul said, 'My God, I bet she's been killed.' At this point I said, 'Come on, Saul. You know Ira wouldn't hurt a fly."

Several hours later, Ira told friends of his and Holly's that he had received a call from Holly, affirming that she was fine and that she wanted to be left alone. She would call in a week, he reported. He asked them to pass this information on to Holly's friend, Saul, in New York.

However, Ira didn't share this purported message with his mother, who he had already upset greatly with his "concern." Nor did he share this information two weeks later when he described his worry and concern about Holly's whereabouts to two friends. Sitting in a restaurant, Ira told them that Holly had left, and that *he had never heard a word* from her since!

So, which was it? Was Ira upset and worried? Or was he in touch with Holly and assured in his own mind that she was fine? It seemed that it was only the people who were concerned about Holly and inclined to try to find her who were treated to the story that she had called and was okay, so as to put them off from looking. Those who were willing to accept her disappearance as just the way things were, were treated to the story that Ira was concerned and in need of sympathy.

That autumn of 1977, Paul Herre and Ron Gelzer, the occupants of the apartment beneath Ira's, were about to start their senior year at university. Paul went away on September 20 to attend a wedding, and it was that weekend that Ron Gelzer first noticed the odor. It came from a closet in the kitchen. Gelzer, a biology student, reported that his first impression was that it "smelled like blood." He tried to track the source of the odor, and lifting a transom in the ceiling in the closet, saw what looked like water. He went outside to view the upper story and try to determine what was directly above the closet that could be possibly be leaking. It was Ira Einhorn's porch. Gelzer went to Ira and asked him if he knew of anything that could be leaking on his porch. An unruffled Ira said that there was nothing as far as he knew.

On September 26, after Paul Herre had returned from his trip, he instantly was assailed by the "gross smell" that pervaded his and Gelzer's apartment when he walked in the door. It was sickening and overwhelming in the kitchen. He described is as being much stronger and a lot less tolerable than the smell of human excrement.

Unable to long endure the stench, Herre went to the elderly property managers who lived in an apartment behind the building to complain. They told him that they were suffering from the smell also. It was so bad they couldn't even eat in their own kitchen. They suggested that if Herre could find the source and clean it up, they would reimburse him.

Herre, Gelzer and a friend, Stephanie DeMarco armed themselves with cleaning supplies and ventured into the kitchen on a search and destroy mission. By now, even going in there was an act that required a strong stomach. DeMarco joked around that it smelled like a dead body, but Paul Herre scoffed at the silliness of such an idea. Herre was elected to enter the closet. He looked up at the transom and pulled it open. In the space between the floors he saw a brownish stain that had come through a crack in the ceiling plaster. It was almost dry. He determined that it was the source of the odor and he attacked it with ammonia. After everything had dried, the odor was still there. He tried pure Lysol, straight chlorine bleach, and nothing killed the odor. Finally, he painted the area to try to seal the odor. That didn't work either. The trio realized that they needed to get creative, and so they just simply stuffed the area between the floors with Odor Eaters. Problem solved for the time being.

A month later, a heavy rain fell and the ceiling began to leak. The odor returned, but by now they had the drill down. Cleaning, painting, stuffing with Odor Eaters. Eventually, after many months, the odor faded, even if it was still faintly detectable in or near the closet as long as Herre and Gelzer lived there.

The property managers opined that the problem was rotting wood. Herre convinced himself that it must be a squirrel that had been trapped and died between the floors. He even went outside with a tape measure on one occasion to determine exactly where the carcass might be lodged. What he discovered was that a closet on Ira Einhorn's screened porch was exactly above the closet in his kitchen.

The following summer, for the *first time* in Einhorn's *seven years of residence* at this apartment, he *did not sublet the apartment* for the summer when he went on his travels.

In September of 1978, after more complaints about the persistence of the odor, the owner of the building ordered some work done over Ira Einhorn's porch to fix the possible leaks that might be contributing to the problem. The roofer hired to do the work reported that the roof over the porch had originally been tarred in early 1977. It was a year later that he received the complaint from the owner about the odor. He thought it might be stagnant water that had leaked in through a crack in the tar. The firm re-tarred the roof. As soon as he learned that work was to be done, Ira contacted the owner and complained about the repairs that might possibly disturb his things. The owner, accustomed to Einhorn's demands for privacy, didn't think the complaint too unusual. Ira specifically instructed the repairmen *not to go near the closet on his porch*.

Meanwhile, the family of Holly Maddux, worried by the lack of regular contact which had always been her habit, called Ira on October 4, to try to find out what might be wrong. Ira reportedly said: "I was about to call and ask you." Ira told Mrs. Maddux that Holly had been in Philadelphia for a few days in early September, and then just "took off." Holly's mother mentioned that it was her understanding that Holly was going to move into a new apartment, and had given them the address, but she was not responding to the mail that they had sent her there. Ira told Mrs. Maddux that he wished she would stop that because he was "tired of collecting Holly's mail."

Whoa! That was COLD!

Two weeks went by, and other members of Holly's family who normally heard from her regularly began to call the Madduxes to try and find out what was going on, since they had not heard from Holly. On October 20, Liz Maddux called Ira Einhorn again, with a list of questions in hand so that she would not be sidetracked by Ira's diversionary tactics.

Mrs. Maddux: When did she leave Philadelphia?
Ira: Three or four weeks ago. She went to the store and didn't come back. Actually, I was in the bathtub when she left. When she didn't

return, I called everyone we knew and checked with the police and the hospitals, but no one knew anything. I talked to a friend of hers....

Mrs. Maddux: Joyce Petschek?

Ira: Yes. You know, she has houses all over so I had a hard time getting in touch with her, but I finally did. She told me not to worry, that I might not hear from Holly for three or four months. I don't know if she meant Holly might just go off for a while or if she knows where she is and won't tell me. I'm the last one she'd tell.

The Madduxes didn't get much from Ira, but we notice that, again, *the story has morphed*. Now it is not that Holly called Ira, but that he called a third party who told him that *she* had heard from Holly, who said "not to worry." Did he really think that no one would ever check it out and compare these stories? Obviously he didn't consider that as a possibility.

The Madduxes tried several approaches to discovering what happened to Holly, finally contacting one of Holly's friends, Lawrence Wells, in Tyler, Texas, Holly's home town. He was an assistant U.S. attorney, and the one who officially reported Holly missing to the authorities in Philadelphia after a check of the hospitals and morgues produced nothing. He also alerted Interpol that she was missing, and then called Ira Einhorn himself. He was given the same story about the bathtub scene. Wells didn't buy it, and called the Philadelphia police directly and spoke to a detective who promised to look into the matter.

Detective Lane conducted a few interviews, speaking to Holly's doctor (Holly was diabetic), as well as her therapist, Marian Coopersmith, who assured him that Holly was not suicidal. Holly had $21,000.00 dollars in the bank that had been untouched since her disappearance.

The police detective paid a visit to Einhorn, where Ira told him the bathtub story. However, he also told him the "Holly called me" story, stating that she had told him "I'm okay. Don't look for me. I'll call you once a week." He then said that when she did not fulfill this promise, he began to get worried. His explanation to the detective, as to why he did not report her missing to the police, was that he had been told that since she was an adult, she didn't qualify. This was correct, and the police investigation ended.

In January 1978, the Madduxes contacted R.J. Stevens, the former chief of the Tyler FBI bureau, who had just retired to open his own private investigation business. Stevens contacted Joyce Petschek, who replied by letter stating that she did not know where Holly was, and suggested that the Madduxes contact Marshall Lever, about whom she said: "He is a transmedium and perhaps could do a transmission for you as to Holly's whereabouts."

Yeah, right.

The Madduxes did not want to leave any stone unturned, so they did contact Lever. He told them that he was about to go on a trip and he would get back in touch. But, he never did. Some "transmedium."

Stevens decided that he needed somebody in Philadelphia, so he contacted another former FBI guy who also was now working as a private investigator: J. Robert Pearce. Pearce did not normally deal with missing persons cases, but as a favor to another former FBI guy, he agreed to take the Holly Maddux case.

Stevens flew to Philadelphia in March of 1978, and the two of them reviewed what they had and made plans to interview Ira face to face. They called to make an appointment. Ira declined. He was too busy organizing Sun Day, an environmental "event." Stevens was persistent and Einhorn just told him that he had no love for Holly's parents — and that, in fact, getting away from her parents was the main reason she had disappeared! (Standard psychopathic trick! Blame the victims.)

Stevens pointed out to Einhorn that Holly had not just cut off contact with her parents, but with all her friends *and Ira himself.* Ira acknowledged this, and said that Holly simply wanted to transform her life by cutting all her ties.

Stevens argued that her parents did not intend to interfere in her life, they merely wanted to know that she was alright. Ira replied that he still would not cooperate since Holly did not want to be found. However, he did tell Stevens the bathtub story with the "Holly called me and said she was alright" variation.

Stevens and Pearce, experienced "G-men," discussed Ira's evasiveness and his unwillingness to assist. They could see no reason for someone who had been so close to Holly, *someone who claimed to love her*, to be so reluctant to help an investigation for the purpose of merely assuring her parents that she was alright. *The fact that Ira refused to help, even while admitting that he knew no one who was in touch with Holly, including himself, was highly suspicious.* At that moment, as Steven Levy recounts in his book *The Unicorn's Secret*, Ira Einhorn became a suspect.

Stevens had to return to Texas, so R.J. Pearce continued the investigation in Philadelphia. He decided to approach the whole matter as if it were a "fugitive" case, develop contacts with people who knew her, and find out who saw her last and when and under what circumstances. After interviewing Ira's parents, *who were baffled that Ira would not cooperate with helping to establish that Holly was alright*, Pearce interviewed the property managers at 3411 Race Street. They loved Holly, but didn't like Ira, saying that he was strange. They weren't sure what he did for a living, saying that he was a guru/consultant about UFOs or something. When Pearce asked if they thought that Einhorn would harm Holly, *they admitted that they had thought of this and discussed it*, but had discarded it since Ira had such a widespread reputation as an advocate of nonviolence.

Walking back to the front of Ira's building, Pearce rang Ira's bell and was buzzed in. Before he had climbed the stairs, Ira came to the door and wanted to know who the visitor was. Pearce described the encounter in a report reproduced in Steven Levy's book:

> He appeared to be dressed in a kind of silk kimono affair. He didn't completely open the door, and I couldn't see how he was attired completely. His eyes are noticeably blue, he has a stocky build, full beard, light brown in color, and rather long hair matching in color. He appeared calm. From what little I could see into the apartment, I didn't think it was heavily furnished, but it did appear orderly. Ira Einhorn was emphatic that he would not "help Holly's parents." He did say he did not know where Holly is but claimed that if he did know, he would tell [me]. (Levy, 1988)

Pearce next spoke to a number of Philadelphia officials in law enforcement, one of whom was George Fencl, who knew Einhorn and had recently been in contact with him as a consequence of the Sun Day shindig. Pearce asked him to have a chat with Ira, and Fencl did so. The result of this talk was that Ira again made it clear that he had no desire to help anyone find Holly. *He also commented to Fencl that this matter was getting an unusual amount of attention for a routine missing persons case — a detective, and ex-FBI man, and now a police inspector.* The implication was, of course, that the attention was focused on Ira, and not simply because Holly was missing. Fencl asked him to just help straighten it out and Ira promised he would. A few weeks later he called Fencl and told him that, after discussing it with all his friends, it seemed that Holly was "out of the country." He said he would do some checking himself when he was abroad later in the upcoming weeks.

George Fencl relayed this promising information to Pearce who then asked him: "Would Ira Einhorn harm Holly Maddux?" Fencl admitted that Ira was strange. "He would write you a letter to set up a meeting and a week later just walk in your office unannounced. But everyone knew Ira Einhorn advocated and practiced nonviolence and the police inspector had no evidence to the contrary."

As Pearce developed leads, he talked to an acquaintance of Holly's who had chatted with her while she was at Joyce Petschek's house on Fire Island, in the weeks before she disappeared. In late August of 1978, this individual told Pearce that Holly had told him that she had broken up with Einhorn permanently and there was a new man in her life, Saul. He then told Pearce about Andrija Puharich.

R.J. Pearce, *an ex-FBI guy*, knew nothing about the "colorful history" of Puharich. As it happened, Puharich's three-story house in Ossining, New York, the headquarters for his "mind-blowing experiments of the Space Kids," had just recently burned down. The police had ruled the case arson.

Pearce dispatched another former FBI agent, Clyde Olver, to investigate the burning of Puharich's house, in order to determine if there might be any connection between that and Holly's disappearance. Puharich was rumored to have fled to Mexico, but the neighbors had plenty to say about the "strange goings on" at Turkey Farm. They said there were people coming from all over the world for "unspecified, and possibly unnatural experiments."

Olver tracked down three of the "Space Kids" who had been living at Puharich's before the house burned down. They were, apparently, feeling pretty badly treated by Puharich because he had just up and disappeared, abandoning them. However, they did know Ira. In fact, they mentioned that they had seen him quite recently: *A day or two after the fire*, Ira had appeared out of nowhere, promising to recruit an investigative reporter to look into the fire.

I must admit that reading this item raised a huge question in my mind. I started to wonder exactly *when* Ira Einhorn began to propose the idea that someone was out to get him, that the work he was doing was dangerous? When, exactly, did the assembly of obviously gullible people with whom Ira hung out, begin to think that they were all playing some kind of exciting and dangerous "cat and mouse" game with the "shadow government?" Who came up with that idea? Who promoted it? It occurred to me that it would be *very* handy for *all* of that group who hung around with Puharich to think that somebody was really after them — which they would naturally think if Puharich's house had been torched. That idea, of course, would lend great credibility to Einhorn's own claims that he was "framed." It would be a very convincing emotional shock to all his friends and would incline them to believe Ira and support him.

In fact, that is a standard ploy of the psychopath: To actually instigate sabotage of some sort so as to create the image of being attacked, stalked, or in some kind of danger, so that they can engender sympathy and support by their "brave and stoic endurance" of their troubles. But again, the clue is that the words and actions are out of sync.

Did Ira Einhorn burn down Andrija Puharich's house in order to put Puharich's supporters in a shocked and confused frame of mind, so that they would be more susceptible to believing his claims about being framed? After all, it was in March that Ira was first approached by Stevens and Pearce. It was after that when Fencl talked to Ira and Einhorn made his comment about the unusual amount of interest Holly's missing status was attracting. Did he then begin to build up the tension of the idea of being stalked by nefarious groups, and did he then begin making plans as to how he would "confirm" this impression? Was the burning of Andrija Puharich's house part of his plan to convince the people of "wealth and taste" who had supported their "paranormal" interests, that the experiments undertaken by Puharich were "dangerous," and that Ira was also in danger? Could Ira have burned down Puharich's house to create the impression that he was "next on the list" of those to be stalked and dealt with? Could he have been influential in convincing Puharich himself of these ideas, so that he would promote them to his supporters? Was this also designed to set their backs up against the authorities who might later wish to ask questions about Holly's disappearance? Such as "ex FBI agents" who Ira already knew were asking questions? And after the burning of Puharich's house, Ira shows up *one or two days later* and promises to recruit an investigative reporter?

The rumors about Puharich and his possible government-connections-turned-sour had run rampant for years. But was any of that really true?

> Clyde Olver contacted the arson investigator for the insurance company with the policy on Puharich's house. The investigator had concluded that a would-be psychic researcher, spurned by Phuarich, was the main suspect. This suspect had once left a statement on Puharich's answering machine such as, "I don't know what's happening to me! I'm going crazy!" He had appeared at the Turkey Farm one day and demanded that everyone listen to his personal problems. One of these problems was, as he explained it, harassment by extraterrestrials. (Levy, 1988)

The insurance investigator had interviewed Puharich on August 15. He confirmed that Puharich was, indeed, a "medical doctor who had authored several books, an expert in paranormal phenomena." Puharich had also provided to the insurance company the necessary information about his income sources: a $10,000 grant from a Baron DePauli, $15,000 from Dell Books as advance on a book about Tesla; $50,000 for the rights to a movie about Uri Geller, and between 50 and 70 grand for his scheduled lectures.

The insurance investigator thought that Puharich was a nutzoid. "It is noteworthy to mention that Puharich stated he has observed numerous UFOs and has communicated with extraterrestrial beings." Puharich was in Los Angeles when the fire occurred, but he suggested to the insurance investigator that the strange behavior of the guy ranting about being harassed by UFOs was not unusual; instead, he pointed out that the fire was very likely started by the CIA as a "warning" to him because he (via Einhorn) had been circulating evidence of Soviet experiments in psychic warfare.

Now, let's stop and think about this for a minute. I want to get this straight: Puharich, Einhorn, and many others since, have promoted the idea that the CIA is after them because they are circulating information about the Russian threat in terms of psychotronic warfare. Here we have Puharich talking to an insurance investigator, for God's sake, who might just as likely have thought that Puharich himself was crazy and burned down his own house based on such loose talk. Why in the world would anybody who was truly rational talk to an insurance investigator this way?

It reminds me of the time I had an insurance appraiser come to the house after we finished some construction work. I was having the fidgets worrying about this person coming in my back room where the psychomantium was. How was I going to explain a black tent to an insurance appraiser? What if they decided we were so weird that we had to pay higher insurance premiums? After all, nobody really knows all the secrets of how actuaries come up with insurance rates. Everybody who has had experience with insurance companies *knows* that you want to seem as "normal" and "in the middle" as possible. Well, since I also used the room for my microfiche viewer, I just moved it to a prominent position in the middle of the room and didn't say anything. The appraiser asked what kind of stuff I was viewing, and I just told the truth — genealogy records. *That* was "normal."

So, back to Puharich: Here he is talking to an insurance investigator, the person who has the power to authorize or withhold payment on his claim, and he was saying things like *that*? I'm sorry. Puharich may have been a genius, but this single incident makes me question his grip on reality.

I'm having a hard time with another thing here: What happened to the fact that, during that period of time, the US government was pretty busy promoting the idea of the "Russian Threat?" If that is the case — and it *is*, historically speaking — why in the world would the CIA object to the circulation of ideas that the Russians were doing us dirty by turning our brains to jello with their psychic experiments and ELF waves? I mean, after all, that would fit right in with their own view. Why would Ira Einhorn claim that the CIA *and* the Russians were mad at him? Oh, of course. Because they would not be expected to show up at his trial and deny the charges. With psychopathic lies, everything has to be so double and triple reversed-covert that nobody will ever be able to get to the truth of the matter. You know, "This tape will self-destruct in five seconds, if you are caught, the State Department will deny all knowledge of your mission" type of thing.

How handy.

Well, just some things to think about. Now, back to Ira and Holly.

After interviewing many witnesses and writing up the report, Pearce was convinced that Holly Maddux had been murdered by Ira Einhorn. From January 3 of 1979 until March, Pearce tried to get action from the Philadelphia police. The detective assigned to the case, Kenneth Curcio, was singularly unresponsive. J. R. Pearce went over his head and spoke directly with the Police Commissioner on March 7, 1979. A new detective was then assigned to the case, Michael J. Chitwood.

Chitwood was a controversial "Dirty Harry" kind of guy who never even carried a gun, yet had been castigated as a "brutal inquisitor of homicide suspects" by a journalist for the *Philadelphia Inquirer*, resulting in his removal from the homicide division. In a plot right out of a movie, the Einhorn case was his first homicide assignment after being reinstated in the good graces of the department, because of his work in recent hostage negotiations. Chitwood read Pearce's reports and also knew: Ira Einhorn was a murderer. He went to work to verify the information so he could draw up a search warrant.

Chitwood visited the city coroner and described to him the problem of the odor coming from Einhorn's closet. The coroner, Halbert Fillinger, told Chitwood, "When you go to Einhorn's apartment, you're going to find a body there." Chitwood simply could not believe such an idea. Nobody *in his right mind* would keep the body on the premises. He was convinced that blood had been spilled, that it had soaked through the floor during a short period of time when the body may have been temporarily concealed in the closet, but that surely the body would be gone by now! After all, it had been 18 months. All he wanted was a search warrant to go in and take out the floor and submit it for testing to determine the presence of human blood and proteins. That was the most he was hoping for in the way of evidence.

On March 28, three weeks after being handed Pearce's findings, Michael Chitwood, Captain Patterson, three men from the Mobile

Crime Unit, and two techs from the chemical lab, armed with crowbars, power tools, cameras, and a warrant, rang Ira Einhorn's buzzer at 3411 Race Street.

It was ten minutes to nine o'clock and Ira Einhorn was still sleeping. He grabbed a robe and pressed the button to unlock the outside door. Chitwood and company were still climbing the stairs when Einhorn opened the door to his apartment and peered into the vestibule. He stood there in his opened robe, naked and exposed, exuding the repellant body odor for which he was notorious. (Ira believed he was a "godlike" being, and therefore, mere mortals ought to drink in, savor, and appreciate his bodily secretions.)

Detective Chitwood identified himself and told Ira that he had a search-and-seizure warrant. Einhorn laughed. "Search what?"

Chitwood was meticulous. There were going to be no mistakes here that could later invalidate the search warrant. He handed it to Einhorn and asked him to please read it carefully. It was 35-pages long. It basically said that the police had permission to search Einhorn's apartment for any evidence relating to the disappearance of Helen (Holly) Maddux.

Einhorn told the policemen his much repeated story, that he hadn't seen her since September 1977, when she went out to do some shopping and didn't return. Einhorn was very calm and just asked if he could get dressed. "Certainly," he was politely told by Michael Chitwood.

A few days later Einhorn told a reporter:

My reaction was, what is this all about. And of course, I have very good control of myself. I immediately gave myself an autohypnotic command — just ... cool it. Cool it and watch this as carefully as possible, which is what I did. Which to them translated into as my being nonchalant, but I was just being totally observant. Because I've been through [tough situations], I've faced guns, I've faced the whole thing. So if you don't act quickly, you can make a mess. I was not about to do anything. So I observed, literally, when they marched to the closet on the back porch.

The team of policemen had no real interest in the rest of the apartment, though Chitwood noted he had never seen a place with so many books. They went straight for the closet. The closet had a thick Master padlock on it and the detective asked Einhorn if he had a key to the closet. Ira said he didn't know where it was.

"Well, I'm going to have to break it [the lock]," Chitwood said.

"Well, You're going to have to break it," Ira responded.

The photographers from the Mobile Crime Unit photographed the locked door. Then Chitwood broke the lock with a crowbar, and the door was photographed again.

The closet was 4.5-feet wide, 8-feet high, and just under 3-feet deep. The two-foot-wide shelves were jammed with boxes — some of which were marked "Maddux" — bags, shoes and other odds and ends. On the floor was a suitcase with the name "Holly Maddux" on it. Behind the suitcase was a black steamer trunk.

The interior of the closet was photographed before Detective Chitwood began removing the items one by one. As they were taken out, they were individually photographed and then examined. The boxes contained such things as kitchen items, clothing, schoolbooks, papers, etc. The suitcase contained clothing, and several letters to Holly Maddux that were more than two years old. Holly's handbag was in a box sitting on the trunk. Inside the bag was her driver's license and social-security card.

Detective Chitwood noted an unpleasant odor as he continued to remove the items from the closet one by one, pausing for the photographers to record every move, every article. Ira had been walking back and forth between the main room of the apartment and the closet repeatedly while all this was taking place, and Chitwood later said that he thought he could sense fear rising in Einhorn. Einhorn himself would later say that he was merely in a "meditative state" — observing — and that he was coming to the idea that this intrusion was connected to his "efforts to disseminate crucial information about top secret things like psychotronic weapons."

At this point, Michael Chitwood was ready to open the trunk. It was sitting on a piece of dirty, folded-up carpet. It measured 4.5-feet long, 2.5-feet wide, and 2.5-feet deep. It was locked. Again, Ira was asked for the key, and again he said he didn't have one. The trunk was photographed before and after the lock was broken. The odor assailed Chitwood as he opened the lid of the trunk. He asked for rubber gloves.

On top, inside the trunk, were newspapers. The latest date on them was September 15, 1977. Underneath the newspapers was shredded foam-rubber packing material and wadded-up plastic shopping-bags. Chitwood began to scoop out the shredded foam. After three scoops he saw something — a wrist and five fingers — shriveled and dark like rawhide.

The coroner had been right. Detective Chitwood followed the hand down the arm to a cuff of a plaid flannel shirt, and then he backed away from the trunk. Pulling off his rubber gloves, he instructed one of his men to call the medical examiner. He went to the kitchen to wash his hands, and standing there was Ira Einhorn, maintaining his cool.

"We found the body. It looks like Holly's body," he told Ira.

"You found what you found," said Einhorn.

At that moment, Detective Chitwood noticed some keys hanging from hooks on the wall in plain view. "What are these for?" he asked Ira. Ira said, "Maybe they fit the locks that you just broke." Chitwood took them and tried them in the locks. They fit.

Going back to the kitchen, Chitwood asked Ira, "Do you want to tell me about it?"

"No."

Chitwood read Ira his Miranda rights. When he came to the part about the right to remain silent, Ira said, "Yes, I want to remain silent."

By this time, the medical examiner, an assistant district attorney, more homicide detectives, and J.R. Pearce had arrived. Cameras were flashing, power tools were being revved up to cut out the floor where the trunk had stood, and more search warrants were arriving authorizing more areas to be searched and more evidence to be retrieved. Through it all, Ira Einhorn was in a state that has been described as "eerie languor." He gave no resistance.

In deference to his prominence as a public figure, he was not handcuffed before he was escorted to the official car for transportation to the holding facility, where he was booked for the murder of Helen (Holly) Maddux, aged 31-years old.

The medical examiner described the cause of death as "cranio-cerebral injuries to the brain and skull. There are at least ten or twelve fractures and maybe more." Her skull was broken under the left eye-socket; there were a series of breaks in the skull on the left side, and several broken and depressed broken places in front of her right ear; the right frontal bone of her forehead was smashed; and there were more breaks around the orbit of the right eye. Holly's lower jaw was broken so badly that part of it was driven into her mouth. According to the medical examiner, "the holes in the skull are so big you can't define how many times one area had been struck," presumably with a blunt object such as a lamp or a bottle. Six was the minimum number of blows, but there were likely more than twice that many.

In other words, Holly Maddux was probably already dead while Ira continued to vent his rage on her, like a crazed 270-pound gorilla stomping on a rival and beating its chest. (No insult to gorillas intended. They are probably far more civilized than Ira Einhorn.)

The reader, having read the facts of the case — the report of the thud and scream, the attempt by Ira to dispose of a trunk, the terrible odor coming from his apartment for many months, his contradictory claims and statements to friends and family, his concerns about the closet where the body was found (of which there were many other notable examples in Levy's book), his failure to follow his habit to sub-let his apartment in the summer, even his little obstructive lies about the keys, and finally, the fact that the body was in *his* closet — *knows* that Ira Einhorn murdered Holly Maddux, stuffed her body in a trunk and, when one feeble attempt to get rid of it failed, returned it to the closet no later than September 15.

Anybody with two neurons in contact with one another would come to the same conclusion.

But, that is *not* what happened. Droves of people — public figures, people of wealth and taste, intellectuals — came to support Ira and declare that it was utterly impossible to even consider that he would harm a fly, much less Holly, whom he loved with all his heart.

Nobody seemed to pay attention to the curious fact that *Ira was not the least upset that the body in the closet might be Holly's*. Interviewed in jail by Howard Shapiro for the *Philadelphia Inquirer*, Einhorn went on the record saying:

"I have been outspoken all my life, but never have I been violent. I want to be very direct about this. I did not kill whoever was supposed to be in there. I am not a killer. I do not know if a body got in there — if it was a body."

Then he declared that he "still loved Holly."

I want the reader to stop for a moment at this point and think about all of the above. Imagine that there is someone you love, someone who you have said is the "love of your life." Maybe even imagine it is your child, if it is easier to get the proper "feeling" that way. Now, imagine that this person has just disappeared. Keep imagining. Imagine that you *are* innocent of any harm to the person. You can even imagine that you are convinced that certain nefarious groups may be "after you" for some reason, so you already have the idea that you could be a target for any number of "arranged problems." Forget any other details about Ira's case, the scream, the horrible smell of rotting flesh, the odd behavior about the trunk and the closet. In your scenario, none of that happened. You are a target of evil organizations and your loved one just disappeared, and you are worried. Really worried. What would you do? What actions would naturally result from such a situation?

If you suspected that you had been targeted by some secret gang, what kinds of actions would you take to discover if harm had come to your loved one? Would you refuse to cooperate with the investigator hired by the family of the one you love? Even if you didn't like the family? In such a situation, even people who don't really like each other generally will unite in the common goal of assuring the well-being of someone they both love. What kind of investigations would you undertake on your own?

I bet you can think of all kinds of things that Ira never did. In fact, there are a number of movies and stories about people in exactly such situations who have made heroic efforts to find a loved one at great risk to their personal safety and reputation. Because of love.

Well, for whatever reason, in spite of the fact of his declared great love for Holly, *Ira did nothing*. Oh, sure, he claimed he took certain steps, but no one supported his claims. None of the people who he would have been expected to contact to find out Holly's whereabouts were *ever* asked by him if they had any ideas or had heard anything. His story about a call from Holly was only used to obfuscate those who suspected foul play. He said he loved her, but *his actions did not match his words*.

But, getting back to our hypothetical scenario: Imagine that you are just sleeping in your bed, after two years of worry and searching for your lost loved one, and the police show up at your door suggesting that the solution to the problem might be in your closet. (Yeah, I know, this is stretching our imagination a bit, but keep trying.) Anybody who *really* loves somebody, who has *really* worried about that person, even if they are incredulous at the mere suggestion that their own closet may hold the answer, is certainly not going to obfuscate the issue of the keys. You are going to want to know *why* anybody would think that the answer is in the closet. Heck, maybe you haven't looked in the closet for years. And, as Ira claimed, maybe you *do* suspect that something might be arranged by shadowy groups to set you up because of your "radical work." Is that going to diminish the feeling you have for your loved one? Is that going to interfere with your desire to leave no stone unturned that may lead to a solution of where the loved one may be found? Of course not.

Okay, next imagine that the trunk has been opened and — to your complete astonishment — a body has been found. A trustworthy individual who would be considered to know what he was saying when he announced that a body has been found, has just told you that a body is there, and that it looks like your beloved one. Keep in mind, the suggestion has been made that this is someone you *love* with all your heart. Someone that you have been trying desperately to find for almost two years. Keep in mind that you suspect that someone may have been after you, to harm you, and it may be that this harm was done to your loved one in order to get to you. As much as you don't want to think that your loved one is dead, you *have* to know! Are you going to just stand there after a detective has said, "We found a body, it may be your loved one," and just answer "You found what you found"?!

Then, even if you are not allowed to rush madly to the closet to *see* with your own eyes, but instead, are arrested for murder, are you going to just say, "Yes, I want to remain silent?"

Now that your beloved has at last been found, are you going to immediately grant an interview with a journalist in which you tell the journalist that you *know* who murdered your loved one, but that you aren't going to say who did it to "set you up." Are you going to follow this with: "I did not kill whoever was supposed to be in there. *I do not know if a body got in there — if it was a body*"?!

Excuuuuse me!

What we are looking at is a classic display of the psychopathic personality. He claimed to "love" Holly, yet in all the material I have reviewed about the case, other than repeating that he "loved Holly," *there is not a single expression of grief or anguish.* He knew the word "love," but clearly did not know what it meant. Oh, sure, he knew how to "act out" love in ordinary ways, but he had never had anyone to observe who was in the situation he found himself in, so he had no "model" to *ape* in order to produce the appropriate responses. This is the weakness of the psychopath. All their "emotions" are "acting," and they can only act based on what they learn from others. Even if they have produced, momentarily, an appropriate display, something about the way their mind works makes them unable to sustain it. They are so focused on themselves, that they simply cannot abide attention being diverted to anyone or anything else, and will unconsciously give themselves away.

Oh, indeed, others noted Ira's coldness, and Ira explained it as being his great "control" — that he had faced so many difficult situations that he gave himself an "auto-hypnotic suggestion" so he would be calm. But that was only *after* it was pointed out how odd his behavior appeared to others.

Those who doubted him had their doubts thrown back in their face on the platform of his public persona. The very idea that they would doubt him became an implied *moral failing* on their part!

"This is a time of testing. *I try to learn from every situation.* Now I'm going to see if my friends believe in me. I think that they do. I know that many people will help. I know that many will refuse to believe what's being said. These people know me. They'll stand by me. I'm sure of it."

And, of course, with such a challenge thrown out there, they did.

Curiously, Einhorn's choice of legal counsel was Arlen Specter, the strategist who conceived the controversial "single-bullet/lone assassin" theory regarding the death of John F. Kennedy. Einhorn was a very vocal critic of the Warren Report and held the single-bullet theory in complete contempt. A mutual friend of Einhorn and Specter suggested Specter as counsel, and Einhorn tossed aside his previous scruples and said "Yes!" If Specter could convince the world that a single assassin had murdered JFK, he could convince the world that Holly got into Einhorn's closet along with all her things, completely unbeknownst to Ira.

The first objective was to secure a low bail. Bail hearings are held to determine whether or not the suspect should be considered reliable enough to be released on bail until tried. Witnesses are brought in who will attest to the fact that the accused is of such good character that they will abide by the law and show up for their trial.

The first character witness for Ira Einhorn was Stephen J. Harmelin, an attorney with a prestigious law firm. He had known Ira since high school, and he described his character and reputation as "excellent."

The assistant district attorney, Joseph Murray hoped to uncover indications that Einhorn might leave the country rather than face a trial. His first concern regarding this was Einhorn's income. There had been years of speculation about just how Einhorn *did* make his money, even though it was obvious that he lived very frugally and didn't need much. He did, however, travel a great deal and engaged in activities that might be thought to require goodly sums from time to time. Where was it coming from?

"What does he do for a living?" Murray asked.

His long time friend from high school wasn't precisely sure. "My understanding is that he acts as a consultant for various institutions in Philadelphia."

"Like what?"

"My recollection is that he was doing some consulting work for the Bell Telephone Company. Mr. Einhorn seemed to be able to bring to the attention of corporate executives a kind of broad spectrum of information which they otherwise wouldn't generally have available to them."

"Like what?"

Edward Mahler, a vice-president of personnel relations at Bell, was called upon to explain the relationship:

"We wish to be responsive to the needs of the community. In that context I was introduced by a friend to Ira, and we started our relationship. I became, as it were, the contact with Mr. Einhorn. I think

I would dare to say we would meet two to three times a month.... We talk fundamentally about things related to the community. I would discuss with him things we were proposing, and he would respond to them. Then we'd talk, of course, about myriad things that had nothing to do with the telephone company."

"And did Mr. Einhorn have contacts with other key executives at Bell Telephone?"

"Yes, he did from time to time."

"Such as?"

"Such as our president and former president."

"How much was he paid as a consultant?"

"Nothing in cash."

"What? In check?"

"No, nothing like that. No money changes hands."

The judge, William M. Marutani, was so curious about this bizarre series of remarks that he proceeded to question the witness himself.

"Do I understand, Mr. Mahler, that you meet Mr. Einhorn say, two or three times a month?"

"Yes."

"And what would these be — you would happen to bump into each other?"

"No, no, no. Excuse me, Your Honor. Ira represented a group in the community that we did not hear from. Where it seemed to us he could represent their point of view. It was not one we could [otherwise] be aware of.... And at least that was the reason for maintaining regular contact."

"Was this a public relations venture by Bell Telephone?" the judge asked.

"Oh, okay, community relations, Judge."

"And Bell Telephone, with all its wealth, never gave him a dime?" The judge was obviously incredulous that the phone company even took Ira seriously.

"No sir, no sir. We did do *one thing* for him. I feel I should mention it to explain in part what went on. It seems that Ira represented the people in the community, like the professor you heard, the lawyer, the physicist, the futurist, and so on. [We] would reproduce articles that Ira would find, or others would find, and *mail them to members of the group*.... It was a service we performed, you could say, for Ira. He was very appreciative of it. But again, it was in line with *our communicating with this broader network*, Judge.

"What would be in it for people such as Ira to go to all this trouble?"

There was no answer that was obvious. Mahler knew that Ira had never wanted money. It was his declared purpose to just be a "networker." Money did not motivate him. He claimed that he operated on a higher and grander system; he worked without pay for Planet Earth! That's who Ira was.

Mahler was excused.

The parade of character-witnesses was long and illustrious, but the prosecutor pointed out that none of their protestations of Ira Einhorn's "good character" made any difference. "This is a very serious case. A case that indicates planning, intention, all the earmarks of a very serious first-degree murder case," he said. He then stated that Einhorn's contacts, habits, access to money, and many travels, all might suggest that Ira would jump bail and disappear. He wanted bail set at $100,000, of which only ten percent needed to be posted.

Arlen Specter argued that this amount was excessive, claiming that the State had no direct evidence that Ira had killed Holly.

Say what?

The judge made a classic series of remarks:

"Isn't it a little unusual to have a dead body... in a trunk in one's own residence? Doesn't that raise some eyebrows...? Remember, I am not determining his guilt — indeed, I want to be candid with you. I think the *Inquirer* quoted him as saying he was framed. He well may have been as far as I am concerned. I mean, he's not in his apartment twenty-four hours a day and somebody... who knows, strange things happen in life... you know, I wouldn't want somebody to find my wife's body in a trunk in my home, particularly if I lived alone. I think I'd be in hot water."

That was the point. Anybody else, under just about any other circumstances, would have really been in the soup if a dead body were found in their closet, and the circumstances we have recounted were present. It would have been a sure thing that everybody would have thought they were guilty. In fact, there have been judges, doctors, attorneys, politicians, millionaires, and high-society ladies of the very type who were testifying to Ira Einhorn's good character, found in far less incriminating circumstances who were *not* presumed innocent either by their friends, peers, or the public.

So why was Ira?

Why did Barbara Bronfman, wife of Seagram heir Charles Bronfman, step forward with the paltry four-grand necessary to post the remainder of Ira's bail, after Judge Marutani had finally set it at $40,000? (Ira's parents assumed the liability for the remainder.)

This is a very crucial question. We might assume that all of the witnesses to Ira's character had a few neurons firing there in the old brain pan. We will allow that they probably did not know all the details that came out in the investigation. After all, the case had not yet been tried. All they really knew was that Holly's body had been found in the trunk in Ira's closet, and he was claiming he had been framed because his networking had made enemies.

I should like to point out that this "explanation" — that Holly had been killed to frame him — completely obviates Ira's former claim that he had received a phone call from Holly saying that she didn't want to be found. It also begs the question: If he really thought that there were "agents" out to get him, and he really was innocent, *why didn't he suspect that they might be responsible for Holly's disappearance?* And, if he thought that, *why didn't he make it his business to find out what had happened to her?* After all, he loved her so much that two nights before she disappeared, he was on the phone to her repeatedly trying to get her to reconcile with him. When she told him she didn't want to see him again, he threatened to throw all her possessions in the street. It was only in order to collect her things that Holly even went to see Ira.

Another question is: If there were agents of some dark conspiracy who were after Ira, and they were able to kill Holly, plant her body in his apartment under the noses of the neighbors in order to frame him, and do it so easily — why didn't they just snuff Ira? After all, it was six of one and half-dozen of the other for him to be martyred by his own death, or martyred by imprisonment because he was framed. Surely, such agents — if they existed — didn't think he would be quiet about being framed — *if he was framed* — and surely, if he was *really* framed, such agents would have assumed that he would blab whatever he knew that they didn't want spread around, and therefore, framing him would have been seen as too costly. Because surely, if such agents really existed, they knew that by framing Ira, whatever information he was supposed to have possessed would have become a matter of public record in an ensuing trial.

The "Ira was framed to silence him" story just doesn't cut it no matter which way you look at it. We already know, from our look at other cases, that when it is desirable to shut somebody up, it can be done so quick, so clean, so completely, that all this nonsense about "the CIA burned my house down," or "the CIA tried to kill me 16 times," or "the CIA killed someone else and planted her body in my closet to frame me so I wouldn't talk about what they are doing," is a complete load of hooey.

But this cock-and-bull story was believed by many, and everybody's ethics were put on the line with Ira announcing to the press that "now he would know who his friends really are." They showed up and got in line, and did the Einhorn dance, and forked over the bucks Ira needed without a single question.

So why? Why did seemingly intelligent and articulate people find it so impossible to think that Ira Einhorn could murder Holly Maddux?

Could it be, as some suggest, that his defenders were *also* part of some sort of conspiracy? Did they all know that Ira murdered Holly, but they were being instructed to get him out of jail and out of the country?

Nice try. But that would make a plain old liar and murderer out of Ira, which would then negate his "frame up" story, based on the whole deal about being involved in things that others wanted hushed up enough to frame him! Again, simple logic cuts through the confusion spread by Ira.

No, I am afraid the answer, on the psychological level, is far more prosaic than that: Ira Einhorn was quite simply a psychopath, a con-artist, a grifter, and all those people who lined up to cast their lot in support of him, intelligent and articulate though they might have been, were simply duped. They were suckers. They were victims of a psychopath.

Remember the guy described by Dr. Robert Hare, the psychopath who was on his way to public prominence and was then exposed as a fraud? The amazing thing about that case, which has elements that exactly mesh with the Einhorn case, was the fact that *the very people he was duping jumped to his defense*, one of them declaring that "I assess his genuineness, integrity, and devotion to duty to rank right alongside of President Abraham Lincoln."

What many people do not seem to fully grasp is the fact that psychopaths are such *good* impostors that they can live their entire lives as phonies and *never* get caught. I am reminded of several cases where the evidence of the "secret" of the psychopath wasn't revealed until after their deaths, in their papers and effects. Naturally, the friends and families were devastated and the repercussions weren't too savory either. But, for the most part, except for a few people who have had certain experiences with the psychopath, who have seen the "man behind the curtain," a psychopath dupes nearly everyone — and does it successfully.

Another important thing is the fact that a psychopath may spend years building up to a "coup." Some of them display almost preternatural cunning in the way they stalk their objective. They can certainly make a big display of working very hard. They can be meticulous and scrupulous about fulfilling certain obligations, "keeping their word," doing all the things that create a certain "image" of benevolence or kindness, or generosity. But there is *always* a pay-back. The smarter ones have learned to delay the "pay-back." Some of them don't ask for an "obvious" pay-back, rather they will use a connection to make another "hit." They will associate with one "target" in order to get to another. When we say that psychopaths make their way by conning people into doing things for them, obtaining money, prestige, power for them, or *even standing up for them when others try to expose them*, it must be emphasized that this activity can be so subtle that only the most acute observer can detect the "glitches" in the program, so to say. They always have a "secret life" that is hidden from the public, and maybe even hidden from every single person in their "regular life." In some cases, they may have revealed themselves only when they were very young, before they "learned" that their way of being is "different."

The man in the case cited by Hare said: "These trusting people will stand behind me. A good liar is a good judge of people."

Ira Einhorn said: "This is a time of testing. I try to learn from every situation. Now I'm going to see if my friends believe in me. I think that they do. I know that many people will help. I know that many will refuse to believe what's being said. These people know me. They'll stand by me. I'm sure of it."

In both cases, and thousands upon thousands of others, exactly the same conditions prevail. A psychopath lies, and people believe him — on a scale that is almost impossible to imagine. The reason they are able to do this is very, very significant: They are able to do it because *most people are gullible, with an unshakable belief in the inherent goodness of man.*

But it goes even deeper. Not only do people believe them and take them at their word, when another person points out the "glitches in the program," the believer will not just refuse to acknowledge it, they will refuse to investigate on their own! If the proofs are handed to them, and all they have to do is read them and *think*, they *won't do it!* In some cases, they will read them, but will trust the psychopath who is lying when he is saying that the proof is all lies.

Just to give a concrete example of this fascinating phenomenon — a real-life example — let me describe this curious aspect as it has been observed in our own experience. In the case of Vincent Bridges and his coterie of followers, it has not occurred to them, for example, that six people who all know Frank Scott personally — *some of them for as long as ten years* — all have the same opinion of him. None of the

Bridges gang have ever met him and have no long experience upon which to base an opinion. Nevertheless, they believe Bridges' opinion, his claims that all of these six people are lying, because only *he* can possibly know the truth of the situation! Why? Because he has so repeatedly declared himself an "expert" that they *must* actually believe he is! The fact that every single one of his "expert credentials" was proven to be false makes absolutely no difference in the world. What is even more shocking is the extent to which the defenders of the psychopath will go, even committing illegal and unethical acts in his defense. In fact, this item might even be a key to identifying the psychopath: The extent to which people will go, acting in opposition to their own welfare, their own professed standards of ethical behavior, and even to their own detriment, in order to defend a psychopath. This may be a clue that the psychopathic reality is "communicable," like some sort of disease.

A concrete example of this behavior is the fact that the very person who was obsessively checking Bridges' credentials, Dr. Nellie Oleson (pseudonym) — the one who first brought to our attention the fact that there were serious questions to be asked about Vincent Bridges's background — is now one of his staunchest supporters!

The fact is that even after we had expressed no real desire to pursue any serious questions about his background — we merely wished to have the free will to "not associate," and that was that — Dr. Oleson *repeatedly brought it up*, adding one find to another, in what now can be seen as an obvious attempt to "stir the pot." Not only that, but she was writing private emails to other members of our e-group, casting doubt on Bridges as an "expert" of any kind, seemingly in an effort to put pressure on us to — what? To investigate him? Or simply to be more and more antagonistic? It's hard to tell. Nevertheless, what is crazier still, as soon as we had arrived at the opinion that Bridges was doing some serious "credential padding," and we began to discuss Bridges privately in terms of Dr. Oleson's findings about his lack of credentials; as soon as she was assured that we all had the idea firmly planted in our minds that Bridges was a liar and a cheat, she *then* flipped over to Bridges' side and started doing the exact same thing with *him*!

What power did he exert over her? Because the fact is, Dr. Nellie Oleson did some things that not just violated our rights, and the rights of our group, but which were also ethically self-destructive and legally actionable!

Dr. Oleson, as a member of the Perseus Foundation, *signed a confidentiality agreement* that anything relating to any of our investigations that was discussed or shared in person, by phone, or email, that was designated "private," was *private*. This was a clear necessity for a foundation that is set up to investigate backgrounds of other people and possible frauds on the internet, etc. Journalists cannot work without confidentiality. *A confidential source has to be assured that they are dealing with a group that will hold the information private, even including their identity.* This is a legal right of investigative journalism based on the necessity of being able to assure that information will be available even in cases when the informant has reason to fear the release of that information.

Now, keep in mind that Dr. Nellie Oleson is a professor of English at the University of Wisconsin, Platteville, and would certainly be expected to be very familiar with how journalists work. However, after being ejected from our e-group for disruptive and aggressive behavior, *to garner favor with Vincent Bridges*, Oleson forwarded to him numerous private emails that had been copied to her *from a confidential informant*, as part of the beginning investigation into the Bridges matter. The informant had given his permission for her to read them, since she was a member of the Foundation with a confidentiality agreement on file. When they had been forwarded to her, they were marked as "private."

However, not only did Dr. Nellie Oleson violate the confidentiality of a third-party confidential informant, she violated her signed confidentiality agreement with the Perseus Foundation. She also betrayed over 200 other people in our e-group and research team.

As an instructor at an institution of higher learning in this country, who not only claims to be informed about such matters, but is also hired to teach them — to *set an example* — we find that *her behavior was not only irresponsible, it was highly unethical and illegal.* I certainly would not want such an individual (especially considering other aspects of her "story") teaching one of my children. What is more, in the world in which she proposes to operate — the world of journalism — this act has put her beyond the pale of serious journalists and writers. In short, she has shown that she cannot be trusted in a field where trust is paramount. Without journalistic confidentiality, no truth would ever be known.

She did this to "defend" Vincent Bridges. We see here a real live example of how people will even do things that are illegal under such an influence. If nothing else, the fact that this woman, with so much to lose, would commit so unethical, irresponsible, and morally reprehensible a series of actions proves the power of the psychopath to confuse and influence.

This brings us to another aspect of the psychopath: The obsessive need *to force their delusions on others* and to force association. Mr. Bridges literally is consumed with the fact that we have refused to associate with him. He spends days and nights obsessing about us, our website, and every single thing we do or say. I've never seen anything like it! It's like he has no life at all except the life he draws from his followers in their endless feeding frenzy on each other's opinions of the Cassiopaea Experiment, about which they know almost nothing! Bridges and Dr. Nellie Oleson will produce endless pages of text about their analysis of me, or about what we say or do, using either my own writings or Tom French's article to diagnose my "psychosis," or "megalomania," or whatever is the *personality glitch du jour*. He intimates that he has had long and friendly talks with Tom, and suggests that they are "hermanos," that they "both know" how "dangerous" I am. What he never mentions is the remark about Frank (and what may have been behind it) that Tom made in his article: "He was very excited, and when F*** got excited, he could be a little pushy. Laura did not mind. She adored F***; besides, the unwritten contract of their relationship allowed great leeway for pushiness."

It certainly was that way. Pushy. *Great leeway* for pushiness, by the way. Noted in print by Tom French, who could not fail to note it, and who also could not fail to note that I was most definitely tolerating a *lot* from Frank. I truly adored Frank, and that was and is the *only* reason he continued with the group for as long as he did. This factor has also been noted in remarks by the six other active members of the group, some of which have been made public, others of which are reserved for legal and/or privacy reasons.

But this comment of Tom's is ignored by Vincent Bridges, all the while he "uses" Tom based on a phone call during which conversation he attempted to convince Tom that I am a raving megalomaniac trying to start a cult — *because we simply do not want to associate with him*. Bridges tried to elicit from Tom French a "new opinion" of me, *after* filling him in with the sordid details of my life according to Vincent the "expert." I guess he thinks Tom was born yesterday.

As it happens, Tom usually calls me on my birthday, as he has for the past seven years. We had a most interesting discussion this last birthday, you may be sure.

So, on the one hand, we have six respectable people, none of whom has ever even "fudged" their credentials[30] offering information that is

---

[30] I freely admit that I have none. Well, that's not entirely true. I have certification in hypnosis; until I quit in 1996, I was listed in the directories of The American Counselors Society and The National Society of Clinical Hypnotherapists, neither of which is any

direct and factual — along with supporting evidence — about a situation, and the individuals involved.

On the other hand, we have a proven liar and manipulator declaring that everything we say and do is directed towards some sinister, nefarious purpose. He has nothing on which to base his statements other than just his say-so. So what we see is a very strange phenomenon that is quite similar to the situation surrounding Ira Einhorn.[31] Not only is our "sin" the same as Holly's — she refused further association with Einhorn — many of the people around them believed *him* and not the evidence. People didn't just refuse to look at or consider the facts, they were quite simply unable to *believe* them. For Barbara Bronfman, it wasn't until years later — and after she had given many, many thousands of dollars to support Ira in exile — that she suddenly was able to "see" him, and recoiled in horror at how she had been used and manipulated and deceived. So it seems that there is some possibility that some victims *do* eventually wake up. But many don't, and I would like to know why.

Over and over again I think about those remarks of Leon Festinger about cognitive dissonance:

> Suppose an individual believes something with his whole heart; suppose further that he has a commitment to this belief, that he has taken irrevocable actions because of it; finally, *suppose that he is presented with evidence, unequivocal and undeniable evidence, that his belief is wrong*: what will happen? The individual will frequently emerge, not only unshaken, but even more convinced of the truth of his beliefs than ever before. Indeed, he may even show a new fervor about convincing and converting other people to his view. (Festinger, 1956; this author's emphasis)

But essentially, what we are talking about here is not anything like religious belief, or belief in Santa Claus or the Tooth Fairy — or are we? What is it about the psychopath that makes them able to *induce belief* so strongly that people will even undertake to do illegal things for them, to lie for them, steal for them, join them in libel, and cruel

---

kind of major endorsement. I also am an ordained minister, but I usually don't tell anybody because I don't do anything with it, and it's so easy to be ordained it doesn't count for anything in my opinion. In fact, it was only because my Reiki master said that in some states doing Reiki is only legal if you do it under the aegis of religious practice that I even bothered with it.

[31] What's more, Mr. Bridges has even likened himself to Ira Einhorn in his long, fantastic autobiographical tale (www.cassiopaea.com/archive/most5.htm) made up of moonshine and wishful thinking! He tells us so self-deprecatingly that Andrija Puharich liked him because he reminded him of Ira. My guess is that, just like his implied camaraderie with Tom French, Mr. Bridges is just doing what he does best here again.

and hurtful activities against others? Is it a psychological weakness that drives people to prefer lies over truth? Or is it *some power in the psychopath*? Is there something going on at levels we do not perceive?

In the previous chapter we presented some of the ideas of psychology as to why such a phenomenon manifests, most of which relate in some way to the idea that people will go along with and support a psychopath, in the face of evidence that they have and *are* being conned, because their own ego structure depends on being right, and to admit an error of judgment would destroy their carefully constructed image of themselves.

I don't think that's necessarily the best explanation. I think that it has more to do with Hare's remark that most people are gullible with an *unshakable belief* in the inherent goodness of man, and it is that which makes them vulnerable. Because they don't lie, they cannot conceive of the level to which a psychopath will go to lie and create support for his lies. They have feelings and would never do such things themselves, and cannot conceive of a creature that does not have feelings. They just don't get it that the psychopath is not a real human being.

But it *has* to go deeper than that, because not only do such people continue to support the psychopath, they become quite active in doing so. Even criminally so. This brings us to that remark of a victim of psychopathy we previously quoted: "The essential feature of Psychopaths is a Pervasive, Obsessive-Compulsive desire *to force their delusions on others*. Psychopaths completely disregard and violate the Rights of others, *particularly the Freedom of Association which includes the right not to associate*."

The person must always, always keep in mind that "forcing" a delusion in the context of the psychopath may have nothing at all to do with what we consider to be force. For the psychopath, the issue is to get the person to believe a lie by using his cleverness, his inborn ability to "sniff out" weaknesses and play on them to gradually turn his victim around just like a masterful horseman. This brings us to the crux of the matter: Psychopaths view any social exchange as a "feeding opportunity," a contest or a test of wills in which there can be only one winner. Here we have found a clue.

Not all victims of psychopaths suffer as Holly Maddux did. Not all "secondary" victims suffer as Holly's family has. But as long as psychopaths continue to operate unrecognized, there is going to be suffering; if not you, then maybe someone you love. On what would have been Holly Maddux's 50th birthday, her sister wrote:

"It has been a very long and painful 20 years since Holly's disappearance and death, an emotional place that is not frequently visited despite the multitude of ways that my life was forever changed with Ira's simple, selfish act. My siblings, their families and I miss Holly. Even though they can hear stories about her, Holly's nephews and nieces will never get to know her as this absolutely wonderful and caring person. I hope no one else has to explain to children how and why someone much loved in the family is dead- trust me, it's not easy and it forces young ones to face fear more intimately than they would ever dream of, and so early in their lives.

"I am now much older than Holly ever was, and I have no older sister to compare notes on life and growing up with, like we had just started to do when she was killed. My own philosophy on life, something to do with the grace of God and the beauty of life and the peace that comes with all of that, is continuously challenged when I hear of a similar situation — don't think for a minute that the OJ Simpson trial was easy to be around! Bottom line, someone is killed and their beautiful life is a memory that it is now up to you — and the hundreds of others whose lives that one special person touched — to remember. Something is just not fair here."

Holly Maddux had made a decision to no longer associate with Ira Einhorn. She said, "No thanks, I don't want to hang out with you anymore." She refused his delusion, and he completely disregarded and violated her right to choose who she would or would not associate with. He furiously and repeatedly bashed away at her head — the seat of the idea of refusal.

But such behavior does not just appear out of nowhere. If Ira was guilty of this, unless he just had a sudden fit of temporary madness, there *must* have been signs. How come nobody noticed? If there were signs, how come they were not remembered at the crucial moment when Holly's body was found in Ira's closet? What was the power Ira had over the minds of other people? How did it manifest? Again and again we come to that question: What is it about seemingly intelligent and articulate people that makes them subject to the delusions of the psychopath? Is it something in the person being duped, or is it some power the psychopath has that other people don't?

We are going to look at that question. We want to know why. As Einhorn himself said during a 1971 mayoral campaign speech, "Psychopaths like myself emerge when societies are about to change."

Chapter 61
Ira's Inner Cesspool

September 11, 2001: Attack on America. I am sure that it is not lost on the reader that it was exactly 24 years to the day, on 9/11/77, that Holly Maddux was brutally murdered. Ira Einhorn was tried *in absentia*, all the evidence was presented, and a jury of his peers decided that Ira did it. It seems clear from the medical examiner's report that Ira must have been in such a frenzy that, even though the first one or two blows probably killed her, he continued slamming her delicate skull over and over again until there was nothing left recognizable of her former beauty or identity as a human being.

Forensic psychologists tell us that when a person is murdered by repeated bashing of the head, it is invariably done by someone who knows them and is enraged with them, who seeks to destroy the source of their thinking and words which the killer cannot tolerate. As the Cassiopaeans said: "Maybe [Einhorn] did what he did, and then convinced himself that he did not, *a la* O.J. Simpson."

In short, Holly Maddux was sacrificed on the altar of Ira Einhorn's ego. He had lost control of Holly, and Ira Einhorn was all about control — control of others.

Many of the principals involved in the case are also now dead. Holly's father, unable to come to terms with the brutal death of his beloved little girl, committed suicide in 1988. Her mother died two years later of emphysema — no longer able to breathe the air of a world in which something so horrible could occur without any hope of justice. R. J. Pearce, the former FBI agent turned private-detective who doggedly pursued the "Unicorn," has also passed on, along with Ira's buddy from his heydays, Jerry Rubin.

Right up to the present moment, Ira Einhorn insists that he did not murder Holly Maddux, that he was "framed" by shadowy organizations because of his "studies" and associations with those doing experiments in psychotronics (see "Einhorn" section of bibliography). Andrija

Puharich is also dead, and most interestingly, before he died he had come to some very different ideas about the entire situation: He scoffed at the idea that studies and information about psychotronics could ever lead to a plot to frame Ira for Holly's death. It is remarkable that so conspiracy-minded an individual as Puharich eventually evaluated what Ira was doing as *not sufficiently important to attract attention from any intelligence agency,* whether CIA, KGB or some other nefarious strike force. And, of all people, Puharich was most familiar with the nature of the information Einhorn was dealing with since, for the most part, *it came from Puharich himself.* This also suggests a significant change in Puharich's mind about what he himself was doing, that he might now believe that the information about "psychic warfare" was false, or that it was disinformation.

In the months following his arrest, Ira Einhorn looked many people straight in the eye and said "I did not kill Holly." The famous intensity of his gaze engaged the doubter and, for the moment he was focused on them, their doubts were cast aside.

> Do you know me as a violent person?
> Why would I kill Holly Maddux, a woman I loved?
> Even if I did kill her — and I didn't — would I really be so stupid as to leave the body in the trunk in my apartment for so long?

For many of Ira's friends, this last point was compelling evidence that he was framed. Somebody who would keep a corpse in his closet, just a few feet from his bed, was not the Ira they knew, who was clear-thinking and practical. Ira pushed the point by reminding everyone that all during the time that Holly's body was "supposedly" stuffed in his closet, he was entertaining friends, women, and so forth. If he really had a body in the closet, would he be so stupid?

His friends kept saying: This is *not* "classic Ira."

Ira would go through the list of contradictions between the actual situation and what other people knew about him based on his "public persona"; and the way he presented those "facts," it always seemed weighted in his favor. He was so cunning that, in his presence, listening to him talk, the majority of people would come away believing he was innocent.

There were a couple of people who had seen Holly with bruises that she admitted were a result of fights with Ira. But Arlen Specter argued that no one had witnessed Ira striking Holly, so the testimony that Ira had ever previously struck Holly was "hearsay." The judge did not agree.

Jerry Rubin told Ira that he could do a great deed for the world by admitting he had killed Holly, and then proclaim himself an example of an overdose of male domination, one of the problems in our world. Then, after paying his debt to society, Rubin went on, Ira could found an institute to study the problem of male violence.

Ira said it was an "interesting idea but irrelevant," since he didn't kill Holly.

Arlen Specter thought that Ira's best hope was in an insanity defense. But Ira flatly refused. There was nothing wrong with *his* mind, and there was no way he was going to go along with any such idea. To him, claiming insanity was as much a disaster as admitting killing Holly — both actions would "discredit his lifetime work." His credibility would be in ashes, and nobody would believe that his ideas were legitimate ever again. The psychopath knows that admitting a lie is the end. Once they have done so, they have lost the one power over others on which they depend: The ability to lie with such flair that gullible, well-meaning people can always be found to believe them.

As the weeks went by, Arlen Specter realized the negative political implications of being associated with Einhorn, and turned the defense over to Norris Gelman. Ira kept boasting that he was going to expose the frame-up. His attorneys certainly hoped he had something along that line to produce, but apparently he didn't. "You think the Russians came... and threw a beam on him?" Gelman, who was privy to Ira's best information, mocked in derision of the very idea.

Gelman didn't urge Ira to cop insanity. "The most brilliant defendant that ever hit City Hall, and I'm going to claim he was insane? No way!" Instead, Gelman's hopes were pinned on just a careful refuting of the evidence, point by point, and minimizing what could be minimized in that which was out-and-out damning. If he could just establish a "reasonable doubt," that was the best that could be expected.

Needless to say, Ira wasn't happy with that plan.

Ma Bell dropped Ira like a hot potato. Many of Ira's friends (see bibliography), as they learned more about the case, began to drop him and withdraw. One friend who did maintain ties walked down the street with him one day, and Ira noticed that people no longer came up to him to hug him, happy to be seen in his presence. Instead, they avoided eye contact, or even crossed the street to avoid him. "I'm not going to be able to be Ira Einhorn now", he complained. "And I realized he was a selfish, arrogant bastard," the friend reported.

Ira decided that hanging out in Pennsylvania while waiting for his trial was a bummer, so he went to California. He met a friend there and spent some time on her houseboat in Sausalito. Then he wandered to

Esalen, where he had spent so much time in the sixties. He met psychic Jenny O'Connor there, a woman who channels "The Nine." Nothing significant came up.

He also met a young woman while there who, while high on MDA, claims that she looked into his soul and knew that "this was an extraordinary being, scandalously charged with a crime he did not commit." So much for MDA insights. He had lunch with Jacques Vallee, a conference with Mike Rossman, and then went on to visit physicist Jack Sarfatti. Over and over again he was telling his story and finding sympathetic listeners. Sarfatti even organized a public meeting for Ira. Sarfatti marveled at how calm and "together" Ira was. Saul-Paul Sirag, another physicist, agrees. "The thing that impressed me was his incredible nonchalance, considering the enormity of what he'd been charged with... he seemed to be in great spirits."

Jack Sarfatti and Saul-Paul Sirag were unknowing witnesses to the amazing lack of conscience of the psychopath.

However, an odd event occurred that confused Sirag: "I had gone to the meeting with my girlfriend at the time... and we walked off in one direction and he went in another. I guess Barbara and I were arguing mildly, and Ira turned back and heard us bickering, and he said — as if this were a joke, but it was still weird — 'Beat her up.'"

Ira returned to Philadelphia fully charged from all his energy gathering in California. He had hoped that his lawyer's motion to declare the search warrant invalid would succeed and he would be able to retrieve his journals. The judge ruled that it was valid. Michael Chitwood had made sure of that. But, when the trial was postponed for other reasons, Ira went traveling again. Ira was spreading his "rap" everywhere he went.

According to Ira, he was doing stuff that 'They' didn't want done. They — the CIA — didn't want Ira to connect together Tesla, psychic discoveries behind the Iron Curtain, and remote viewing.

Even though his passport had been taken when the things in his apartment had been confiscated, Ira managed to get another and flew to England to drum up support there. But the folks in England were already having doubts due to press coverage. Playwright Heathcote Huffer asked Ira if he did it. "He looked me straight in the eye and said no." Yet again, people were troubled by Einhorn's confidence and nonchalance. "I got the feeling that the whole thing was an inconvenience to him. That the work was important and this was kind of a nuisance."

After his return to Philadelphia, Einhorn took a new approach. He backed away from his hammering of the KGB-CIA theory and began to suggest that there was some sort of headquarters of the Nazi party in Tyler, Texas and that Holly's father was involved in the frame-up!

Oh, yes indeedy! Blame the victim! Classic psychopath.

The means of setting this idea in motion included asking his friend, George Andrews, to check out the possibility that Frank Maddux was a high officer in the American Nazi Party, and that the Nazis were somehow involved in planting Holly's body in his closet. Andrews was understandably shocked at Ira's implication, but he dutifully set about checking it out. He was unable to establish any validity to such ideas.

As time passed, the "blame game" became more and more vague. After awhile, he just would refuse to address the issue or answer any penetrating questions at all, saying only that he knew who the guilty parties were, and would reveal them at the proper time.

In November 1979, the FBI lab report on the floorboards from Ira's closet were disclosed. The tests showed no blood and/or human protein. Ira crowed that this proved that Holly's body had not been in his closet since September of 1977. He began to declaim loudly and vociferously that he was *now* vindicated!

In an interview with Claude Lewis of the *Bulletin*, Ira said:

There's no blood in the apartment, they found no blood in the entire apartment! How can you fracture somebody's skull twelve, thirteen times, and no have any blood? It's crazy.... The skull was supposedly fractured, six to twelve — multiple times. You can't tell me blood didn't squirt all over the place. And no matter how careful you are, you wouldn't be able to get it all up! [...] With this new data, the ball is in my court. [...] What this has given me is a new lease on life. I feel totally free. I feel like a citizen of the world. I can go settle anyplace. Because I'm going to do a book on all this." (Levy, 1988)

Yes, indeedy. Ira was gonna write a book. His book proposal said: "I intend to write a book directed to a mass audience, for it is obvious that my case has enormous mass appeal. [...] An inside view of someone who lives on the edge of thought, but I will never become overly abstract or philosophical."

The proposal was rejected because Ira made it clear that, even though he would talk about the murder, he would not speculate on who actually did kill Holly. "This ambiguity will not detract from the value of the book, for so much of the public interest in this case hangs upon the ambiguity of my present public persona."

In short, the book was going to be all about Ira, and how Ira so heroically dealt with a nasty murder charge. Holly's life and experiences, who may have killed Holly and why, was not important.

The only thing that was important was "the public interest... [which]... hangs upon ... my public persona."

Ira wasted no time sending the FBI results around to everyone in his network. He wrote:

> It took a court order to [get it]. You may draw your own conclusions from this behavior. It is what I have struggled against during the last 8 and 1/2 months of difficult uphill battle.
>
> I was conspired against in the most hellish way; so many of my friends could not grasp the conspiracy for it is beyond normal ken. [Read: Of course I understood, but that is because I am so speshul!] For this there can be no blame or recriminations as the press played right into the hands of those who wished to silence me. [Read: All of you who doubted me, I forgive you, I will let you kiss my foot if you ask nicely!] This is not the time to name them who are guilty of murder — that will come.

And so on. It was basically an appeal for money to pay his lawyers to defend him against the "hellish conspiracy" that was now an established fact, in his mind, because he had the evidence of the FBI report. Ira promoted this item as "pivotal," and continues to do so to this very day. What he doesn't talk about — because psychopaths never address the things that cannot be addressed without exposing them for what they are — is the fact that a negative result on the floorboards and plaster did *not* preclude the possibility that blood or human protein might have been in the materials at an earlier point, having dissipated in the 18-month interval between the murder and the discovery of the body.

Ira also continues to avoid the fact that a *second* test, using newer and more comprehensive methods, later showed that there *was* human protein in the floorboards and plaster! Those who knew Ira said that he took this information very hard and that this was what prompted him to flee. He *knew* he was nailed.

As of April 17, 2000, Ira posted a statement that somehow managed to mention the two FBI chemical analyses as though they *both* said that there was no evidence of blood or human protein:

> After almost two years of pre-trial manoeuvring, which included much judicial misconduct: Foreshortening of Xeroxed pages so that the damaging pages of a now unnumbered report could be removed from the information due to us; information that included a public sighting of my former girl friend by three bank employees six months after I am supposed to have killed her; and most significantly, a failure to accept two lab reports, issued by the FBI and a nationally known laboratory, National Medical Services, which did all the work on the O.J. Simpson case; reports made at the instigation of the prosecution, which indicated that there was no blood or human protein in the supposed leakage from

the body and that material found outside the trunk did not correspond to material found inside the trunk. This was a stiff blow to the prosecution's contention about the murder; the response: The most prestigious local magazine published an article with blood on every page. Such behavior was repeated no matter where I turned, and fearing the death penalty, I left Philadelphia for a life underground.

Oddly, in the above remarks, a single person, a bank employee, who thought she saw Holly in the bank (identified by a photo) after the time the murder was supposed to have been committed, has now become *three* people. This identification has always been regarded as mistaken because, in fact, there is no record of any bank transaction by Holly *after* September 11, 1977. Why would she be in the bank if she did no banking while there?

Einhorn also quotes, as proof that he knew of a building threat, a letter from Stafford Beer, about which Ira tells us, "This conversation quoted verbatim from Stafford's letter took place in the summer of 1977 about two months before the murdered woman, Holly Maddux, disappeared."

One day I looked up from my desk and saw that someone was approaching down the path. This was most unusual, because hardly anyone knew yet of my whereabouts, which I was keeping virtually secret, and the place is eight miles from the nearest village. I could hardly believe my eyes: it was Ira. But yes — he would have been one of the very few who would know my new address, because I wanted to keep up the flow of packages from Bell.

Ira and I became locked in a fascinating discussion of very sensitive matters. They concerned monumentally important scientific discoveries, and their possible impact on human life and society. I have not to this day disclosed what Ira told me, and I do not know whether what he told me can be substantiated. I am sure that Ira believed what he said, and I could without difficulty accept that it might be the case (that is, "no alien life forms"). What followed is indelibly fixed in my mind, and I shall get as near as I can to the *ipsissima verba*:

Ira: "I am making a special visit to you, and to a few other friends who have the knowledge to understand what I have found out, because my situation is dangerous."

Stafford: 'I can believe it. Are you worried about your own government or 'the competition'?"

Ira: "Your call."

Stafford: "Well, are you saying that you think you might be bumped off?"

Ira : "The trouble is, that wouldn't do. It would provoke a whole lot of investigation, and the truth might get out. No, I think that I have to be in some way discredited."

Stafford: "Aren't you in some way discredited already? Plenty of people think you are a nut case. And plenty think you are immoral — a bad influence. That lot got Socrates after all."

Ira: "And the ideas survived. Just my point. No, it has to be a lot stronger to count."

Stafford: "Any Ideas?"

Ira: "None. I don't know what I need to protect myself against."

I can swear to this testimony. That's exactly what happened. It made me apprehensive for my friend. (Levy, 1988)

Aside from the fact that Stafford Beer is interesting in his own right, the fact is, Holly Maddux left Ira Einhorn in Europe in the midsummer of 1977. When she left, it was already clear that Holly would be moving into her own place. In other words, at that point in time, Ira already knew that Holly had rejected him, and at that point in time, he began to plan her demise. At that point in time, he began to plant the seeds of his "frame-up theory" and his first "target" of this campaign was Stafford Beers.

His later "morphing of the story" was more elaborate, but again, casts the blame on Holly for engineering her own death. This variation was given to journalist Russ Baker, who chronicles his experience with Einhorn in *Esquire* (see bibliography). After much back-and-forth maneuvering, Einhorn finally lays out the new theory as follows:

It is a tale that begins in the mid-1970s, with Ira penetrating deeper and deeper into an understanding of the evil work of his government, chiefly regarding psychic warfare and UFOs, and continues with Ira's growing determination to expose the truth as he saw it.

This tale involves a former CIA man, a current CIA man, a prominent ufologist, shadowy figures, and psychic, Uri Geller. This tale involves the scurrilous rumor that the former CIA official, who at the time was head of the Weird Desk at the CIA, had an affair with Holly Maddux. It is Ira who calls this rumor scurrilous, but he brings it up repeatedly, and this rumored affair is clearly the basis of Ira's theory of the crime.

[According to Ira], [T]here was a great, roiling debate within the secret agency, it seems, that centered on Ira. One faction favored the release of top-secret data on the UFOs, the other did not. Ira was the loose cannon who was going to blow the lid off the story. The CIA had to frame Ira for Holly's murder, and to facilitate this, one of its men arranged an amorous liaison with her. And Ira says that the former official's successor at the agency is in "constant e-mail contact" with him, confirming parts of this story....

"They're using my case to fight over the CIA stuff. I'm convenient."

Ira is the teller of the story, and *Ira is the wronged hero of it, too*. It is a story that takes you on a tour of the interior of Ira's head. [According to Ira], [T]his story is the defense that Ira might have offered at trial had he been there. It is impossible to tell whether he has actually brought himself to believe it. (Russ Baker, "A Touch of Eden," *Esquire*, December 1999, Contributors Page)

Russ Baker was aware that Ira was "different" when he went to interview him. However, what Russ didn't realize, and what so many, many people do not understand, is the power of the psychopath. Russ comments on this in some amazement:

> I could not have imagined this a couple days ago, but after almost thirty hours of Ira, I have to get away from him. Although I'm constantly aware, on some level, of his manipulation, I also know that he is exceptionally good at it. I suppose that *I was not entirely prepared for this, at least not to this extent.* This is humbling. I do not know what a clinical diagnosis of him would be, but *he is masterful at manipulation.* He is a professional. There's a reason that he succeeded so spectacularly on the lam for so long. *I realize that I'm fighting to maintain perspective.* (Baker, 1999; this author's emphases)

It is, indeed, humbling, to realize that with all our powers of observation, all our intellect, all our knowledge and awareness, we can still be captured and manipulated. Any of us who think that we have a real handle on it, that we cannot be fooled, or we assume that we are *not* being deceived by a psychopath, better think again, apply some tests, and watch closely.

In the end, the safest thing to do once one has the suspicion or feeling that something is not quite right, that something is confusing, that something just doesn't mesh, is to just get away! There is *no* way you can interact with or converse with or debate with a psychopath without getting slimed. None. Zip. Zilch. Nada.

From our perspective, there is no doubt that Ira Einhorn killed Holly Maddux, but that then leads us back to the problem of why, for a full year and a half, did he make no effort to dispose of her body and the other evidence that would ultimately indict him? This, of course, is a clue to the deep nature of the psychopath as a truly predatory being, whose behavior *can only be understood in terms of comparison to the hunting behavior of beasts.*

We believe that Einhorn's dissembling after the murder, his cruel cunning employed in hiding his crime, demonstrates a specific type of cold-blooded, conscious thinking consistent with the psychopath, and that it also suggests that Ira was planning for *months* in advance to murder Holly Maddux. During these months, he carefully and cunningly prepared the ground of the minds of his followers and associates, planted clues about conspiracies, shopped for a trunk with the express intention of using it to store Holly's remains, *for the same reason that a predator in the wild keeps the carcass of his kill nearby — in order to savor the kill again and again.*

At the same time, there are very deep implications here in terms of the Theological Reality. Ira Einhorn was closely identified with helping establish what has become known as the New Age Movement. With the benefits of education and a brilliant mind, he had created a sphere of influence among people who stand above the common man. His world was a reality of scientists, educators, intellectuals, occultists, and some folks with seriously big bucks. He brought them all together and almost single handedly, through the cognitive dissonance resulting from the murder of Holly Maddux, *created a mythos that still grips the minds of modern day researchers in those fields.*

Am I suggesting that there is no conspiracy? Well, the readers know me better than that! Indeed, we cannot but notice the terrible significance of the dates: September 11, and the fact that the brutal barrage of blows to Holly Maddux's beautiful head occurred precisely 24 years before the brutal barrage of blows to the beautiful World Trade Center towers, the "head" of the illusory American way of life and sovereignty. Both cases signaled a significant change in the perceptions of the American people. In both cases, huge disinformation projects were launched as a result of the events, and both cases, the events are what the CIA and other intelligence agencies term a "sideshow." They create a distraction away from what is really going on. The ideas born from these sideshows are being vigorously promoted right up to the present day. Tom Bearden, in a recitation of the various persecutions suffered by "truth tellers" of his acquaintance, wrote that:

> Ira was suddenly confronted with a decomposed body in a trunk in his apartment. He is now awaiting trial, charged with the alleged murder. And that is so totally incredible, so out of character with the high consciousness and attunement of my friend, that I for one don't believe it for an instant. Particularly since he was working on Tesla material, directly in contact with the Yugoslavian government, and trying to get information out to the public on Tesla-weapon effects. At any rate, it appears suspiciously as if the psychotronics investigators/researchers are slowly being eliminated or nullified. (Levy, 1988)

Bearden then went on to claim that a whole series of "unusual deaths" were related to this particular cause: psychotronics and remote viewing, etc. He listed the arson of Puharich's house; the plane-crash death of Itzhak Bentov, author of *Stalking the Wild Pendulum* (also a crony of Einhorn); and the fatal-stroke death of Wilbur Franklin, an

Ohio scientist studying Uri Geller. At the present time, the list is expanded by "theatrical physicist" Jack Sarfatti:

> Was Jan Brewer telling the truth about the Fourth Reich using Arica to influence the New Age? Brewer was part of the original Esalen group of forty that went to Chile for the first Arica training with Oscar Ichazo. Arica was big at Esalen at the same time that the Soviets were soaking in the hot tubs. Was I pulled out of the operation by George Koopman because in his opinion I was unpredictable and uncontrollable? Or is the truth still even stranger than even I can imagine? Was Michael Murphy a brilliant Puppet Master or merely a lucky charming "useful idiot," a Forrest Gump character like me? Was Ira Einhorn framed? Was Jean Nadal murdered? Was Francois Trauffaut murdered? Was Harold Chipman murdered? Was George Koopman murdered? Is this all my paranoid exaggeration? What do you think?[32]

Please note that the one person who seems to have stood by Ira Einhorn all the way down the line is Tom Bearden, who shared many of the Unicorn's conspiracy theories regarding the use of psychotronics to monitor and control human behavior and modify weather patterns. This, along with the fact that Tom Bearden is closely connected to Richard Hoagland and the Stargate Conspiracy, ought to give us pause.

It's all a sideshow. People are really being killed in this theater of the macabre. As we have seen in our discussion of the deaths of Morris K. Jessup, Phil Schneider, and Stefan Marinov (discussed in Book Six, *Facing the Unknown*), such events are engineered to create these sideshows, and generally the ideas and beliefs that emerge from them are disinformation. Just as Marinov may have been programmed to jump off the fire escape of the library at the University at Graz, so it is likely that Ira Einhorn was "programmed" to kill Holly Maddux, so that in the ensuing media circus, certain disinformation or "memes" could be promulgated.

One crucial difference to consider is the fact that many of those who are dead were either murdered or driven to take their own lives. Einhorn, on the other hand, brutally took the life of another. That point is significant. It seems that programming can only be done *within the parameters of the basic consciousness frequency resonance*. That is to say: The innate tendencies of the individual can be amplified. A positive person may not be programmable to harm another, but they can certainly be programmed to self-destruct. No, indeed! A *very* special kind of person is needed to do what Ira was "assigned" to do, whether he was conscious of the assignment or not.

---

[32] *MindNet Journal*, Vol. 2, No. 2A, http:// www.stardrive.org/Sarmail8-16-01.shtml

More importantly, it doesn't even have to be in the context of some super-duper-secret government laboratory where white-coated Fourth Reich scientists wearing coke-bottle glasses gleefully hook electrodes to their helplessly drugged subjects, and then giggle fiendishly while they pull levers and turn dials to send torturous electrical currents into brains and genitals. No, indeed, it can be done far more globally and subtly than that, as we have already noted. We have already noticed the curious thread of connections that, when we begin to tug on them, lead us back to Princeton; and we are going to find that Ira Einhorn is also connected, even if indirectly. What do we think these connections indicate? Well, certainly not that which has been widely promulgated in the "Sideshow of the New Age," led by such ringmasters as Tom Bearden, *et al.*

Dick Farley, whom we have met before, writes:

By its nature and by design of those on Earth who have been and still are involved in these murky areas of "almost science," the applications of "Anomalous Phenomena" for political, psychosexual and biowar non-lethal spacewarfare paranoia-seeding counterintelligence for global fascism, a battle for the minds (some say the "souls") of humanity and our future on our home planet (if it really IS our home planet ;-) is underway. Billy Graham, a friend of Mr. Rockefeller's, who has funded much of what is described in [*The Stargate Conspiracy*], is "in the loop." Likewise some former presidents, world leaders and "who knows who" now.

What is going on is a battle for "control of the bridge of Spaceship Earth." At the level where this stuff may be or may not be "real," as we objectively may define reality, it's all just an academic exercise. But beliefs kill, as events of September 11 have shown us.

Reliance on yet another set of "revealed truths" from "hidden sky gods" is unlikely to advance humanity's collective intelligence sufficiently or in time for us to avoid, or resist, the planned depopulation and global takeover the oligarchs who have funded this clearly have planned for "the rest of us."

Mr. Rockefeller is said to believe (because one of the channelers of the Nine told him so) that he has been born again on Earth, reincarnated into his position of wealth and global influence, because he was once a Pharaoh, and before that a ruler of Atlantis. The "search" underway for artifacts of ancient Egypt, which these people believe is descended from the Lost Continent of Atlantis, which reputedly had contacts with ETs and in some fashion may have been caught up in a "War in the Heavens," as Billy Graham and others have called it (in print, variously), is the oligarchs' quest for "legitimacy." They believe they were sent "back down here" to help Earth correct and avoid the mistakes of the long distant past, revealed to us now only in the fogs of myth and legends. Shamanically, using hallucinogens secreted from Nazi Germany before the end of the Second World War, these men (mostly men, at the beginning) believed that Hitler was on the right path, but his own megalomania corrupted and sunk him in his efforts to impose a Teutonic resurgence to world rule. This is what THEY believe!

To them, having stumbled into their "nest" and subsequently researched their own literature, doctrine and in literally thousands of hours of research and conversation with them, I have said. "Your plan will not work. You will have to 'show your talons' in order to impose your rule over humanity, because of core traits in human character. We love freedom; without it we die. But not all of us. You will never conquer us all."

And to them, and to you, I add this: "The degree of tyranny necessary to govern Earth in the future is inversely proportional to how effective we (and "They") can be as teachers." By their fruits, shall ye know them. Let's be careful "Out There."

As we have noted again and again, this is precisely the conspiracy that the Cassiopaeans have repeatedly pointed out. We can certainly surmise that The Nine have an agenda, and that this whole gang of Big Bucks Oligarchs are just the tools of the Hyperdimensional Control System, who have been pumped up to think that they are doing some grand and glorious and heroic work to "correct" something. But we notice the big differences in this perspective and the Cassiopaeans' perspective; they have made it pretty clear that the present situation is a replay of the Atlantean deal. The Atlanteans tried to take over the world and were defeated by the "Athenians." We see from Farley's astute analysis that these people are trying to "correct and avoid the mistakes of the long distant past." That is to say, they want to set things up so that the Atlanteans will *win*, and *rule the world*.

We also notice the use of drugs and hallucinogens as a characteristic of the negative agenda. As the Cassiopaeans have repeatedly suggested, and even said outright:

Accessing the higher levels of psychical awareness through such processes [as mescaline, peyote, LSD, etc.] is harmful to the balance levels of the prime chakra. This is because it alters the natural rhythms of psychic development by causing reliance on the part of the subject, thus subjugating the learning process. It is a form of self-imposed abridging of free will. [...] The other substances mentioned are, at least in part, synthetic, with the exception of peyote. But even that is not a natural ingredient of the human physiological being.

Of course, Farley also points out that these folks are really "looking for something." This something is supposed to be able to tip the balance in their favor, help them "correct the mistake" of the past, when they lost the war and lost control of the Earth.

So, how does Ira Einhorn fit in here? The "medium is the message," as Einhorn himself proclaimed, and Ira Einhorn was the medium *and* the message. In an attempt to come to some understanding of what that message was, what Einhorn was promoting, what changes occurred in our society as a result of his presence and his "spider web of contacts," again we have to look behind the veil to the man behind the curtain, pulling the levers and creating the entire sideshow of smoke and mirrors.

How did Ira Einhorn become such a truly efficient alien reaction machine?

Ira: "My mother lavished a tremendous amount of attention on me before I started school. She fed my curiosity and was teaching me at the third-grade level before I went to kindergarten at five. Aside from giving me a strong ego, this tutoring just made me a freak in my milieu when I entered school."

Bea Einhorn: "He read incessantly. He'd come to dinner, a book would be in his hand. He'd go on his vacation, he'd take so many books, you wouldn't know what would be up. I don't think he slept more than three, four hours a night, ever. Five would be the top. I would get up in the morning to call him, and he would be reading."

Because he was intellectually advanced, Ira was bored in school. Going to school was so repellent to him that he threw up every morning before starting off to his classes. Once he was in class, he was disruptive, yelling out in a rude and taunting way and refusing to stay in his seat. He made good grades academically, but had consistently poor marks for conduct.

After transferring to another school where he was told that he would not get good marks in his class work if he did not also get good grades for conduct, he apparently modified his behavior. His mother notes, however, that the new school made an effort to keep him busy. "They realized what they had in his mentality," she says proudly. One of his friends remembers him, at the age of 13, playing with rhymes and strange words. "What are you doing?" Ira would answer, "I'm practicing." The friend remarked: "His flow of language was exquisite. *He could mesmerize people through language.* He had the most incredible vocabulary that ever came down the pike. Ira could talk constantly, without hesitation, without pause."

Because Ira made top grades in junior high school and had high scores in the standardized intelligence tests, he was admitted to a prestigious college prep high school. His IQ, according to Ira, was "upwards of 140."

He was said to be an independent thinker by the time he got to high school, having so strong an ego that he dared to violate the standard dress of his peers, wearing shorts when no one else did. He also continued his disruptive behavior. He was reported to have repeatedly challenged his instructors, breaking up assemblies with loud behavior and indulging in general-nuisance activity. But, because he continued to get good grades, this behavior was tolerated.

Regarding his personal relations at this time, one of his friends from high school reported: "Ira was really a force to be reckoned with. He would attempt to be dominant in conversations, yet when he was in the presence of somebody whom he knew to have more knowledge than he, he would listen intently and respectfully, and find out where it was they got that information." He would then seek it out on his own. Ira apparently fantasized about himself as a member of the European, intellectual, philosophic elite, and not just a suburban Jewish kid in America.

Steven Levy, author of *The Unicorn's Secret*, reports a curious item: One of Ira's high school friends said that quite often when he would visit Ira at home, Ira's mother would direct him to the bathroom, where he would find Ira *ensconced in the bathtub* with an open book in front of him. The friend thought this was extremely odd, but Ira was oblivious to the discomfort of his visitor. He would basically "hold court" in the bathtub, discoursing on whatever he was reading at the time. The reason I find this interesting is the fact that Ira claimed to have been *in the bathtub* when Holly Maddux supposedly left his apartment for the last time, never to be seen again. Because there has been some conjecture that Ira may have drained Holly's body of blood and fluids for several days after changing his mind about disposing of it, I cannot help but wonder if his story may hold a partial truth: that he *was* in the bathtub when "Holly left," in the sense that the image of Ira bathing in Holly's blood flashed unbidden into my mind. Such an act, as the reader will see, would be entirely in keeping with the psychopathic personality.

But, back to Little Ira. He was a larcenous lad, no doubt. His friend reported that he would repeatedly send away for batches of books given away in book-club advertisements in magazines. When asked how he was going to pay for them, or the required subsequent orders, Ira would reply: "Don't worry! They're not going to get me — I'm a minor. If they're dumb enough to send it to me, they're not going to prosecute me." His friend noted wryly that Ira saw himself as a sort of Robin Hood, robbing the world to give to a good cause — Ira.

At one point in his high school career, Ira spent the summer lifting weights and doing push-ups to bulk up his body. He became, in the words of his friends, a "hulk," with an accompanying macho attitude. Unfortunately, he was never able to do anything about his spindly legs, and he was plagued by acne.

One very curious item is the fact that Ira boasted to his friends that he didn't feel pain. He demanded that they test him by stubbing a cigarette out on his hand and, sure enough, Ira held his hand steady and never flinched.

Ira started smoking marijuana in 1956, when it was still quite rare in American high schools. He graduated in 1957 with a scholarship to the University of Pennsylvania. Interestingly, he refused to let his picture be included in the yearbook, didn't want a high-school ring, and attended the prom in jeans. This was so shocking that the high school threatened not to graduate him, but a sympathetic English teacher intervened saying, "I ought to wring that boy's neck, but I cannot deny the world Ira."

In college, Ira was contemptuous of the idea that one could learn by attending classes. He did, of course, continue to read, but generally, he refused to fulfill the requirements of his classes. Again, his behavior was tolerated because of his evident intellect. Also, Ira's professors probably passed him just to get rid of him, because he would intimidate and challenge them by citing obscure works that contradicted their own opinions.

So far, in general, we have described a very bright, very precocious, very independent kid. Except for vaguely troubling items that could be insignificant, we are not alarmed. I bet that most of the readers can identify with Ira, most especially his insistence on his own way of being, his resistance to authority.

I can identify with Ira myself. I was reading when I was three; I was often, though not always, bored in school; and in later years I was somewhat resistant to authority that I perceived as unfair or unreasonable, most particularly my mother. I also resisted the authority at school a few times. I remember one occasion when I deliberately baited a teacher who I had seen use her authority cruelly on another student. I reduced her to tears in front of the class. At my 30-year class reunion, I was reminded of that incident, much to my embarrassment. What was a surprise to me was that it was seen as sort of heroic. At the time, and in my own recollection, I was just simply being obnoxious like any other kid who is too full of themselves. I deeply regretted having made another person cry. Even if some still perceive it as justified, I don't. I don't excuse her behavior, and I don't excuse my own.

I am certain that many readers also remember little "tests of courage" or games of "chicken" played with other children. (The only one I ever remember playing was one in which the object was not to blink, or some such thing.)

I was similarly not interested in the minutiae of classwork. However, unlike Ira, I often made poor overall grades because, even though I scored very high on all the tests, I rarely turned in the daily homework because I was too busy reading. In fact, I generally was hiding the book I was reading inside the textbook so that I could read undisturbed through classes. I thought that was a much better use of my time than listening to some dry old lecture about something that I already had learned through my reading. I also was generally too involved in my reading to "act up" in classes. I was called down for talking to the pupils on either side of me — whispering and not paying attention — and was even once sent out to stand in the hall by the classroom door for this. Heck, if I'm going to confess, yes, I smoked in the bathroom, and once instigated a conspiracy to shoot jelly beans at the Spanish instructor while the classroom was in darkness during one of her endless slide presentations of her summer vacations. She flipped the light switch on and glared at all of us — and informed us that the slide presentation would *not* continue until the guilty parties confessed. With a sigh of relief, the entire class became amnesiac.

Yes, I read under the covers with a flashlight because I was required to turn off the light in my room at a certain hour. I would often read all night and feel like death warmed over when I had to get up and go to school. But, except for the fact that I was something of a social semi-geek, I didn't mind going to school until about 11th grade. At that point, I was just simply tired of it. It *was* boring.

The point is: Regarding Ira as a child, we haven't really seen anything in the record assembled by Steven Levy from his interviews with all the principals of Ira's early life, that would indicate anything truly abnormal, except perhaps, his conscious, premeditated larceny, his lack of consideration for the feelings of others, and maybe the "I feel no pain" episode.

But now we come to that most interesting of times in Ira's life, when new players enter the stage. One of these was Morse Peckham, "the prize and pariah of Penn's English department." Morse Peckham was a "Renaissance Man." He was a polymath whose depth of knowledge was matched by its breadth.

For Peckham, the life of the mind was the only life. This had been the case since childhood. He has described his parents as imbued in nineteenth-century culture; his mother read Tennyson to him before his naps. At ten, he was using chess pieces to emulate the stage movements of Shakespeare's characters. He was the first University of Rochester student to take graduate English work at Princeton, where he earned a doctorate, but not before serving in World War II, where he spent his European tour writing the official history of the Ninth Bomber Command. [...] [He was a] large man, more than six feet tall, with fine features and a beard. [...]

A lifelong bachelor, Peckham dressed elegantly, and smoked cigarettes in a long white holder. [...] The thrust of his work was transdisciplinary scholarship. [...] He saw the culmination of [the romantic era] in the writings of Friedrich Nietzsche, and his book on romanticism, *Beyond the Tragic Vision*, would be hailed in academic circles as a masterpiece. By the early sixties, Peckham was starting a more ambitious project that would use his cultural knowledge to go beyond criticism of art, music, and literature and *probe the essence of humanity itself*. [...]

At the time Ira Einhorn found his way into one of Morse Peckham's classes, Peckham was working in virtual isolation, living alone, sharing his intellectual theories and discoveries on a daily basis with no one. [...] [T]here was some heavy intellectual bonding between Ira Einhorn and Professor Peckham. While most of the students were gasping for breath at Peckham's hairpin intellectual turns, Ira would ostentatiously be keeping pace with the master, providing verbal footnotes or suggesting esoteric comparisons to the point under discussion. It was no secret that this mental jam session continued outside of class as well. [...] Inevitably, some of Ira's peers wondered how close the relationship really was. [...] There is no reason to surmise that the speculation of homosexuality was in any way founded. [...]

[Peckham] considered his mental life intense and thrilling, but it precluded any emotional life outside of the pursuit of ideas. "In Ira," he says, "I found someone whom I could try these ideas on. Because I didn't have anybody else." (Levy, 1988; this author's emphasis)

What were Peckham's "ideas"? Some of his early work includes a study of various editions of Charles Darwin's *Origin of the Species*. As already noted, he was interested in romanticism. In 1951 he published *Towards a Theory of Romanticism*, in which he wrote:

Shift away from thinking of the universe as a static mechanism, like a clock, to thinking of it as a dynamic organism, like a growing tree. [...] For those who make the shift, the values of static mechanism — reason, order, permanence, and the like — are replaced by their counterparts in an organic universe — instinct or intuition, freedom, and change.

Romantic thought is relativistic and pluralistic; it rejects absolute values, formal classifications, and exclusive judgments; it welcomes novelty, originality, and variety. It is less interested in distinctions than in relationships, particularly in the organic relationship which it posits between man and nature, or the universe, and (less often) between the individual and society. The great chain of being is replaced by an indefinitely extended and complicated live network of connecting filaments, as in the vascular system of a plant or in a mass of animal nerve tissue, by which every phenomenon is tied by countless direct and indirect contacts to every other.

When a new fact appears, it is not just another link in the chain or cog in the machine; it is an evidence of organic growth and development, and its emergence changes every previously existing aspect of the universe. A new characteristic is evidence of a totally new and different world. Therefore a romantic artist will strive, not to imitate an ideal perfection of form which has always existed, but to originate a form which has never existed before and which will uniquely express what he alone feels and knows. To do so, he will rely more on imagination than on logic, more on symbols than on signs or allegories, more on unconscious than on conscious powers. He will believe that he is creating a genuinely new thing and thereby changing and renewing the whole of his organic universe. (Peckham, 1951)

He also wrote *Explanation and Power: The Control of Human Behaviour* in 1986, wherein his Darwinistic approach to cultural development is made clear:

For human beings, the world consists of signs, and it is impossible for human beings to consider the world, or themselves, from a meta-semiotic point of view or position. The world is an immense tapestry of innumerable threads, emerging and disappearing in the presentation and evanishment of indefinably innumerable designs, and human beings themselves form some of those same threads and patterns. We are figures in the tapestry we observe, and respond to, and manipulate. The old notion that the world is an illusion is sound, for no sign (configuration) dictates our responses. But it is sound only up to a point, because the physical character of the world limits the range of our responses. We can do lots of things with water, but as yet we have no way to build a skyscraper out of it, though the possibility has its charms; nor can we walk on it without doing something either to ourselves or to the water. Or to use another notion, the world is Idea, our Idea, but it is also Reality, Actuality, Factuality. The mind transcends the world, but then it does not transcend the world. Plato's *demiourgos* did not create the reality he set about ordering; he set about ordering a chaos, a recognition that *human behavior works on material that is really there*. Or, to put it in somewhat newer terms, the world is object, and man is subject, and the subject is different from the object but, nevertheless, somehow the same. (Peckham, 1986; this author's emphasis)

Morse Peckham theorized that it was only through "cultural vandalism" — *the aggressive undermining of established values through random, mindless acts of destruction* — that social innovation was stimulated. He theorized that humans needed to push themselves to such disruptive extremes; otherwise there was no hope of matching the insects' astonishing ability to adaptively alter their physiology and behavior in a relatively brief time. Peckham theorized that our mammalian talents for memory and self-reflection serve largely to oppress us with the dead weight of the past. Unburdened by mammalian scruples, insects effortlessly practice the Nietzschean virtue of active forgetting: The adult fly doesn't remember anything the maggot once knew.

In short, Peckham was glorifying psychopathy, and in Ira Einhorn we see Peckham's glorified psychopath in action. About Ira Einhorn, Morse Peckham said: "Ira stood out because of his really wide reading and his ability to understand what he read."

However, after spending some time *out* of Ira's direct presence, Peckham began to realize that something was wrong in the interaction. He had the odd feeling that Ira was parroting his own words back at him. "I was still very interested in him and very friendly with him, but I began to feel that talking to him was like being in an echo chamber, just my own ideas being fed back to me without any modification or any thought on his part."

Just like Ross Baker, Morse Peckham had fallen under the sway of the psychopath. But he had also analyzed the problem, and in his analysis he put his finger on one of the clues to identifying the psychopath. They are parrots, apes, echo chambers. But, as Baker pointed out, it was humbling to realize that, after a period in Einhorn's presence, he was having difficulty with his mental clarity. Morse Peckham, as brilliant as he was, took some time to come to this realization because he was, indeed, dealing with a brilliant psychopath.

As sympathetic as we may be for Morse Peckham and the fact that Einhorn duped him, there is something else crucially interesting about Morse. Let's go back to that most interesting remark about Morse Peckham: He did his Ph.D. where? At Princeton. When? Oh, in the same general time period as when Nash was there. Peckham was, as some have described him, an "intellectual raider." He advocated that in order to be a "cultural historian," one had to "know everything." He would read so extensively in a field that he soon could think in the way the professionals in that discipline thought. From looking at his work, we suspect that Morse Peckham was powerfully influenced by Game Theory.

What do we conclude? That Morse was part of a conspiracy? That he consciously was interacting with Ira, preparing him for his future role? Or do we think that Morse was just simply who he was, and Ira was who he was, and maybe there was some "tinkering" with the Matrix to ensure that the two of them would come together so as to pump all those theoretical ideas into Einhorn's head, with the surety that he would put his own spin on them?

There is nothing simple about any of this. When you start pulling on these threads, you just never know what is going to spring out of the closet. What we discovered is a connection linking Peckham to the telephone company which later "utilized" Ira Einhorn as described by the Bell executive at Einhorn's bail hearing. "AT&T's Experiment In Humanistic Education, 1953-1960," by Mark D. Bowles (*The Historian*, vol. 61) suggests that Ira's "network," was designed to counteract a previous experiment in social engineering that hadn't turned out quite the way the experimenters wanted it:

> The unexpected Soviet detonation of an atomic bomb in 1949 triggered a wave of paranoia and anxiety in the United States. As historian Vincent LaFeber wrote, "Few American officials had expected the Soviet test this early." The result was a new era of "nuclear fear" that spread throughout the culture....
>
> One of the central reasons for instituting liberal arts training was to preserve the American way during the Cold War, yet the Viteles data indicate that the participants became more tolerant of non-capitalistic political ideologies. After training, the number of participants who believed that liberty and justice were possible in socialist countries nearly tripled (Question 1), and significantly fewer participants believed that democracy was dependent upon free business enterprise (Question 3). Clearly, this represented a threat to AT&T's corporate leaders; no longer could they continue to support a training program that might undermine America's own economic system.... English professor Morse Peckham designed the program. (Bowles, 1998)

Again, economics rears its ugly head. The one thing this report tells us is this: Those guys in charge of all this aren't omniscient. But it was clear that, at the point in time when Ira Einhorn was in close association with Morse Peckham, the program that Peckham had designed, obviously with a particular agenda that supported the economic theories that were being developed around the work of John von Neumann and John Nash, was now known to be a failure. Plan B was obviously going into effect, and Ira Einhorn was central to this plan: Restore paranoia! Restore belief in Russian superiority or Russian evil experiments on mankind!

So we wonder just what kinds of cerebral jamming Ira was doing with Morse Peckham?

In the fall of 1959, when Ira was a junior, he met Michael Hoffman. Hoffman was just entering grad school in English and the two soon became close friends. Hoffman soon married and had a child, and Ira regularly urged him to toss the "normal life," to ditch his wife and child and really "live." This gives us some clue to the effect Morse was having on Ira.

Ira had already decided that earning a living was not for him, so when Morse Peckham urged him to attend grad school, even offering to pay his tuition (!) he thought that was as good a way to develop an occupation for himself that was more to his liking. However, since he had been the resident "Falstaffian figure" at Penn, and had spent most of his time in reading, talking, sex, and doing drugs, as well as traveling extensively, he had to really scramble to convince his professors to accept his final papers as proof that he had successfully mastered his courses. Most of them did, but one gave him a failing grade, which meant that Ira would have to repeat the course. He refused.

Ira's friends and family campaigned vigorously to get him to change his mind. His mother went to talk to the professor, who said, "Look, Mrs. Einhorn, I don't even know what your son looks like — how can I pass him?"

Levy says that Ira finally changed his mind, suggesting that Ira could see that his own well-being was enough reason to bend to the dominance of the institution. He fulfilled his obligation under duress and with many complaints, and received his degree in 1961. However, considering the influence of Morse Peckham, we wonder about this uncharacteristic change of mind.

In 1962, Morse Peckham was in Europe traveling, and Ira's friend Michael Hoffman took a teaching job in Maryland. Ira wrote to him regularly, so there is something of a record of his thinking through this period. He also was writing in his journals. On December 14, 1999, 2:53 p.m., Ira Einhorn wrote in an email about these journals:

On the morning of March 28, 1979 ... a slew of Philadelphia police armed with a search warrant entered my small apartment. When they left, they carried away evidence, a partially decayed body, that effectively ended my life, as a social activist. They also carried away, for no apparent reason, all of my papers, including 63 volumes of personal journals and all the information that I had been collecting and distributing for years on my international information network.

I have never seen any of this material again. I probably never will. The private diaries were later turned over, as mentioned above, to a journalist who quoted from them extensively and often out of context in a book that painted me in totally black terms as a murderer. This use of my private work is of course illegal, but my situation does not allow me to do anything about it. It is part of the pattern that has characterized all official action in my case. Action that continues to this day.

When reading the above remarks, I was so astonished I practically choked. Notice that, the finding of the "partially decayed body" of Holly Maddux, the woman he was supposed to have loved dearly, was described by Ira as having inconvenienced him, as she "effectively ended my life as a social activist"!! Not one thought for the fact that the "partially decayed body" was once a living, breathing woman whose life as anything at all was effectively ended in a way that Ira simply does not grasp. No pity for Holly, nothing but self-pity for Ira.

He then goes on to feel sorry for the loss of his diaries. Oh, lord! How sad it is! "For no apparent reason," too! Never mind that the dead body of a woman he claimed to love was found in his house, there was "no reason" to take Ira's journals away from him. Poor widdle Ira. He's gonna cry now! More so, because of what was in those journals, which he complains are illegal to use because he is just so innocent and pathetic and abused and mistreated by those nasty people who think that it's not nice to bash people's heads in!

Levy quotes from these journals and from letters Ira wrote extensively. The reader may wish to obtain the book *The Unicorn's Secret* and read it a time or two to get the full impact of Ira Einhorn in all his psychopathic glory. But the book is far more than just the story about Einhorn, it's a history of the politics and pop-culture of the 1960s and 1970s, the platform on which the New Age was constructed. There is no way to understand what is going on now without understanding what preceded it, and the part Ira Einhorn has played in creating the great "Sideshow of the New Age" — to distract attention away from the *real* Stargate Conspiracy.

Getting back to Ira and psychopaths, and Ira whining about how the police (for just no reason!) stole his journals and how Levy was "plagiarizing" them to write about him; he has given us some interesting clues to the perceptions of the psychopath:

Eibl-Eibesfeldt (1970) and Konrad Lorenz (1966) proposed mechanisms that limit aggression in social animals. They noted that in animals such as dogs, who bare their throats when attacked by a stronger opponent, this display of submission results in a termination of the attack. James Blair, in 1995, proposed a model of psychopathy based on this idea: that there is a functionally analogous mechanism in humans: a violence inhibition mechanism (VIM) that is activated by nonverbal communications of distress — emotions, for example, or expressions of pain or suffering. Blair suggests that having *a violence inhibition mechanism is a prerequisite for the development of three aspects of morality*:

1. the moral emotions (such as sympathy, guilt, remorse and empathy),
2. the inhibition of violent action, and
3. the moral/conventional distinction.

Blair proposes that psychopaths lack a functional VIM and this is why they are not affected by distress cues from others. Blair made some predictions based on his model:

> (1) that psychopaths will not make a distinction between moral and conventional rules;
>
> (2) that psychopaths will treat moral rules as if they were conventional; that is, under permission conditions, the psychopaths will say that moral as well as conventional transgressions are OK to do;
>
> (3) that psychopaths will be less likely to make references to the pain or discomfort of victims than the non-psychopath controls. [Blair & Morton, 1995, p. 13]

Using subjects identified by Hare's *Psychopathy Checklist* Blair's research demonstrated that:

> ...while the non-psychopaths made the moral/conventional distinction, the psychopaths did not; secondly, and *in contrast with predictions*, that *psychopaths treated conventional transgressions like moral transgressions* rather than treating moral transgressions like conventional transgressions; and thirdly, and in line with predictions, that psychopaths were much less likely to justify their items with reference to victim's welfare.
>
> (Pitchford, 2001; this author's emphases)

Now, this little discovery of Blair's may be very significant: Psychopaths treat conventional transgressions like moral transgressions. In short, what this may reveal is that the psychopath perceives something that another person does to them that they don't like as a "moral transgression." They may even see a disagreement with another person as a "moral" reason to cause them harm. This then leads to the idea that the psychopath perceives their own wants and desires as being "godlike," so to say. Anything done to the psychopath — for whatever conventional reason — any disagreement with the psychopath, is a "sin," so to say, and their responses to that "sin," are to complain about it as though something terribly and immorally wrong has been done to them. This gives us a clue that the psychopath will seek to justify their truly immoral behavior as "moral," or on "moral grounds, all the while unable to see any moral justification of the other at all, in any way, shape, form or fashion.

This then leads us to the issue of ego. As Steven Levy wrote, Ira Einhorn had the Gibralter of self images. The root of this ego is easy to trace: his mother. She had instilled a tremendously strong self-image in her son by her pride, her boasting of his mental prowess, her constant attention to developing his "superior" mentality, and her protecting him from consequences of his behavior *because* of his "genius."

So, as a genius, as a "mythic, godlike being," Ira Einhorn could do no wrong in his own eyes. And so, he is not even embarrassed at what his journals reveal about him, even though it makes a lie of everything he ever claimed to stand for in his carefully nurtured public image. As long as Ira himself never admits a lie, he can continue to maintain his image, completely unaware of the effect of utter amazement he is having on those who *know* he is lying! The psychopathic liar also knows that there are plenty of people who will believe lies over truth, even against the evidence; and he will cling, to the very end, to that group, that source of "food," knowing instinctively that if he ever admits a lie, he has lost his position as the "alpha male," that he believes himself to be. Loss of that position represents annihilation. And, as we have already seen, playing "semantic games" is the psychopath's solution to answering direct questions.

So, what was in the journals?

The years in question are described by Ira as simply "two years of continuous reading." This was his version of what was going on in the strange writings of that period. He also said that he was taking a bath when Holly Maddux left, never to be seen again.

Steven Levy, the journalist Einhorn described above as having "quoted from [my journals] extensively and often out of context in a book that painted me in totally black terms as a murderer," *didn't just quote from the journals*. He talked to the people described in the journals. So, in addition to letters to Hoffman, the journal entries, there is actual witness (or should I say "victim"?) testimony. Ira doesn't have a *leg* to stand on. He's a liar, pure and simple. But like all psychopaths, he is not embarrassed to lie. He is not embarrassed by the facts that prove he is lying. He continues to lie out of an ego the size of Gibralter, and simply does not feel the shock and horror of those who *know* that he is lying and who would, themselves, be mortified to be caught in such lies, and because they are capable of embarrassment and mortification, have a conscience, and therefore do *not* lie!

So, on to 1962: Ira was doing a lot of traveling. He went to Ohio, Chicago, New York, and goodness knows where else. He also spent time in Bennington, Vermont, with a girlfriend he'd met while an undergrad. Around this time he wrote to Hoffman:

I'm trying to resolve my future, but at present all I see is chaos — boy do I love it. [...] At present I can't even dream of teaching before thirty. The more I read about the background of great scholars in various fields the more I realize the great importance of long years of careful preparation.... I must learn to wait and be patient — the moment will arrive and then I will be able to bring to bear all I have read on the problem that confronts me. Americans just don't know how to ripen — they all want to produce, produce, etc. We must learn to wait and be silent. (Levy, 1988)

Then he wrote the exact opposite to the above ambitions to be silent: "I've talked a blue streak of late. All have listened — few understand." He then told Hoffman a story about a woman who was so shaken by what he had to say that she returned to her dormitory "hysterical and on the verge of a breakdown."

The question is, was the above an "Ira version" of something quite different? Is this similar to his story: "I was taking a bath when Holly left, and she never came back?" Or "I spent two years reading"?

This brings us to something of the crux of the matter. Ira Einhorn was at the center of promoting ideas of personal freedom, resistance to repressive government, ecological action, and bringing attention to the paranormal and UFOs.

Many have suggested that a giant ego ought not to reflect badly on the message. Just because the guy is a psychopath, can't he still do good — even if by accident? Didn't Ira do a lot of good? The only problem is, as Ira so repeatedly promoted from the works of Marshall McLuhan: "The medium *is* the message."

William Irwin Thompson said: "There are some people whose auras I don't like, and I didn't like to be around him." Another critic called Ira a "social psychopath... I think the definition of that is someone who believes that he can make up his own rules. His idea about society was that he had a special role to play. He probably would manipulate anyone for anything. He just happened to find the right niche at the right time."

There were always a few people like Thompson who saw through Ira and considered him to be just a left-wing con man. The important thing that a lot of people miss is the fact that, even though he rose to prominence on the platform of the antiestablishment counterculture, his main accomplishment was to promote the very paranoia that is counterproductive to a movement for peace. If we consider the fact that the ultimate message that Ira Einhorn was promoting was to pit one group of humans against another, and to suggest that some alien gods — i.e. "The Nine" — were the only ones who could come in and straighten out the mess we are in (by their manipulations, I might add), then we need to look at his message in a different light.

In short, "the medium *is* the message" says more than we might suspect. Since Ira *was* the medium, to know Ira is to know the forces inside and behind him.

The cognitive dissonance produced by Ira's "predicament" over the murder of Holly Maddux was related to not only his private life, which obviously was a complete contradiction of his public persona, but also the earthquake it produced in a general sense: The visual shock of seeing someone who was a representative of the idealistic "make love, not war" shtick being exposed as a complete phony. Ira was a visual image of a fraud, mouthing the platitudes of the sixties while secretly flouting every single ideal of that generation. It was almost as shocking as the realization that a sitting president could be assassinated in broad daylight in front of his people, and nobody could stop it or do a damn thing about it. It produced a feeling of helplessness, the loss of the father, so to say.

Ira's murder of Holly Maddux produced similar sensations of helplessness. Who can you trust? You can't even trust people who claim to be dedicated to love and ecology? After all, in a world where our Big Brother Ira Einhorn murdered Holly, and Elvis Presley died of drug use, what was the counterculture really about? The parade of activism was stampeded over a cliff into irrational paranoia, and in reaction, the era of the 1980s — the mature stage of the post-WWII baby boom — withdrew into a cocoon to metamorphose. As noted, many of them produced babies that were conceived as a result of the "free love" ideas of the times; and that is certainly a condition in which those who are the most aggressively sexual and rampantly promiscuous will spread their genes like wildfire — psychopaths, in fact.

In 1964 Ira was described as a "collector of books and women, somewhat more a connoisseur in the first instance… [he] professes to believe that it is normal for boys of his age to have slept with 200 or 300 different women." The interviewer commented: "Whether this is rationalization for a compulsive Casanova syndrome, whether he is trying to prove something to himself, I cannot say, but I am certain [he is] not average. He is exceptional both in intellectual capacity and the capacity for what I would be tempted to term sexual excesses."

For Ira Einhorn, it was one woman after another. He participated in "gang bangs" though his tastes did not run in that direction. Too much dividing of attention, no doubt. Even though Ira gave lip service to having outgrown his exploitation of women, his descriptions of his "techniques" and methods of seduction give evidence of actions that don't match the words. He remarks, at one point, "Because of our heritage of a two-faced attitude towards sex, very few people are capable of healthy promiscuity, enjoying sex as you would a good meal and then forgetting about it." This is classic psychopathy; it implies, of course, that Ira is among those elevated few who *can* enjoy "healthy promiscuity."

Reading Levy's book, recounting Ira's own words and actions, paints a portrait of not merely a raging sexist, but a sexual predator of the worst kind. How can it be that always being on the prowl for quick and anonymous sex, "enjoying sex as you would a good meal, and then forgetting it," is an "unselfish" act?

Though some of Ira's partners in these types of relationships claim that they were satisfied, many of the hundreds of women which he claimed he had conquered and satisfied, gave quite a different version of the story. According to them, Ira's style was perfunctory and disappointing. "To hear Ira talk about sex, he's made their elbows have orgasms. When it came right down to it, the four or five women I knew [who had slept with him] didn't think so. You didn't really get the feeling that a sexual experience was taking place. It was like there was nobody home there. It was something like territory conquered and then done with."

One supporter of Ira's reputation, having heard these stories being circulated, decided to set up a test. The test was that he called on a very attractive woman who "agreed to meet Ira for the experiment." Here's the clincher: The friend was in the bedroom as an observer and reported that Ira performed "royally."

Well, you idiot! What did you *think*? The psychopath *is* a performer. You didn't think he was going to show his true self to an audience, did you?

One of the clues by which one can identify a psychopath is revealed by a woman who knew Ira for years. When they first met, she had an affair with him.

> He wanted to know me, and he wanted to know how I felt about everything, but I never remember him ever talking about his own feelings. He's an extremely controlled person, everything he did was thought out and calculated. He liked having the upper hand and doing something for others, and not feeling like he was dependent on anybody.

207

I thought that got to be weird. I even said, *Don't you ever get angry, or sad, or express any kind of emotions? Don't you really have feelings for things you do and for people?* He didn't even want to talk about that. He was very, very closed. (Levy, 1988; this author's emphasis)

The fact is, Ira *did* have four "intense relationships" that amounted to more than idle dalliance or one-on-one anthropological interviews. In each of these four relationships, Ira fancied himself "in love." The truth is, except for having a woman to serve him, his only interest in them seemed to occur when he became obsessively fixated on a woman who would not submit her will to him, and who eventually rejected him. For Ira, "love" was "lust for power over another." In three of those instances, when Ira did not achieve mental and emotional submission from his "objects," Ira responded to this rejection with violence. One of those women was Holly Maddux. In his journals, Ira Einhorn is exposed for the stalking beast of prey that he is.

Ira Einhorn met "Rita Siegal" in early 1962. Their relationship was more a function of what he imagined it to be than what it actually was.

Rita was a Bennington girl from Long Island, smart and direct in her speech. She was a dancer.... At the time she met Ira... she was having trouble with her self-esteem.... "If anyone would pay me a good word, or pat me on the head, I would just lap it up," she recalls now, with the perspective of a quarter-century of reflection. "So I was real needy. So that's where the relationship came from, he patted me on the head and showed me some interest. And then I respected his intelligence, he seemed very intelligent to me." But never, according to Rita Siegal, did she consider it a towering relationship, the key union in her life. This contrasts to Ira Einhorn's perception at the time:

"I hope that the beauty of Rita's love will be able to sustain me in all the agonies of those restless days ahead. When I'm with her, all disappears — and no matter how she acts the calm of 'knowing' descends upon me — all becomes joy and light. Long live love!" (Levy, 1988)

Apparently, Rita didn't feel the same. Ira was doing what the psychopath does so well: viewing his reality as some fictional construct, projecting himself onto the others. His literary-inspired romantic ideas were imposed on Rita, and he thought he could control the entire drama the way a writer controls the development of a novel. Rita explained: "Eventually he got into all these fantasies. But I never really took it very personally somehow, because he was so far removed from reality. The, quote, great romance of the century, was what was in his head and he was fantasizing, but that wasn't what was happening at all. I never got the feeling that this was a love relationship. I got the feeling it was a sick relationship."

In short, the words didn't match the action — the clue to the psychopathic personality. Like the rest of us might be, Rita was probably reluctant to end the relationship because she was afraid she would "hurt" Ira's feelings. After all, when somebody declares that the Sun and Moon rise and set in your eyes, you sort of think that they have real emotions, right? With all his reading, Ira undoubtedly had some really good lines!

Rita allowed the relationship to continue through the spring and into the summer. She took a summer job in Hanover, New Hampshire, and even though she already had the idea that Ira was a really strange person, she allowed him to move in with her.

Ira's lack of direct contact with Peckham or Hoffman seems to have released him from any restraints that might have modified his behavior, and his reading was gradually taking over his entire reality. He was reading Nietzsche, Lawrence, Henry Miller, Marquis de Sade — and some of his ideas frightened Rita. She said: "It was as if somebody were talking about something, and dreaming about something, and fantasizing about something, and all of a sudden he would *become the thing itself.* And he would start glaring at you, and leering at you."

Rita reports that at such times she felt that Ira would actually do harm to her if she didn't get away from him. "I could see when he'd be clicking into crazy, and I'd just run, physically get out of there. I was fast... he couldn't catch me. Who knows what he would have done if he'd caught me. It was like that."

As the days passed, Ira continued poring over texts, and Rita became more and more uncomfortable with his presence, though she was also afraid to leave. At the point he began to experiment with torturing cats, she realized that he was really quite capable of terrible things.

Yes indeed, our esteemed guru of "make love, not war," our guide to "save the planet," the guy who keeps saying "I did not kill Holly," was torturing helpless animals. Here we have found the last piece of the psychopathic puzzle. Of course, Ira is outraged that his journals were handed over to Levy, because what Ira was writing in his journals in late June of 1962 happens to correspond to Rita's descriptions of her experiences, with the added factor that we begin to see just how sick this turkey really is:

Sadism — sounds nice — run it over your tongue — contemplate with joy the pains of others as you expire with an excruciating satisfaction. Project outward the vision of inward darkness. Let no cesspool of inner meaning be concealed. Reveal the filth that you are. Know the animal is always there.... Beauty and innocence must be violated for they can't be possessed. *The sacred mystery of another must be preserved — only death can do that.* (Quoted in Levy, 1988; this author's emphasis)

Whoa! You don't have to be Siggy Freud to figure that one out! But what is crucial is that Ira has told us here why he kept Holly's body in a trunk. That was who he really was. The attempt to dispose of it was merely a thin veneer of humanness that was quickly overridden with the small frustration of failure. What is more, Ira has told us more than we wanted to know about what went on at, or after, the time Holly was beaten to death. I'm not sure if people who get their cookies from hurting other people are also likely to be necrophiliacs, but I've read a few cases that would turn your stomach. I keep hearing Ira say: "I was in the bathtub when Holly left," and the image that brings is a nightmare. Not only that, but the date of the last newspaper in the trunk where Holly's body was found was four days *after* the murder. What was Ira doing with Holly for four days?

Back to 1962: Rita figured that her summer job would soon be over and she would return to school. She hoped that this would create the opportunity to extricate herself from Ira's clutches gracefully. She had tried to be honest and deal with him like a normal human being. She made it clear that when September came, she did not wish to continue the relationship. What is interesting is that her expression of her wish to break up took on the form of a "moral failure" in Ira's mind. He, of course, was perfect and long-suffering. On July 28, he wrote in his journal:

Rita and I have come to an impasse — I can no longer tolerate either her selfishness or lack of faith. To give and give some more is my desire, but not to one so unthinking as her. My dreams are realizable and will not be snuffed out by the fear of anyone — I too have a right to a life of my own and to that I will dedicate myself. She lacks faith and the ability to respect another: Without these qualities, no matter what she is, she is as nothing. Come September and all is over. The pieces will be picked up and all started anew. The progress of my soul must not be crushed by the failings of a selfish young woman. So good-bye my love and good luck to my replacement — may he be more willing to be taken advantage of. (quoted in Levy, 1988)

In the above remarks, what catches my eye is Ira's remark, "my dreams are realizable and will not be snuffed out by the fear of anyone." Just exactly what "dreams" did he wish to realize that Rita was afraid of? Why did he later relate this to the "progress of my soul" that was "crushed by the failings of a selfish" woman? What, exactly, did he want to *do* that she refused to participate in? What "progress" of soul was he thinking about since he was also writing, "let no cesspool of inner meaning be concealed. Reveal the filth that you are. Know that the animal is always there...."?

On July 30 Ira had just read *Venus in Furs*, a sadomasochistic classic. As soon as he finished the book, he wrote in his journal:

> We so carefully hide the blackness of our soul from all those around us (even ourselves) we forget so easily the impulses of power which unconsciously control so many of our actions! A book like *Venus in Furs* reminds us of what we are — blackness and light. To beat a woman — what joy — to bite her breasts and ass — how delightful — to have her return the favor in our sensitive areas. How is life to be lived? That is what the book asks unknowingly. Should we subjugate or be subjugated. Realize our darkness or at least become aware of it. Can I love Rita as she is or must I break her spirit. Does she provide me with what I want. Often I think not. Investigate — plunge deeply — leave no stone unturned. You are one of the rare free spirits do not be saddled by one who isn't. Life to be lived at its full must be lived freely. Let nothing stand in your way to getting what you can — not even the illusion of love which you know to be so transitory. (quoted in Levy, 1988)

Were there marks on Holly's body that Ira knew could identify him? Is that a reason he decided it was too risky to just dispose of it?

Rita had returned to Bennington at this point. Ira went after her and found her in her dormitory room, and she made it clear that he ought to leave, that she did not want him there. Here we find what may have been the model of the story about the girl who fled to her dormitory in terror of what Ira had said to her:

> "I probably said, 'We need to end this thing.'" she says. It was then that she perceived that silent click that told her that Ira had shifted into a darkly determined being. The shift was difficult to explain. It wasn't like he lost his reality," she says. "He totally knew what he was doing. He just went over and locked the door. It was quiet, premeditated. It wasn't a rational buildup of temper at all. It was almost like, you watch one of those supernatural movies on television, and eyes change. Like a werewolf. It was like that truly. And so I knew then when that happened I was in the room with a madman."

> The room was not large, and there was only one door. "I could have gotten out if I had been a more assertive person," says Rita, thinking back to when she was nineteen years old and young Ira Einhorn was approaching her with madness in his eyes. "I could have screamed. But I didn't. I suppose I could have jumped out the window, but I was afraid of getting cut. The man was strong."

> Ira Einhorn moved steadily toward her. He did not rush his movements. For a brief while, Rita Siegal tried to fight him off, but then

she let go. Ira Einhorn's hands were around her neck, choking her. And then she passed out.

On July 31, 1962, Ira Einhorn wrote: "To kill what you love when you can't have it seems so natural that strangling Rita last night seemed so right." (Levy, 1988)

Apparently, Ira shared the incident with his friend, Michael Hoffman. "He talked about how, watching the color of her face change, something clicked at the last minute and he looked up, he let go."

Ira later tried to convince Rita that he made sure she was still alive. She awoke to find his finger-marks still visible on her throat and spent the night at the school infirmary. She did not press charges against Ira, but she did talk with school authorities, who banned Ira from the campus.

Steven Levy comments about Ira's "strange" reaction to the incident. After all, this event could have ended his career before it even started. He suggests that Ira "tried to will it to insignificance." But we know better by now. The psychopath does not have to "will" anything to insignificance. Everything *but* their own will *is* insignificant. His reaction to it also reminds us of Nash and his arrest in California for trying to come on to a cop in a public restroom. Sylvia Nasar was similarly puzzled by Nash's "non-reaction."

> Others may have sought help after almost killing a woman, but Ira seemed to regard it as a step in a struggle for self-realization. He did not even seem to think that it need affect his relationship with Rita, and a full month after the incident he wrote that "I want to love Rita (my entire being cries and needs the love we could have) but it is so difficult to anticipate the shifting of her unstable sands. Afraid, trapped, unsure, insecure... but beautiful in the desire which entirely grips her (even though she can't admit it to herself) to realize what she has...." (Levy, 1988)

Say what? The woman told him to get lost. He tried to kill her. Now he is saying that her fear of him is the "shifting of her unstable sands"?! She is gripped by desire for Ira, only she just doesn't know it?

In your dreams, buddy!

What really bugs me is the fact that Ira's friend, Michael Hoffman, knew about this for *years*.

Ira later said, "I think people should be able to hit each other occasionally within the context of a structured situation where the violence is understood and accepted."

The end of Ira's one-sided relationship with Rita was a thorn in his flesh. He wrote to Michael Hoffman: "The melancholy that had plagued me like a silent specter ever since Rita left for good... I still love Rita but my spirit does not brood over lost glory or rage about with a fine frenzy; it has learned to sit and wait — to care and not to care and to be assured of others who will pursue the mystic union of flesh with the same ardor as the one I lost."

Is the reader thoroughly sick of this guy yet? Is your stomach churning? Do you feel like you don't want to eat anything at the thought of somebody like that being out there in New Age Land, pretending to be an all-wise, all-knowing, caring and huggy-bear type of guru?

Remember what Gurdjieff said about the activities of these "automatons," and our reactions to them; also, what Reich said about the intolerance and cruelty with which irrationality and illusion are expressed and our relationship with cosmic energy.

This brings us back to Alternative 3, of which one of the main premises is the "joint US/USSR space program that is concealed by political posturing and maneuvers." Keeping in mind that disinformation is generally composed of truth wrapping a lie, we can suppose that there is, indeed, a "One World Government" at some level, where the US and Russia are unified in some effort. But it seems sort of transparently obvious that the statement is promoting the idea that it is a *human* government. The statement seems designed to distract our attention from the extremely advanced technological capabilities of *hyperdimensional existence and focus it on a somewhat advanced human technology.*

We have asked the question: What kinds of minds develop and operate within these "rules." We have looked at John Nash and Ira Einhorn.

Game Theory is all about gaining control of the free will of others. In Game Theory, the best way to know what knowledge or beliefs the players have is most easily controlled by *creating* the beliefs that assist in the covering up of the information that would assist the player in formulating a winning strategy.

Are you offended by reading about Ira Einhorn? Do you feel that it doesn't belong on a site that deals with the development of knowledge and skills designed to set man free?

Please allow me to remind you that, based on the research I have read, the importance of psychopathy in the present day cannot be overstated. Simply put, it is a growing phenomenon and it is going to impact every single one of us individually and collectively in the not-too-distant future. It is also extremely important to understand psychopathy in order to fully understand Game Theory, and how it is the underlying dynamic being used at the present time to move all the pieces into place for the Secret Games of the Gods. As the Cassiopaeans say, "knowledge protects," and the knowledge of the functional modes of the psychopath could save your life.

In order to understand why it is necessary to learn these things about our reality — unpleasant though they be — consider the fact that a physician must study diseases in order to be able to heal them. In this way, Ira Einhorn is a gift. Indeed, it is true that many, if not most, psychopaths never commit heinous murders. But Ira Einhorn did, and the fact that he did, all the while presenting a carefully nurtured public image as a teacher of "spiritual principles" is a goldmine of knowledge. Because Ira also kept journals, which have given us a rare look inside the mind of the psychopath.

As we have noted, psychopaths have a more detrimental effect on our society than all other psychiatric disorders *combined*. So few people are even aware of this fact. They may know all about schizophrenia, or bipolar disorders, or ADHD, or whatever, because all of those things can be medicated and controlled to one extent or another. Also, they are disabling to the individual. Conversely, the chief thing about psychopathy is that it is not disabling to the individual unless certain other factors are present. In general, psychopaths always manage to do very well for themselves. People ask, "Isn't psychopathy maladaptive?" The terrifying answer is: *It may be maladaptive for society, but it is adaptive for the psychopaths themselves.*

Why does it seem that we have a veritable epidemic of psychopaths? Sociobiologists are suggesting that increasing psychopathy is an expression of a particular genetically based reproductive strategy. Simply put, most people have a couple of children and devote a lot of time and effort to their care. Psychopaths systematically mate with and abandon large numbers of women. They waste little of their energy raising children, and in this way, psychopathic genes are being propagated like wildfire. The sociobiologists aren't saying that the sexual behavior of people is consciously directed, only that "nature" has made them a certain way so that it will happen effectively.

We have come full circle. Over and over again we come up against that little problem: religion and belief systems that have to be defended against objective evidence or the beliefs of others. We have to ask ourselves, "Where did these belief systems come from that so evidentially are catastrophic?" Then, we have to think about the fact that now, in the present day, when many of these systems are breaking down and being replaced by others that similarly divert our attention away from what *is*, it becomes necessary to "enforce" a certain mode of thinking. That is what psychopaths do best.

Psychopaths dominate and set the standard for behavior in our society. We live in a world based on a psychopathic, energy-stealing food chain, because that's just the way things are. Most people are so damaged they no longer have the capacity to even imagine a different system based on a symbiotic network.

With the brief review we have examined, we are acutely aware that this is *not* a phenomenon confined to our present "time." It is a trans-millennial program that, step by step, has brought us to our present position. What emerges in the present day is just Machiavellian diversion that focuses the attention of those who are easily deceived. This is reinforced by the "clappers" in the audience, and there seems to be an entire army of psychopaths among us whose job it is to act as vectors of attention and direction.

"Contemporary culture requires automatons. ... One thing alone is certain, that man's slavery grows and increases. Man is becoming a willing slave. He no longer needs chains. He begins to grow fond of his slavery, to be proud of it. And this is the most terrible thing that can happen to a man." (Ouspensky, 1949)

Intolerance and cruelty are *needed* to guarantee the "cover-up." A certain kind of "human being" acts on behalf of this cover-up. In this sense, psychopaths, as alien reaction machines, are the playing pieces in the Secret Games of the Gods.

Chapter 62
Secret Games at Princeton

Let's now talk about Game Theory. I don't profess to be an expert, and I have to admit some difficulty in dealing with the "jargon," but after wading through a number of texts on the subject, I have enough of a grasp to realize just how these ideas are being used on every one of us here and now, on a global level.

Some of you have wondered why I have lingered so long over the subject of psychopaths. Well, there is a reason for it, and it isn't just personal. You see, it has been through interactions with them, while being guided by the Cassiopaeans, that I have been able, step by step, to come to some understanding of psychopathy. The way the entire dynamic unfolded made it clear to me that there was a deep lesson here. Of course, I didn't get it all at once. It was only when I was able to put the John Nash piece into the puzzle that I comprehended that this lesson is a microcosmic view of the entire STS reality.

It was with deep shock that I understood that the basic mode of the psychopath is that of Nash's contribution to Game Theory, and that Game Theory is the means by which a trap is being closed on humanity. It was then that I came to the idea that if we can really get this one, if we can really wrap our minds around the psychopath, we can then use that knowledge to "see through" the games at the many different levels. I also came to the idea that, by understanding Game Theory, we could learn the strategies, and learn to avoid being entrapped.

Aside from the prosaic reason of wanting to avoid the pain and suffering that is the ultimate end of any relationship with a psychopath, there is also a deeper reason: *conservation of energy for transition*. See Appendix B for some excerpts from the Cassiopaeans that will make this objective clear.

There are a couple of remarks from the provided excerpts that particularly struck me. The first ones are the two remarks about dancing, which I would like to compare: "In fact, remember, it takes two to tango, and if you are both tangoing when the dance hall bursts into flames, you both get burned!!!" and "You are dancing on the 3rd density ballroom floor. Alice likes to go through the looking glass at the Crystal Palace. Atlantean reincarnation surge brings on the urge to have a repeat performance."

The first remark was given in the context of what could be described as personal interactions or relations with psychopaths, or other 4th density STS "conduits of attack." The second remark was made in reference to the global control grid, which suggests that our present state is a mirror image of the conditions that existed at the time of the destruction of Atlantis. In both cases, it is suggested that failure to identify and disengage from the dynamic is a *de facto* choice *for* the STS dynamic. And, as noted by the Cassiopaeans: When you play in the dirt, you get dirty.

On other occasions, the Cassiopaeans have mentioned that much of the "secret activity" — i.e. the "Alternative 3" ideas — that is really going on is related to the implementation of the Atlantean crystal principles — i.e. total control of humanity as we have described in Chapters 57 and 58. That it might be essential for certain "automatons" or "psychopaths" to be sprinkled liberally throughout the population in order to "play the game" is a logical conclusion.

05-04-96

Q: (L) You once said that HAARP was something that was to be used to "transfer perimeters." I am assuming that this means to manipulate space, time and density.

A: Yes.

Q: Is it possible that they are planning to use this to bring up the Atlantean crystals to utilize.

A: Not so much to "bring up," as to utilize.

06-09-96

Q: (L) Once before you said that the HAARP assembly was a continuation of the Montauk project, and was being used to "transfer perimeters." I guess this meant space/time travel, correct?

A: Yes. And resurrect Atlantean crystal principle.

Q: (L) Do they plan to actually attempt to bring up the Atlantean crystals?

A: No.

Q: (L) Do they plan to use this for mind control?

A: And other uses.

In the very first session with the Cassiopaeans on July 16, 1994, the seriousness of our situation was described, and we were appalled. After all the many "love and light" messages of other sources, being told that we were in a real fix here was disconcerting, to say the least. I was horrified and asked:

Q: Why is this happening to us?
A: Karma.
Q: (L) What kind of Karma could bring this?
A: Atlantis. [...]
Q: (L) What can protect us?
A: Knowledge.
Q: (L) How do we get this knowledge?
A: You are being given it now.
Q: (L) What knowledge do you mean?
A: You have it. [...]
Q: (L) How do we get this knowledge?
A: Deep subconscious.
Q: (L) When did we get it?
A: Before birth.
Q: (L) Is there anything else we can do for protection?
A: Learn, meditate, read.
Q: (L) Are we doing what we need to be doing at the present?
A: So far. Need awaken.

The item about Atlantean karma being the dynamic that is being currently replayed in our time-loop existence has been discussed already to some extent. The factor that many people do not realize is that the United States is the "Reincarnation" of Atlantis. According to Edgar Cayce, Atlantis was destroyed in "three stages." The Cassiopaeans have connected the Nazi agenda to some future event, and we can even suggest that WWI and WWII were the dynamic replays of the first two "stages" of the destruction of Atlantis. We are now approaching the third stage, and psychopaths are major players in the drama.

07-25-98

Q: (L) I read the new book [*The Threat*] by Dr. David Jacobs, professor of History at Temple University, concerning his extensive research into the alien abduction phenomenon. [Dr. Jacobs wrote his Ph.D. thesis on the history of the UFOs.] Dr. Jacobs says that now, after all of these years of somewhat rigorous research, that he *knows* what the aliens are here for and he is afraid. David Jacobs says that producing offspring is the primary objective behind the abduction phenomenon. Is this, in fact, the case?

A: Part, but not "the whole thing."

Q: (L) Is there another dominant reason?

A: Replacement.

Q: (L) Replacement of what?

A: You.

Q: (L) How do you mean? Creating a race to replace human beings, or abducting specific humans to replace them with a clone or whatever?

A: Mainly the former. You see, if one desires to create a new race, what better way than to mass hybridize, then mass reincarnate. Especially when the host species is so forever ignorant, controlled, and anthropocentric. What a lovely environment for total destruction and conquest and replacement... see?

Q: (L) Well, that answered my other question about the objective. Well, here in the book, Dr. Jacobs says that there is ongoing abductions through particular families. I quote: "Beyond protecting the fetus, there are other reasons for secrecy. If abductions are, as all the evidence clearly indicates, an intergenerational phenomenon in which the children of abductees are themselves abductees, then one of the aliens' goals is the generation of more abductees. Are all children of abductees incorporated into the phenomenon? The evidence suggests that the answer is yes. *If an abductee has children with a non-abductee, the chances are that all their descendants will be abductees.* This means that through normal population increase, divorce, remarriage and so on, the abductee population will increase quickly throughout the generations. When those children grow and marry and have children of their own, all of their children, whether they marry an abductee or non-abductee, will be abductees. To protect the intergenerational nature of the breeding program, it must be kept secret from the abductees so that they will continue to have children. If the abductees *knew* that the program was intergenerational, they might elect not to have children. This would bring a critical part of the program to a halt, which the aliens cannot allow. *The final reason for secrecy is to expand the breeding program, to integrate laterally in society, the aliens must make sure that abductees mate with non-abductees and produce abductee children.*"

A: We have told you before: The Nazi experience was a "trial run," and by now you see the similarities, do you not? [...] Now, we have also told you that the experience of the "Native Americans" *vis a vis* the Europeans *may be a precursor in microcosm.* Also, what Earthian 3rd density does to Terran 2nd density should offer "food for thought." In other words, thou are not so special, despiteth thy perspective, eh? And we have also warned that after conversion of Earth humans to 4th density, the Orion 4th density and their allies hope to control you "there." Now put this all together and what have you? At least you should by now know that it is the soul that matters, not the body. Others have genetically, spiritually and psychically manipulated/engineered you to be bodycentric. Interesting, as despite all efforts by 4th through 6th density STO, this "veil remains unbroken."

But back to the items that attracted my attention in the excerpts given above: The comments about dancing, given in terms of personal and global connections, suggest to us that "engagement" with those who are acting as STS conduits is designed to drain our energy. The Cassiopaeans clearly remarked to my brother that interaction with such persons drains one's energy and prevents the accomplishment of what is important.

The next item was, of course, that which relates to Frequency Resonance Vibration (FRV). The Cassiopaeans said: "Frequency resonance vibration! Very important." Putting this together with my questions about STO "ascension," the Cassiopaeans said that "STO tends to do the process within the natural flow of things. STS seeks to alter creation processes to fit their ends." Putting that together with the all-important clue...

Q: (A) Are there some particular DNA sequences that facilitate transmission between densities?
A: Addition of strands.
Q: (L) How do you get added strands?
A: You don't get, you receive.
Q: (L) Where are they received from?
A: Interaction with upcoming wave, if vibration is aligned.

...we come to the realization that identifying and disconnecting from psychopaths and other conduits of attack is crucial to our Frequency Resonance Vibration. As Don Juan pointed out, we have to "grow" our awareness, and this is not an easy task. Discerning the predator's mind in ourselves and others and refusing to be its food is the all-important task of work on the self. It is, in fact, the path to enlightenment. *If we cannot accomplish this, we cannot achieve sufficient awareness to be able to even see the next step on the journey.* Which reminds us of that curious series of remarks:

08-16-97
Q: Well, I am lost in a sea of puzzle pieces and I have not even begun to try to assemble them!
A: Step by step.
Q: We have the phoenix, cranes, herons, doves, ravens, and all are related somehow to speech or writing. Why are all these birds related this way?
A: Pass the test.
Q: What do you mean "pass the test"?

A: Discover.

Q: Well, writing is related to the words for cutting and inscribing and even shearing and sharks. You called the Etruscans "Penitent Avian Lords," who were also "Templar Carriers." Is this related to these bird images? Then related to speech, writing and shearing?

A: Pass the test.

Q: So, if you are writing, and you pass the test, then you can be a phoenix, dove or whatever?

A: Discover.

Those of you who are familiar with alchemical literature would have instantly understood the above reference to the *Conference of the Birds* by Farid Ud-Din Attar. At the time of the questions, I was not acquainted with this work. But I *was* familiar with a previous "test" that had been identified by the Cassiopaeans, though I was not yet fully aware of all the implications. As it happened, it was through the activity of "SV" that Frank first "revealed" himself, as I have already described in an earlier volume. When the time came to ask questions about this, the deeper clue about Frank himself just went right over my head:

03-23-96

A: SV is storehouse of vital information, clue for you was in name, but you failed to notice! [...] This is why the frustration is for you; nothing of value comes without a price!!

Q: (L) Number one, SV has lied to us. Number two, it seems that she began to demonstrate emotional affect, after we had discussed the fact that there was a serious lack of emotional affect, after you had told us that these robot people are people who spend a lot of time alone and have....

A: The price, my dear, continues....

Q: (V) By continuing the relationship?

A: The Nordic Covenant was a duality.

The remark "the price, my dear, continues" puzzled the heck out of me, because I had pretty much made up my mind that I would terminate any relations with SV. The Cassiopaeans pulled me up short by saying:

A: All persons of Nordic heritage hold secret power centers, can be of darkness, or of light.... SV is of Teutonic bloodline leading directly to such super power source, such as Thule Society and others, and she is aware of her powers and mission. It is of positive orientation. However, you are being tested by 4th through 6th density forces to determine if you have the strength and wisdom for continuance! [...]

Q: (L) Is there any significance to the fact that SV spent all of those years living with the Outlaws motorcycle gang and this covenant?

A: Yes, and that is what has led and is leading to the destruction of the "Outlaws," a group associated directly with 4th density STS.

Q: (V) Her presence there caused them to break up? This was a good thing. Is this what is meant here? That her presence was uplifting to them? (L) Well, it's not uplifting them, it's breaking them; they are all going to jail!

A: Yes and because of circumstances planted by "Agent SV." This is why the perceived lack of emotion connected with that whole situation. Vitale is the bravest human you have ever known! All evidence to the contrary is veil; part of the testing process.

Q: (L) If we're being tested, why are you telling us? (V) So that you do not fail?

A: Yes. […] Vital that you do not fail.

Q: (V) Is there a pivotal word that might break this open to a clearer understanding?

A: Discover.

The "test" in the above case was to determine if I was able to see "through" the words of the Cassiopaeans themselves and to realize that, in order to preserve free will, the choice had to be totally mine, based on my own examination of the data and my own conclusions about what that data implied. They couldn't tell me directly, and so chose the opposite course of contradictory statements to challenge my awareness. For example, connecting SV to the Thule society, which the Cassiopaeans had already identified as part of the STS consortium, and then pronouncing it a "positive" mission, was so contradictory that I was being challenged to think about the whole thing very carefully. I came to the realization that this was what the Cassiopaeans were doing, telling me something deep and important in code:

Q: (L) Well, OK, I'm going to trust you guys, and I'm going to go with the flow, and I'm going to assume that you are right, and I'm going to assume that this is for the best and for the good, and I'm going to stop my knee-jerk reactions, and stop worrying about such things.

A: Suggest you look before you leap. All can be wrong in their quick judgments, whether the result be acceptance or rejection. All is not as it seems.... Remember, those that come into your group, or your circle of influence, can be different than you think. […]

Q: (L) You said we were being tested. Tested for what?

A: Continuance.

Q: (L) Continuance of what?

A: All.

Q: (L) Continuance of all. OK, and we are being tested through SV?

A: Currently.

Q: (L) Are you saying that what we have been considering attacks were just tests?

A: The ones associated with Vitale. And no, all with that name are not of this orientation, but this clue was installed for you.

Of course, the major attack that was "associated with Vitale" was the incident when Frank's true nature was revealed, as described in a previous volume, but which I "shoved under the rug" in favor of my programming to "turn the other cheek." It was after the termination of the session that the realization fully dawned on me about what I was supposed to be "getting." I understood in a flash the "coded" nature of what the Cassiopaeans had been saying, and the resolve formed in my mind that I was not going to accept the identification of SV as a "positive" person, and that I was even being challenged to consider Frank as the individual who needed to be observed and learned about in order to "pass the test." With this resolve in my mind, I asked for the Cassiopaeans to come back and comment on my "internal" questions and conclusions, though I didn't want to voice them as they were formulated in my mind.

Q: Hello. Are you there? I am not comfortable with this information about SV. It seems to be contradictory to everything I can observe and feel.

A: Hiklu Cassiopaea. Worry not further! Discomfort is not necessarily danger, and is indicative of growth and learning. So, proceed and celebrate!!

So, apparently, I passed *that* test. But, of course, the Cassiopaeans did say that I was only "currently" being "tested" through SV. This meant that there were more such "tests" to come. Indeed, that is the case, and it is precisely the type of "testing" that ultimately leads us to the goal: *The Conference of the Birds*.

The Conference of the Birds is a philosophical-religious poem in prose written by a twelfth-century Persian poet and Sufi mystic. The story starts with all the birds of the world, known and unknown, assembled together and discussing the quest for a king. The Hoopoe addresses the gathering of birds, urging them to set out on the quest for the Lord of Creation, the Simurgh. The journey consists of many trials and tests. Of all the thousands of birds who start on the journey, only thirty make it to the end.

It was, of course, only after reading this wonderful little book that I understood the nature of the tests that the Cassiopaeans referred to. It is this that I am trying to convey, share and make available to those who "resonate."

For the individual who wrote to complain that this subject is not going to bring "enlightenment," think again. I hardly think that the Cassiopaeans would have so subtly guided me through these learning experiences, which are often protracted and painful, if it were not for a reason. If I have benefited from it — and I can assure you it has been remarkably freeing to come to this understanding — perhaps I can share it to the extent that when any of readers encounter it who haven't already done so, you will be armed with the prior knowledge that will make the process "smoother," as the Cassiopaeans say. Also, thanks to those who have written of similar successes in becoming free! Your stories are inspiring, and reassuring that our opinion of the importance of this subject is not overstated!

Now, back to games. Game Theory stands on two theorems: von Neumann's "min-max theorem" of 1928, and Nash's equilibrium theorem of 1950. Von Neumann's ideas are the cornerstone of games of pure opposition, or "two-person zero-sum games" as they call them in mathematical terms. As it happens, two-person games have no real relevance in the real world. It wasn't until Nash came along that the distinction between cooperative and non-cooperative games was introduced.

Cooperative games are games in which players can make enforceable agreements with other players. That is to say, as a group they can fully commit themselves to specific strategies. Non-cooperative games posit that collective commitment is impossible. There are no enforceable agreements. By expanding the theory to include games that involve a mix of cooperation and competition, Nash opened the door to applications of Game Theory to economics, political science, sociology and even evolutionary biology. We have noted that Morse Peckham must have been seriously influenced by the ideas of games in his role as a "social historian."

In general, the outcome of a game for any one of the players depends on what all the other players choose to do and vice versa. This means that such games are "interdependent." Games like tic-tac-toe, hangman and chess involve one kind of interdependence, because each of the players moves in turn and has a chance to be aware of the moves of the other and to analyze them before making his or her own move. In such cases, each player will look ahead to possible moves and how they will affect him, he will try to assess the likelihood of these various moves by the other player, and will then reason "back" to his current situation and pick a move based on these analyses. In such games, the players have to anticipate and assume not only the strategy of the other player, but how that other player will respond to his next move, and so on. The player's best strategy can be determined by looking ahead to every possible outcome. In chess, these calculations are too complex, so the players only look ahead a few moves at a time and constantly adjust their strategy based on their experience of the other player.

Games like poker, on the other hand, consist of simultaneous plays wherein the players are ignorant of the other player's current state or possible actions. They are forced to think, "I think he thinks that I think that he thinks that I think..." and so on. Each must figuratively put himself in the place of the other player and try to calculate the outcome, including his own move.

Such games, where there is a lack of information which leads to a logical circle that just goes around and around, are what is dealt with by Nash's concept of equilibrium wherein each player picks his best move based on the idea that each of the other players will also pick a "best move," or will have a "best situation" from which to play.

The problem is: The way the theorem is described is very confusing because of the jargon. Nash defined equilibrium as "a situation in which no player could improve his or her position by choosing an alternative available strategy, without implying that each person's privately held best choice will lead to a collectively optimal result." But the bottom line is this: Nash's equilibrium states that each player ought to assume that the other player is out to screw him royally and is probably in a position to do so, and therefore he must use the strategy that is optimal — which is either to submit completely, because he knows he doesn't have a good position or a good move available and the other guy is going to decimate him, or — assuming his position is such that he just can't lose — to screw the other player firstest and mostest.

Today, Nash's concept of equilibrium from strategic games is one of the basic paradigms in social sciences and biology. He got a Nobel Prize for coming up with the idea.

Nash, Shapley, Shubic and McCarthy, along with another student at Princeton, invented a game involving coalitions and double-crosses. Nash called the game "Fuck You, Buddy." It was later published under the name "So Long, Sucker." Nash and the gang created a complicated set of rules designed to force players to join forces with one another to advance, but ultimately to double-cross each other in order to win. The point of the game was to produce psychological mayhem, and apparently, it worked. Sylvia Nasar records that McCarthy remembers losing his temper after Nash cold-bloodedly dumped him on the second-to-last round, and that Nash was absolutely astonished that

McCarthy could get so emotional. "But I didn't need you anymore," Nash kept saying over and over again.

Keep this game in mind because it is the essence of Nash's ideas: To force cooperation to advance, followed by a big double cross in which only one player is the winner.

Sounds a lot like the current-day craze of "survival" shows, yes? Which of course, leads us to wonder what kind of "programming" or "example" such things are setting up as models for human behavior. More importantly: Why?

Nash's Game Theory was all the buzz at RAND even before he arrived there under contract. RAND had been, prior to Nash's ideas, preoccupied with games of total conflict between two players, as defined by von Neumann, since that seemed to fit the problem of nuclear issues between two superpowers. However, as weapons got ever more destructive, the idea of all-out war was seen as a situation in which both players might have a common interest. Bombing the enemy back to the Stone Age no longer made any sense because it could lead to a war of complete extermination on both sides.

Von Neumann had long believed that RAND ought to focus on "cooperative games." That is, games that ought to be played "sequentially." Such games should involve "moves" based on information, such as in chess or tic-tac-toe. Players ought to communicate and discuss the situation and agree on rational, joint action. In such games, there is cooperation and collaboration, and *an umpire around to enforce the agreement.*

Economists, however, did not like Von Neumann's ideas. They said that it was like saying that our only hope for preventing a dangerous and wasteful arms race lay in appointing a world government with the power to enforce simultaneous disarmament. As it happens, a "One World Government," composed of member nations, was a very popular idea among mathematicians and scientists at the time.

But the social scientists —the economists —were doubtful of the idea that any nation, much less the Russians, would cede sovereignty to such an organization. In other words, in Cooperative Game Theory, who's going to force the other side to cooperate?

But Nash came along and solved the problem. He demonstrated that non-cooperative games *could* have stable solutions. In short, one "player" could have a strategy in which they "force players to join forces with one another to advance, but ultimately to double-cross the other players in order to win."

To put this in practical terms: A One World Government might be advocated by a major player, promoted, set up, and all the other players might follow the rules — but that one player has every intention of *being* the One World Government and overthrowing the powers of all the other players at the last instant.

Now, just what government in the world today seems to be playing Nash's strategy? Take your time. There's no hurry.

Nash's theory inspired the most famous game of strategy of all social scientists, called the "Prisoner's Dilemma," which goes as follows: Imagine that the police arrest two suspects and interrogate them in separate rooms. Each one is given the choice of confessing, implicating the other, or keeping silent.

No matter what the other suspect does, each suspect's outcome — considered alone — would be better if he confessed. If one suspect confesses, the other ought to do the same and thereby avoid the harsher penalty for holding out. If one of them remains silent, the other one can confess, cut a deal for turning state's evidence, and the one who remains silent gets the whammy. Confession — or "cooperation" — is the "dominant strategy." Since each is *aware of the other's incentive to confess*, it is "rational" for both to confess.

Here we come to the realization of the power of the psychopath, and how Game Theory is being "used" against us. You see, the psychopath, having no conscience, does not have the ability to "imagine" the consequences of non-cooperation in terms of being able to "feel" it. Without this ability to imaginatively feel the consequences, he is virtually fearless, and can therefore direct his behavior according to his own fantasized outcome with no regard whatsoever to reality, remembered experiences, the imagined experiences of others and so forth. That is to say, for the psychopath, rationality is determined by virtue of the idea that it is self-serving to the max. "Rationality" is the assumption that everyone else is looking out for Number One, and to hell with everybody else.

Never confessing thus becomes the psychopath's "dominant strategy."

The reader will probably immediately see the dynamic of human relations involving a psychopathic personality and a "normal" human. Psychopaths, having no conscience, always play their dominant strategy, which is totally "rational" without the influence of emotions conjured up by imagination. They do not modify their behavior or choices based on emotion or consideration for the feelings or motivations of others. They will implicate the normal person in the Prisoner's Dilemma, and will refuse to confess their own guilt, because they simply have no ability to perceive hurting another as morally reprehensible. This is the psychopath's "dominant strategy." They will never, in such a situation, consider cooperation.

Normal people, on the other hand, having conscience and emotion, will make choices based on imagination reinforced by emotion. In some cases, in the Prisoner's Dilemma, they will refuse to confess out of loyalty to the other, never realizing that the other might be a psychopath who has not only refused to confess his own guilt, but has undertaken to make a deal for himself by implicating the other. Some people may even confess in order to "save" the other person from suffering pain, never realizing that they have been manipulated into this role by a psychopath who is all the while saying, "Yes, he did it! I am innocent!," and when, in fact, the truth is the exact opposite.

It's easy to see that in any interaction between a psychopath and a normal person with full range of emotions, the psychopath will always "win."

Two of the scientists at RAND set up some experiments using a couple of other scientist-contractors as "guinea pigs." They wondered if real people playing the game would be mysteriously drawn into the "equilibrium strategy." They ran the experiment 100 times. Nash's theory predicted that both players would play their "self-serving" strategies, even though playing their "cooperative" strategies would have left both better off. As it turned out, the results of these trials did not turn out according to Nash's theory. Why? Because the two scientists tended to choose cooperation more often than cheating. Once they had realized that players ought to cooperate to maximize their winnings, that is the strategy they chose.

When Nash learned of the experiment, he wrote: "The flaw in the experiment as a test of equilibrium point theory is that... *there is too much interaction.* [...] One would have thought them more *rational*" (quoted by Nasar, 1998). In short, the players had consciences, and this contributed to their choice of maneuvers.

At RAND, Nash devised a model of negotiation between two parties whose interests neither coincide nor are exactly opposed. It is a classic example of what we see taking place in our world today:

Stage One: Each player chooses a threat and says "This is what I'll be forced to do if our demands are incompatible and we can't make a deal."
Stage Two: The players inform each other of the threats.
Stage Three: Each player chooses a demand that he thinks is worth agreeing for. If the deal doesn't guarantee him that, at least, no deal.
Stage Four: If the deal is made (under threat, mind you), both players get what they want. If not, the threats must be executed. This means, don't

threaten what you really can't deliver, and always deliver what you threaten.

Nash showed that each player has an "optimal threat," or the threat that *ensures the deal* no matter what the other player chooses. Again, do we see this style of play in operation today? Either in terms of politics, or in terms of the relations between government and the people?

Now, coming back to psychopaths: It is fairly easy to see that they often manipulate others to join forces with them in order to help them to advance, but ultimately, when they don't "need them" anymore, they double-cross the others in order to win. The result is deliberate psychological mayhem.

In short, it isn't even necessary for a grand and logistically complex government mind-control program to be in operation in order to produce the conditions necessary to ultimately enforce total controls on humanity. *It is only necessary to have strategically placed psychopaths in the population*, to train and influence selected ones in particular ways through what would be seen on the surface as "ordinary means," and simply calculate the fact that they will always operate with their dominant strategy — serving self.

I expect that the reader is beginning to make all kinds of connections regarding how Game Theory may be being utilized to bring the world to heel.

In December 1994, Vice-President Al Gore announced the opening of the "greatest auction ever." What was being auctioned was "thin air." Billions of dollars were bid for licenses to broadcast airwaves for things that employ wireless communications. (Think about Ma Bell and her connection to Morse Pinkham, Ira Einhorn, Uri Geller and others.) When the auction finally closed in March 1995, the winning bids totaled more than seven-billion bucks, making it the biggest sale in American history. It was, in fact, the sale of public assets. By the late spring, another three-billion dollars had been raised in Washington in similar auctions. The press and the politicians were ecstatic. The corporate giants had been able to protect themselves from competition, and they all called it a "triumph for Game Theory." Governments from Australia to Argentina have used Game Theory to sell scarce public reprocess to buyers "best able to develop them."

Enron.

Now, let's come back to Ira Einhorn. As I noted in the previous chapter, we have a unique situation with Einhorn: *a psychopath who kept a journal*. As we have also previously quoted (but it bears repeating until we all really get it): "In spite of more than a century of clinical study and speculation and several decades of scientific research, the mystery of the psychopath still remains. Some recent developments have provided us with new insights into the nature of this disturbing disorder, and its borders are becoming more defined. But the fact is, compared with other major clinical disorders, little systematic research has been devoted to psychopathy, even though it is responsible for far more social distress and disruption than all other psychiatric disorders combined." (Hare, 1999)

Everyone who has written to me about this subject has repeatedly confirmed that you just simply do *not* know that you are in the clutches of a psychopath until it is almost too late! The individual who wrote complaining that he was not being "enlightened" by studying psychopathy, mentioned that he had been "ripped off" for ten grand and he just "got over it and moved on." So, naturally, he had decided that this was the correct approach. Don't try to learn anything from the experience; don't try to share it so as to help others to avoid future entrapments; just "get over it!"

Well, what if the "psychopath" is a gang of hyperdimensional beings trying to take over the world, rather than just a single guy getting ripped off to the tune of ten grand? What if the "payoff" that you must "get over" is the loss of your soul, or your opportunity to grow beyond the limitations and controls of this reality?

Learning to identify the psychopath by analysis, by finding clues that will aid our tracking, is most definitely an essential tool of enlightenment. As the Hoopoe was described in *The Conference of the Birds*: "On her breast was the ornament which symbolized that she had entered the way of spiritual knowledge; the crest on her head was as the crown of truth, and she had knowledge of both good and evil."

What is even more significant is that the mystic writer of this famous series of clues to the achievement of the ultimate quest of the soul has told us that the first words out of the mouth of the Hoopoe are: "I am one who is engaged in divine warfare, and I am a messenger of the world invisible."

That is, ultimately, the foundation principle of these volumes. We are talking about "Divine Warfare" in practical terms. We are talking about invisible worlds. That means that we must have knowledge of both good and evil. Based on the condition of the world, there is a serious lack of the latter. There is far too much of the "get over it and move on" syndrome, which allows the psychopaths to continue to operate undetected in our reality. We desperately need to find some series of definitive identifiers, some "mark" that we can "see" in the realm of the unseen that will serve as warning. We need to remember that "psychopathy is a personality disorder defined by a distinctive cluster of behaviors and inferred personality traits... [and] diagnosis is based on the accumulation of evidence that an individual satisfies at least the minimal criteria for the disorder." More importantly, once we have some clue, or warning, we have to understand the strategy of the psychopath, and then we must find within the strength to act in favor of our own destiny, realizing full well that the psychopath, as one who seeks to force his delusions on others, will perceive the choice to "not associate" as being "attacked" in a "morally reprehensible" way, and will use every ploy in the book to reestablish his control by these means.

Since we know that the only things that science can tell us about the psychology of psychopaths are those things which are inferred from a "cluster of behaviors," we realize that truly seeing inside the mind of a really good psychopath — as he sees himself and the world around him — is something that is hardly likely to happen.

However, with Ira Einhorn, that is exactly what we are presented with. A gift. A window into the soul of the psychopath.

"You know my method. It is founded upon the observance of trifles," said Sherlock Holmes in Conan Doyle's *The Boscombe Valley Mystery*.

It made me almost physically sick to write about Ira Einhorn. It is making me sicker yet to observe the dynamics of the political/economic scene, and to realize fully what maneuvers have been set into motion and what the ultimate objective is. Game Theory reveals to us a scheme that has been mathematically defined, so that it can be utilized to set up the dominoes to fall into place for the purposes of instigating control over the entire world.

Psychopaths like John Nash, Ira Einhorn and others, give us insight into the inner workings of this scheme, without which we are helpless to defend ourselves or even — possibly — to stand firm as anchors of a different outcome. In short, *the activities of psychopaths are the microcosm of the macrocosmic service-to-self reality and plan.*

As we have noted, part of the disinformation campaign of Alternative 3 includes the idea that "There is a secret joint US/USSR space program that has gone far beyond what the public sees. Astronauts landed on Mars in 1962. It has been discovered that there is other intelligent life in the universe. The earth is dying. We have polluted it beyond repair. The increasing 'greenhouse effect' Cassiopaeans will cause the polar ice caps and glaciers to melt and flood the Earth." And so on.

All of these ideas — most of them emanating from movements of the 1960s and the spiritualist interests of Andrija Puharich and the Stanford Research Institute (SRI) — coalesce in Ira Einhorn's "work." These ideas, having been "adopted" by Einhorn to use in his psychopathic maneuvers, naturally became the keystone for his "I didn't kill Holly" rant. We must suspect that these are precisely the concepts that were deliberately "planted" in his reality to be utilized in his psychopathic "dominant strategy" of fearless self-preservation. What is less obvious, is whether or not someone — or something — had the awareness that he would eventually commit murder?

The Tyler, Texas issue is going to become *most* interesting further along. As we have noted, Vincent Bridges has *identified himself with Ira Einhorn* — and his attempts to further the claims of Einhorn's finger pointing at Tyler, Texas naturally suggests to us that he chooses to believe Ira's rant that he did not kill Holly (see Appendix C).

Nevertheless, Einhorn did commit murder and this act could be seen as a "breakdown" in one sense, though this lapse in his camouflage was the natural outcome of his personality structure. Again, was that understood at some level? Was it part of a deliberate manipulation?

Ira Einhorn was a "sensation junkie" who thrived on chaos. He "came to his full powers" on the destruction of the old way of being. As the traditions of the establishment (which were certainly arbitrary and repressive) crumbled, Ira saw his advantage and seized the day. In summer 1964, Ira divided his time between Berkeley and Palo Alto, where he instantly recognized what was happening in the launching of the counterculture. At a party given by Ken Kesey, Ira dropped acid and wrote: "I was so high and the experience was so strange that I must have more distance."

Obviously, that trip made him a bit nervous for some reason. He later claimed that he had first used LSD in 1959, and that he subsequently took "trips" about twice a month for over two years. He considered acid to be the key to self-exploration. As a side note, his acid use preceded his assault on Rita Siegal by two years. Could there be a connection?

In the fall of 1964, Ira, probably through the influence of Morse Peckham, was hired to teach a literature course at Temple University. He came to class in rumpled clothes and with his tie cut in half. He would have his class spend entire sessions pondering the difference between eroticism and sexuality. He utilized Morse Peckham's technique in his discussions of poetry — the second line could not be discussed until the first was fully understood. He made it clear to his classes that his lifestyle was daring and different and free. Rumors flew that his apartment contained nothing but a bed and books. This was heady stuff to the kids of 1964.

Ira was interviewed at some point, telling a journalist that "I dress like the students. If they ask about marijuana and LSD I give them straight answers about the delights and the dangers. I make no bones about my contempt for the academic world. I'm very popular with the kids. I wouldn't say I was with the administration."

He constantly talked about being a writer, finishing his novel; but the fact is, he simply wasn't a very good writer. His efforts in those directions were repeated failures. Bernie McCormick, writer for *Philadelphia* magazine, said that Ira "was a terrible writer. It was Ira's frustration."

However, in the true style of the psychopath, by virtue of the ignorance of his audience, he was able to engender an illusion of literary proficiency by his raiding of the writings of the best minds of civilization (see Appendix C). That personal charisma was the power that enabled him to mesmerize people and to climb to a sort of counterculture fame and glory. The word "love" became Ira's instrument of aggression.

He began to assemble a small network of "revolutionary seers" and sharpened his vision of "dire apocalypse" and spiritual transformation. He got in touch with his "personal power," and formed a public persona based on his personal charisma.

Just like Hitler, Ira believed that he was the supreme judge of what was good and what was not in just about every field of endeavor. And, like Hitler, he functioned as an "amplifier" of the secret desires, least permissible instincts, resentments over sufferings, personal revolts against pressures to behave responsibly, anger at the hard reality of work and misery, and repressed and suppressed shadow sides of young people in revolt against their parents and the demands placed upon them, giving it voice and permission to exist, as well as justification.

Just as Germany was "ripe" for Hitler, the young people in America were ready to be plucked by Ira Einhorn. The social, economic and political situation engendered by the Vietnam War was a bomb just waiting for a detonator, and Ira sought to provide it. Like all psychopaths, Ira had the ability to sense what people wanted him to give them — a reason to do and be — and he was then able to manipulate these themes so as to arouse their emotions of esteem and adoration for him. He flattered and cajoled, built up straw men and knocked them down, and he always managed to say what the majority of his audience was secretly thinking but could not verbalize — because they were dark, violent and unacceptable thoughts, thoughts

233

and feelings born and reared in the chaos of the psychological traumas of their parents, most of whom had suffered through WWII.

Of course, Ira always presented his ideas in the guise of "daring to speak the truth and to defy the authorities and oppressors of the evolution of mankind, the oppressors of the human spirit!" Ira's inducement to chaos appealed to the most primitive and basest instincts and inclinations, and cloaked them with the image of nobility and high ideals. He was thus able to justify all actions — no matter how revolting and antisocial — as the means to the attainment of these "high ideal goals." Those who knew him said that he was a "goal justifies the means" kind of person.

We should note that psychopaths would not be successful in society if there were not conditions of oppression and inequality in that society to begin with. In this sense, we ought to remember that we are probably dealing with Game Theory applied to social engineering: Machiavellian ploys of producing a deplorable situation, raising up an "enemy" to blame for it, and then producing "saviors" who claim to be able to lead people to the "right solutions."

What is even more important than just our own personal interactions with psychopaths (painful lessons though they may be) is the work of the psychopath in larger arenas — the sideshows of social, spiritual and political activities that grab and hold our attention, form our impressions, and generally mold our awareness into something that has little at all to do with what is really going on. There are, of course, very serious spiritual implications that we will address at some point. But for now, it is crucial to fully understand this dynamic. Because the fact is: There is no possibility of awakening without first becoming aware of what keeps us in sleep, and *how* it is done. And, of course, since we know from the partial information that has been obtained about psychopaths that one of the chief things about them is their ability to ape, to imitate, to camouflage their true nature, discovering the clues that will help us identify them is crucial.

Ira Einhorn is, in fact, a goldmine. We have before us a real live New Age-type guru who manipulated a *lot* of very smart people. Ira Einhorn was a symbol of nonviolence for over a decade. Even the people who were troubled a bit by his rough treatment of Holly overlooked his little "glitches." Ira had created a carefully nurtured public profile that made it almost impossible for others to rationally assess the small clues. His supporters felt that he was an "extraordinary being scandalously charged with a crime he did not commit," because their impressions had been so thoroughly "shaped" by the carefully nurtured public image. What is more significant is the fact that his careful image making extended even to his closest friends. Even they only had brief glimpses of his dark side, so be assured that uncovering a psychopath is an art of great subtlety, necessitating what Don Juan has called "systematic harassment" accompanied by close and careful observation of the "trifles" of the reactions of the individual. Do not think for a moment that they will reveal themselves easily! So, we are indeed handicapped, and all we really have to learn from is "failed" psychopaths. But that's better than nothing.

It is truly difficult to appreciate just how different the functioning of the psychopaths is compared to that of normal people. After killing a waiter who had asked him to leave a restaurant, Jack Abbott denied any remorse because he "hadn't done anything wrong"; because "there was no pain, it was a clean wound" and the victim was "not worth a dime" (Hare, 1999, 42-3). John Wayne Gacy murdered thirty-three young men and boys, but described himself as the victim because he had been "robbed of his childhood." Kenneth Taylor battered his wife to death and then couldn't understand why no one sympathized with him in the tragic loss of his wife! A female psychopath allowed her boyfriend to rape her five-year-old daughter when she was too tired for sex, and then was outraged that social services took the child away! Diane Downs shot her three children, then wounded herself to create "evidence" of an attack by a stranger; asked about her feelings regarding the loss of her children on *The Oprah Winfrey Show* (September 26, 1988), Downs replied, "I couldn't tie my damned shoes for about two months... the scar is going to be there forever. [...] I think my kids were lucky" (Hare, 1999). Hare remarks:

> Another psychopath in our research said that he did not really understand what others meant by "fear." However, "When I rob a bank," he said, "I notice that the teller shakes or becomes tongue tied. One barfed all over the money. She must have been pretty messed up inside, but I don't know why. If someone pointed a gun at me I guess I'd be afraid, but I wouldn't throw up." When asked to describe how he would feel in such a situation, his reply contained no reference to bodily sensations. He said things such as, "I'd give you the money;" "I'd think of ways to get the drop on you;" "I'd try and get my ass out of there." When asked how he would feel, not what he would think or do, he seemed perplexed. Asked if he ever felt his heart pound or his stomach churn, he replied, "Of course! I'm not a robot. I really get pumped up when I have sex or when I get into a fight." (Hare, 1999, 53-4)

One of the truly scary things about psychopaths is the fact that most psychotherapies actually seem to make psychopaths more likely to further violate the rights of others on even grander scales, probably because psychopaths use psychotherapy to hone their skills in psychological manipulation. In no case has it been confirmed that such therapies have ever helped a psychopath, even if they will use the fact that they have had therapy to con people, because inside the psychopath sees no need to change their admirable personalities.

One researcher, Linda Mealey, describes psychopaths in terms of "cheaters." This suggests to us that our study of psychopaths might be helped along by considering them in terms of "card sharks":

> Human cheaters would not be detectable by instruments routinely available to his or her conspecifics... [and] should be very mobile during their lifetimes. The longer a cheater interacts with the same group of conspecifics the more likely they are to recognize the cheater's strategy and to refuse to engage in interactions with him or her. There will be costs of mobility, since the mobile cheater will have to learn a new social environment after a move, and he or she will need to be skilled at it. A third prediction is that human cheaters would be especially facile with words, language, and interpersonal empathy... Human male and female cheaters should exhibit very different patterns of cheating, reflecting the obligate mammalian dimorphism in reproductive strategy and potential. A male cheater should be especially skillful at persuading females to copulate and at deceiving females about his control of resources and about the likelihood of his provisioning future offspring. Females, on the other hand, should feign lack of interest in copulation in order to deceive males about their paternity confidence. They should also exaggerate need and helplessness in order to induce males to provide them with more resources and support than they might otherwise provide. Finally, female cheaters might abandon offspring as soon as they perceived that the chance of offspring survival exceeded some critical value. (Harpending and Sobus, 1987)

Mealey distinguishes between congenital or primary sociopaths who are "born cheaters," and secondary psychopaths who *become* cheaters, in order to enhance their "mating and acquisition" possibilities. Her model suggests that primary psychopaths can be recognized at an early age — as toddlers — and that secondary ones manifest their psychopathic nature somewhat later — possibly around the age of puberty. The primary psychopath seems to be more prevalent among well-to-do, well nourished and well nurtured classes, and the secondary psychopath tends to emerge from disadvantaged backgrounds.

In this sense, I think that the terms psychopath and sociopath might be useful to distinguish the two. Mealey's "secondary psychopaths," which we will refer to as sociopaths, are generally of low socioeconomic status; have low intelligence and poor social skills; experience parental neglect, abuse, inconsistent discipline, and punishment; and their antisocial behavior is a response to social pressures.

According to Mealey, primary sociopaths are "designed for the successful operation of social deception and… are the product of evolutionary pressures which… lead some individuals to pursue a life strategy of manipulative and predatory social interactions" (Mealey, 1995). In short, they are designed to be the vectors of our reality. Game Theory.

Not every indulged child becomes a psychopath. However, we have to wonder at the indulgent, hothouse-nurturing approach that is "prescribed" for the "special Indigo children," which is precisely designed to do precisely that.

The many people interviewed by Steven Levy generally agree that Ira was able to manipulate the emotions of others in such a way as to numb their critical faculties to the point where they were willing to believe anything he said. He was able to manipulate them to believe that he was the noble David of the counterculture against the Goliath of "the establishment." His listeners were ready to believe anything about him that *he* said, because they wanted to — even if the facts indicated the exact opposite. This emotional bond he created in his audience was not easily dissolved, and its chief effect was to rob people of their critical-thinking functions. They did not want to be confused with facts, they did not want to have to think, they only wanted Ira to give voice to their own feelings of anger and resistance.

With his enormous intellect, Ira was able to pull up facts and figures about just any topic you could name or mention, which gave an impression of infallibility. He had a talent for repeating things he had heard in such a way that the listener was led to believe that the ideas and insights were his own, that he was very clever — an intellectual, even a genius. He was a consummate "poseur." His inability to produce or "give" anything truly creative in terms of writing, was evidence that his "powers" had to do strictly with interactions wherein subtle manipulation was geared toward eliciting emotional responses from his audience.

It is said that, in any room, Ira Einhorn commanded attention to himself. He would routinely unleash a dazzling fusillade of powerful or well-known names he was in contact with, inside information he had access to, and the elevated means of understanding that he had attained. He had an odd way of twisting his apocalyptic vision so that he could speak of the world's inhabitants in the first-person plural, yet somehow be personally exempt from the category. Things are happening so fast that people don't know how to deal with the situation, he'd declaim, with the implicit understanding that Ira Einhorn himself had no difficulty comprehending the disorienting complexity of the world around him.

The very unspokenness of this superiority could make it more infuriating, because you could not put your finger on it. Ira had a way of appropriating the high ground for his opinions and attitudes, simply because they were Ira's, and by that measure correct. He would borrow your vacuum cleaner, and if you asked him, months after the loan had passed, if he might see fit to return the vacuum cleaner, he would casually reply that, oh, the vacuum cleaner was broken, and change the subject. If you persisted, tried to elicit at least some clarifying comment on the missing vacuum cleaner, he would regard you with some disappointment — you actually care about a vacuum cleaner? And you, bound in your material possessions, would shrink a little, thinking of course vacuum cleaners are but dust in the great mandala of existence. And if you resented the fact that you were out one vacuum cleaner, you kept it to yourself.

Ira Einhorn handled formal rejection in much the same spirit. It was seldom his failing, but the inadequate qualities of the rejecter that led to those problems. Once, writer William Irwin Thompson refused Einhorn permission to attend sessions at Thompson's New Age conversation pit, Lindesfarne. In fact, as Thompson recalls, "I told him he was full of shit." And what did Einhorn do? "His way was to become patronizing and condescending," Thompson recalls. "To [imply] I was a benighted person with neurotic hang-ups, a gifted person with blocks, who would never amount to much because he had all these strange blockages to the evolutionary momentum of the human race. That I was beyond salvation."

"Ira always lived as if the rules didn't apply to him, and for an extraordinary amount of time he got away with that," says Ira's friend Mike Hoffman. "He got away with it because he convinced people of his very special quality. Which he had. The intensity he put into all the reading and thinking and the willingness to go out there farther than anyone else to follow some train of thought." [...] Ira really was [to himself] the center of the universe, says his friend Ralph Moore, who ran the Christian Association at Penn. "He would have an 'ends justifies the means' attitude that says, 'My agenda is the legitimate one here.'"

"Ira psychologically had no superego," says Stuart Samuels. "One of the reasons that his smell was so bad, despite people literally telling him about it, was that it didn't matter, because as far as he was concerned, he was larger than the world. So it didn't matter if he smelled. Ira was totally egocentric, so anything you said would always turn back on his knowledge of it, his point of view. It was always from his perspective."

"What I saw Ira do most was take over, wherever he was," says Jeff Berner, a [friend of] Ira's. "Dominate every social scene, take over every room use, every environment, and every space fully as his own. He was one of the few people I've allowed to do that in my own home. And I didn't mind because by the time he left, I was richer." Thus friends and associates accepted Ira's ego as part of a package which, on balance, was marvelously entertaining, intellectually provocative, and righteously motivated. Best of all, if you joked about it, Einhorn would be the first to laugh with you. [...] He would rail about macrobiotics for half an hour, and you could, as one friend did, finally interrupt him by saying, "Great, Ira, now let's go and get a hamburger," and Ira would say, "Sure," without missing a beat." (Levy, 1988)

Einhorn assiduously promoted drug use and frequent, promiscuous sex among young people. He would call for "open forums where people will listen, instead of shoving drugs under the rug." Ira passed out DMT and hash to students and friends, and taught classes entitled "Analogues to the LSD Experience." This maneuver was designed to create even more controversy about himself. Even though LSD was being discussed by everyone in private, Ira was the first in Philadelphia who dared to publicly proclaim acid's "virtues." He encouraged and participated in "sexual encounters" (otherwise known as orgies) and his reputation as a "cocksman" was, as we have already noted, legendary, if somewhat misleading. Ira had quantity, not quality, it seems.

Keep in mind, he began his public "ascent" in 1964-66.

At the same time that Ira was creating his public role as a counterculture hero, freeing mankind from the oppression of the establishment which didn't want to allow them unlimited drug-induced and sex-induced spiritual ascent, he met a student at Penn named Judy Lewis and became obsessed with her. Judy was seven years younger than Ira, and was, according to her own report, experiencing some emotional stress. Ira's friend, Michael Hoffman, said: "Ira was so intense that when you got involved with him, he would get inside your head. You would have that kind of relationship — inside of one another's heads. He was particularly that way with women because he needed to dominate. [With Judy], it was not a placid relationship, it was obviously a very passionate one, and at a certain point I think she wanted out. Because I think like a lot of other people, she finally felt mind-fucked by the guy."

Said another observer of the relationship, a friend of Judy's: "She was very interested in him, because of his mind, basically. And he got more and more possessive. [...] She was not allowed to do anything except be with him, and I think that she wasn't the kind of dependent female that he was used to or that he needed. I remember she used to talk about how he would insist on staying up all night long and talking, and if you wanted to go to sleep that was disloyal. She began to get a sense of his intensity, and his emotional violence. He was grasping and tenacious and nuts."

Again, Ira had created a fantasy that he projected onto a woman. He completely ignored the fact that the longer the relationship continued, the less interested the other party was in continuing in it. Ira produced reams of writing about Judy's beauty, her depth, and her selfish refusal to give everything to him: "Joy would erupt if Judy could only learn the simple acceptance of the magic which flows between us."

Here we come to the crux of the matter. Ira wrote in his journals that Judy could, if she could be persuaded to be willing, provide Ira with the "absolute trust my mother's strong relationship imposed on my psyche. Do I wish to master a woman sufficiently so that she will take care of me as my mother did?"

At this point in time, Ira was *suffering from excruciating headaches* (possibly symptomatic of dopamine deficiency). Something was definitely going on. Ira wrote: "I have a strange lightness about the head which is beginning to frighten me. There seems to be a strong possibility that I may eventually be permanently psychotic!"

Just as Rita Siegal had, Judy became terrified of Ira. She also discovered, as Rita had, that leaving Ira was not a simple matter. You didn't just walk away from Ira. It was months before Ira finally "got it" that Judy wanted to end their relationship. He disregarded her words and constructed fantasies where her wishes to not see him were just "vacillation." He would create scenarios in which he perceived her wishes as evidence that she really wanted to continue and expand the relationship. He also created scenarios of what would happen if that turned out to not be true. In November 1965 Ira wrote in his journals:

> The violence that flowed through my being tonight... still awaits that further dark confirmation of its existence which could result in the murder of that which I seem to love so deeply. The repressed is returning to a form that is almost impossible to control.... There is a good chance that I will attempt to kill Judy tomorrow — the rational awareness of this fact brings stark terror into my heart but it must be faced if I wish to go on — I must not allow myself to deviate from the self-knowledge which is in the process of being uncovered! (Quoted by Levy, 1988)

There was no violence the following day, but a week later Judy again tried to become free of Ira. Ira seemed to *know* that he was "deviant." He made a note in his journals to ask his mother about his behavior as an infant, with the comment: "So much of my deviancy could be explained in terms of an impotent, uncompleted rage."

Meanwhile, of course, in public Ira was marching on to fame and glory as *the* manifestation of the benefits of drug and sex-expanded awareness and spiritual superiority. He wrote in his journal in March 1966:

I feel as if things are about to culminate in the creation of an involvement that will allow me to do the work that will enable me to become more of what I am or as a result of this partial madness I will bring my world crashing down about me.

The struggle to deal rationally with what he clearly understood was a "game plan" that was not to his advantage, does nothing to suggest that there was any real emotion involved. To Ira, it was simply "moves" in a game. He was essentially attempting to impose cerebral and strategic rationality on his fundamentally predatory nature. On March 14 he wrote:

How ridiculous the thought of killing Judy appears, yet I held it in my mind just four short hours ago — this particular ability of man is both his horror and his joy. Violence creeps over my body as I reach toward the destruction of Judy, a hopeless victim in this infernal entanglement which seems to be draining the life's blood of both of us... the foolish ambivalence of our desires still tosses us beyond the recall of reason to a point of suspension on which we hang in perilous balance threatening to destroy or be destroyed in an instant or reckless action — we must come together or die.

To that I say: What do you mean *we*?

In any event, three days later Ira's predatory nature overwhelmed his "rational" thinking and the "event" occurred that was recounted by Judy to Detective Michael Chitwood thirteen years later. The situation, as she described it, centered around the fact that Ira had insisted on a meeting. She agreed as long as he just came by for coffee and nothing else. Ira, of course, arrived full of confidence that he could mesmerize Judy with his ideas and words about why the relationship should go on. Their discussion was interrupted when Judy went out briefly to get milk for the coffee and donuts. Ira himself recorded the event in a poem entitled "An Act of Violence." The poem describes Judy returning with the items and serving the coffee; and as she does so, Ira is mustering up the wherewithal to commit some, as yet, unnamed act. He discards the idea and writes that, as he is putting on his jacket to leave, "suddenly it happens."

Judy's back is turned and Ira moves toward her with a coke bottle in hand. "Bottle in hand I strike/ Away at the head...."

The bottle broke, however, and Judy began to bleed. Ira wrestled her down to the floor, holding her by the neck. She hit her head against the table as she fell, and Ira was strangling her. Like Rita Siegal before her, she went limp and lost consciousness. Ira wrote: "In such violence there may be freedom."

Judy recounts that the neighbors had heard the uproar and had come into the room. She told them to call the campus police. By this time, of course, Ira had disappeared. He went home to write in his journal:

> Where am I now after having hit Judy over the head with a coke bottle, blood on my jacket and pants — then making some feeble attempts to choke her. She wanted to live that has been established.... I'll be able, if she does not have me arrested, to go back to living a normal life. Violence always marks the end of a relationship. It is the final barrier over or through which no communication is possible.

As was the case with Rita Siegal, and so many other women who seek only to forget such violence perpetrated against them, Judy did not press charges against Ira. Ira was, however, informed that if the assault were repeated, he would face serious legal action.

Ira admitted that his action was "ridiculous." However, instead of a single instant of remorse, he seemed to think that his action was a "liberating response" to a woman who was "too selfish" to agree to indulge his perversions. He saw his violence as something that contributed to his growth, something that freed him from depression and moping about Judy's wish to leave him. Effectively, the reality that he had again almost killed a woman who simply wished not to associate with him, was completely lost on Ira.

Ira's friend, Michael Hoffman, was again in his confidence. By this time, however, he was appalled and tried to gain some understanding about it from his friend by questioning him at length about the event.

> He would sort of disengage himself from himself when he talked about these things. He didn't take responsibility. He didn't have the same kind of guilt that you or I would have, in that he didn't say "Jesus Christ, how could I have done that terrible thing?" He would talk about how it had grown out of the nature of the relationship. How it's not really possible to have that kind of full, rich, sharing relationship that a man and a woman needed to have. And somehow that would be part of the explanation. He probably had the most elaborate defense structure I've ever seen in anybody. He would literally walk in after doing something like that and want to discuss the reasons for its having been done from a psychoanalytic point of view, a sociological point of view, its place in history... so that he immediately had a very elaborate structure to put it in. (Levy, 1988)

You see, Ira didn't think that he needed help. He thought of his violence as evidence that he was some sort of romantic hero. He knew he was "deviant," but he saw it as just who and what he was, and that his way was *right*. He wrote to Hoffman from California a few months later: "Rita and Judy practically destroyed, Peckham unable to go any farther! I need the confrontation of my monsters lodged in some external being — to meet and see what haunts me — to face it and fight it every day without it disappearing.... I live quietly and calmly with real joy on the edge of a volcano that might explode into nova-like being at any moment... when it happens, beware!"

Ira despaired of finding a woman who could satisfy his lusts in terms of the violence and pain he desired. He wrote:

> I'm slowly beginning to realize the enormity of the problem which my development has created in respect to women. The interaction with Rita is just an example of how difficult, even at that age and with such a magnificent partner, any final linking is to be. Judy provided in her striking beauty a repository for always wandering projections, and the strength of our deathlike struggle is a good indication of how impossible my quest is to be. I refuse to admit the inevitable — that I can live without a woman (my mother). Until I accept this my productivity will be intense, like my countless infatuations, but sporadic. *I'm faced with a hell that is somewhat relieved by my incredible energy which is so capable of constantly creating that joy which is deeper than sorrow.* (quoted by Levy, 1988; this author's emphasis)

What, exactly, did Ira mean by the above? What kind of "hell" was he living in? A "hell" that denied his impulses. What kind of joy was he desirous of "creating" by indulging those impulses? A joy that was "deeper than sorrow"? Take note of his reference to Rita and the fact that she was unable to complete the "final linking." Earlier, about Rita and this desired "linking," Ira had written:

> Sadism — sounds nice — run it over your tongue — contemplate with joy the pains of others as you expire with an excruciating satisfaction. Project outward the vision of inward darkness. Let no cesspool of inner meaning be concealed. Reveal the filth that you are. Know the animal is always there.... Beauty and innocence must be violated for they can't be possessed. The sacred mystery of another must be preserved — only death can do that. [...] My dreams are realizable and will not be snuffed out by the fear of anyone — I too have a right to a life of my own and to that I will dedicate myself. [...] The progress of my soul must not be crushed by the failings of a selfish young woman. [...]
>
> We so carefully hide the blackness of our soul from all those around us (even ourselves), we forget so easily the impulses of power which unconsciously control so many of our actions! [...] We are — blackness and light. To beat a woman — what joy — to bite her breasts and ass — how delightful — to have her return the favor in our sensitive areas. How is life to be lived? [...] You are one of the rare free spirits, do not be saddled by one who isn't. Life to be lived at its full must be lived freely. Let nothing stand in your way to getting what you can, not even the

illusion of love which you know to be so transitory. (quoted by Levy, 1988)

I hope that the reader has noticed the references to his mother in Ira's comments. This is, as noted, a clue — the crux of the matter — as we will eventually see when we examine more closely the Negative Macrocosmic Reality that seeks to overtake and dominate our own reality via the machinations of the psychopath. And, interestingly, it appears in the "breakdown" of John Nash, as we will cover in the next chapter.

Chapter 63
Murdering the Feminine

As the reader might guess, a person who has no fear of consequences because they cannot imagine them — or even if they do have the intellect to do so, has no emotion associated with those consequences which would tend to enforce them as behavioral choices — is the ideal vehicle for the violation of the free will and rights of others.

It also means that such a person is free to choose to do things that are potentially self-destructive, without giving a single indication to another "player" that his or her choice is based entirely on a delusion. Very often, they "win" because the sheer boldness of their action is unrestricted by conscience, which is a construct of emotions. But, interestingly, this also has the potential to leave the psychopath open to total destruction.

It's like a poker player who has absolutely nothing in his hand, but because he is so intent on winning and is so unmoved by the possibility of losing, and because lying produces absolutely no internal emotional reaction of fear of being discovered or the potential shame or disaster inherent in such an event, is able to bluff so convincingly that the other players, any of whom might have a winning hand, fold and walk away — because they are convinced by the psychopath's confidence that he must have the winning hand of all time. Only he doesn't, and this means that the psychopath's strength is also his Achilles Heel. Once he has been spotted, identified and understood, he no longer has the power to bluff. Once knowledge enters the game, the psychopath is exposed, and has no more ability to "con" the other players. The sad part is: He also has no ability to learn from this experience anything other than how to make his bluff better and more convincing next time. The psychopath never gets mad because he is caught in a lie; he is only concerned with "damage control" in terms of his ability to continue to con others.

Such was the case with Ira Einhorn when he boldly and arrogantly decided to keep Holly's body in the trunk in his closet. It wasn't an act of stupidity; it was the act of a psychopath. The plain fact of the matter is, if Holly's family hadn't had enough money to pay a private investigator to keep digging, Ira would have gotten away with it forever.[33]

So it is in our world: Economics very often provides major payoffs to those who are psychopaths, and penalizes those who are not.

Of course, the reader will also easily be able to see how and why "dumbing down" a society is useful to the psychopathic manipulation of that society. A population that does not have knowledge, or the inclination to obtain knowledge (or even the awareness that they can), is much more easily bluffed by the "losing hand" of its leaders. But, there is something far more insidious at play here: Game Theory, law and economics. The earliest applications of economic reasoning about legal structures were done with an eye toward how legal rules affect societal behavior. The simplest of strategic problems will serve here to highlight the current situation.

When two individuals interact with each other, each must decide what to do, without knowledge of what the other is doing. Imagine that the two players are the government and the public. In the following model, each of the players faces only a binary choice: to behave ethically either in making laws or in obeying them.

The assumption is that both players are informed about everything except the level of ethical behavior of the other. They know what it means to act ethically, and they know the consequences of being exposed as unethical. There are three elements to the game. 1) The players, 2) the strategies available to either of them, and 3) the payoff each player receives for each possible combination of strategies.

In a legal regime, one party is obliged to compensate the other for damages under certain conditions, but not under others. We are going to imagine a regime wherein the government is never liable for losses suffered by the public because of its unethical behavior; instead, the public has to pay for the damages inflicted by the government due to its unethical behavior.

The way the payoffs are represented is generally in terms of money. That is, how much investment does each player have to make in ethical behavior, and how much payoff does each player receive for his investment?

---

[33] *Update*: Since the above was written, Einhorn has been extradited to the United States, after which a jury affirmed his murder conviction on October 17, 2002.

In this model, behaving ethically according to standards of social values that are considered the "norm," costs each player $10. When law detrimental to the public is passed, it costs the public $100. We take it as a given that such laws will be passed unless both players behave ethically.

Next, we assume that the likelihood of a detrimental law being passed, while both the public and the government are behaving ethically, is a one-in-ten chance.

In a legal regime in which the government is *never* held responsible for its unethical behavior, and if neither the government nor the public behave ethically, the government enjoys a payoff of $0 and the public is out $100 when a law detrimental to the public is passed.

If both "invest" in ethical behavior, the government has a payoff of minus $10 (the cost of behaving ethically) and the public is out minus $20 which is the $10 invested in being ethical *plus* the $10 of the one-in-ten chance of a $100 loss incurred if a detrimental law is passed.

If the government behaves ethically and the public does not, resulting in the passing of a law detrimental to the populace, the government is out the $10 invested in being ethical and the public is out $100.

If the government does not behave ethically, and the public does, the government has a payoff of $0 and the public is out $110 which is the "cost of being ethical" added to the losses suffered when the government passes detrimental laws. Modeled in a Game Theory bi-matrix, it looks like this (with the two numbers representing the "payoff" to the public being the left number in each pair, and the two numbers representing the "payoff" to the government being the right number in each pair):

		Government	
		No Ethics	Ethical
Society/Public	No Ethics	-100, *0*	-100, *-10*
	Ethical	-110, *0*	-20, *-10*

In short, in this game the government always does better by not being ethical, and we can predict the government's choice of strategy because there is a single strategy (no ethics) that is better for the government no matter what choice the public makes. This is a "strictly dominant strategy," or a strategy that is the best choice for the player no matter what choices are made by the other player.

What is even worse is the fact that the public is *penalized* for behaving ethically. Since we know that the government will never behave ethically, because behaving with "no ethics" is the dominant strategy, we find that ethical behavior on the part of the public actually costs *more* than unethical behavior.

In short, the public is being manipulated to make choices that are unethical.

The public, as you see, cannot even minimize their losses by behaving ethically. It costs them $110 to be ethical, and only $100 to not be ethical.

Now, just substitute "psychopath" in the place of "government" and "non-psychopath" in the place of "public," and you begin to understand why the psychopath will always be a psychopath. If the "payoff" is emotional pain of being hurt, or shame for being exposed, in the world of the psychopath that consequence simply does not exist; just as in the legal regime created above, where the government is never responsible for unethical behavior. The psychopath lives in a world in which it is like a government that is never held responsible for behavior that is detrimental to others. It's that simple. The form game above will tell you why psychopaths in the population, as well as those in government, are able to induce the public to accept laws that are detrimental. It simply isn't worth it to be ethical. If you go along with the psychopath, you lose. If you resist the psychopath, you lose even more.

Erosion of ethics and responsibility are the objective. Economics is the weapon. "I think the essence of being an economist... has to do with learning to turn off all your emotions and say, 'Let's think abstractly about adoption or crime.' We have a name for people who are innately really good at this. They're psychopaths, people who are incapable of feeling guilt, or remorse, or any emotional reaction. So there's a good reason, I guess, why one looks with a degree of nervousness at people who are really good at this activity [economics]" (Romer, 1983).

In America, a great many households are affected by the fact that work, divorce, or both, have removed one or both parents from interaction with their children for much of the day. *This is a consequence of economics.* When the parents are absent, or even when one is present but not in possession of sufficient knowledge or information, children are left to the mercies of their peers, a culture shaped by the media. Armed with joysticks and TV remotes, children are guided by *South Park* and *Jerry Springer,* or *Mortal Kombat* on

Nintendo. Normal kids become desensitized to violence. More susceptible kids — children with a genetic inheritance of psychopathy — are pushed toward a dangerous mental precipice. Meanwhile, the government is regularly passing laws, on the demand of parents and the psychological community, designed to avoid imposing consequences on junior's violent behavior.

As for media violence, few researchers continue to try to dispute that bloodshed on TV and in the movies has an effect on the kids who witness it. Added to the mix now are video games structured around models of hunting and killing. Engaged by graphics, children learn to associate spurts of "blood" with the primal gratification of scoring a "win."

Again, economics controls the reality.

The psychic stresses of our world are right in the home. There they can easily act on any kid who believes that "the world has wronged me" — a sentiment spoken from the reality of existence, a reality created by economic pressures instituted via Game Theory.

Is there a solution?

The obvious solution would be a world in which, at the very least, the psychopath — in government or in society — would be forced to be responsible for unethical behavior. But game-theory modeling demonstrates that *selfishness* is always the most profitable strategy possible for replicating units. It seems that, over centuries, this has been one of the agendas of the hyperdimensional control system — to encourage the reproduction of genetic psychopaths — so that in this day, in this present time, all their pieces are on the board for the Secret Games of the Gods.

Could it ever be an evolutionarily-stable strategy for people to be innately unselfish?

On the whole, a capacity to cheat, compete and lie has proven to be a stupendously successful adaptation. Thus, the idea that selection pressure could ever cause saintliness to spread in a society looks implausible in practice. It doesn't seem feasible to out-compete genes which promote competitiveness. "Nice guys" get eaten or out-bred. Happy people who are unaware get eaten or out-bred. Happiness and niceness today is vanishingly rare, and the misery and suffering of those who are able to truly feel, who are empathic toward other human beings, who have a conscience, is all too common. The hyperdimensional manipulations are designed to make psychopaths of us all.

Nevertheless, a predisposition to conscience and ethics can prevail *if and when* it is also possible to implement the deepest level of altruism

— making the object of its empathy the higher ideal of enhancing free will in the abstract sense, for the sake of others, including our descendants. In short, our "self-interest" ought to be vested in *collectively ensuring that all others are happy* and well-disposed too; and in ensuring that children we bring into the world have the option of being constitutionally happy and benevolent toward one another.

In short, if psychopathy threatens the well-being of the group future, then it can only be dealt with by refusing to allow the self to be dominated by it on an individual, personal basis. Preserving free will for the self in the practical sense, ultimately preserves free will for others. *Protection of our own rights as well as the rights of others, underwrites the free will position and potential for happiness of all.* If mutant psychopaths pose a potential danger then true empathy — true ethics, true conscience — *dictates using prophylactic therapy against psychopaths.*

It seems certain from the evidence that a positive transformation of human nature isn't going to come about through a great spiritual awakening, socioeconomic reforms, or a spontaneous desire among the peoples of the world to be nice to each other. But it's quite possible that, in the long run, the psychopathic program of suffering will lose out because misery is not a stable strategy. In a state of increasing misery, victims will seek to escape it; and this seeking will ultimately lead them to inquire into the true state of their misery, and that may lead to a society of intelligent people who will have the collective capacity to do so.

So it is that identifying the psychopath, ceasing our interaction with them, cutting them off from our society, making ourselves unavailable to them as "food" or objects to be conned and used, is the single most effective strategy that we can play.

There is, of course, a deeper reason, and that brings us back to the clue of "The Mother."

*

In 1958, at a New Year's Eve party, John Nash was apparently being his slightly weird, slightly off-center-from-normal self, known and accepted by his family and friends as just who he was. Many allowances were made for his behavior because he was so brilliant. However, by the last day of February 1959, John Nash would undergo a terrifying and bizarre change.

Just prior to that complete breakdown, Nash appeared at the New Year's party *dressed in a diaper* as the "Infant New Year." Well, maybe that's not so weird. But what made it even stranger was the fact that Nash spent most of the evening *curled up in his wife's lap.* This was very disturbing to the other guests who later commented on the discomfort that this behavior evoked in them. The sensation of being "ill at ease" in the presence of certain people is often just swept under the rug, explained away or ignored by most of us. But in this case, it was most definitely a sign that something was wrong — seriously wrong — with John Nash.

Those who have looked at it in retrospect suggest that long before the New Year's party, Nash had "crossed some sort of threshold." However, the deterioration of his mental state was disregarded because he was a known eccentric to begin with. According to those who knew him, his social discourse had always been odd because he never seemed to know when to speak out or stay quiet. He seemed to be unable to participate in an ordinary give-and-take conversation. He was prone to telling lengthy stories with cryptic or off-center endings.

In the months before the party, Nash had been teaching a course in Game Theory. His students noted that he paced a great deal and fell into trance-like states in the middle of lecturing or answering a question. Just before Thanksgiving that year, Nash confided to his assistant and a student that there were "threats to world peace" and "calls for a world government." He hinted that he was to play a significant role in some upcoming drama. It was after classes resumed following New Year's that Nash asked his assistant to teach a couple of his classes because he needed to go away. Then he disappeared.

As it happened, Paul Cohen, the mathematician who had become famous for solving a logical puzzle posed by Gödel, also disappeared at the same time. After a few days, people began to notice the absence of both of them, though Cohen was eventually found to be visiting his sister. Nash had "driven south," ultimately ending up in Roanoke at his mother's house, but no one knows any other details. It is thought that he may also have gone to Washington, DC.

A couple of weeks later, Nash, holding a copy of the *New York Times* in his hand, walked up to a group in the faculty common room and pointed to a story in the paper, saying that "abstract powers from outer space, or perhaps it was foreign governments," were communicating with him through the paper. He claimed that the messages were meant only for him and were encrypted. Only he could decode them and he was being allowed to share the information with the world.

Nash began to say that radio stations were sending messages to him. He gave one of his students his expired driver's license, telling him that it was an "intergalactic driver's license." He told the student that he was a member of a committee and that he would put the student in charge of Asia.

Nash recalls that period as one of mental exhaustion accompanied by a growing series of images and sense of revelation regarding a secret world that others around him were not privy to. He started noticing men in red ties and thought that this was a signal to him. He thought it had something to do with the communist party, or so he claimed in 1996.

At this point in time, Nash was still working on the Riemann hypothesis. He became paranoid and thought that other people were conspiring against him or stealing stuff from his trash can. In France, mathematician Claude Berge received a letter from Nash, written in four colors, complaining that his career was being ruined by aliens from outer space.

"One day, Nash wandered into someone else's office. He drew 'a set that resembled a large, wavy baked potato. He drew a couple of other smaller shapes to the right.' Then he fixed a long gaze on his audience of one, pointed to the baked potato and said: 'This is the universe. This is the government. This is heaven. And this is Hell.'" Nash began writing strange letters. They were addressed to ambassadors of various countries and Nash attempted to mail them via interdepartmental mail. The department secretary put them aside to show to the department head, Ted Martin. Martin panicked and tried to retrieve the letters (not all of which were addressed, most of which were not stamped) that had been dropped in mailboxes all around the campus.

Now, we really need to stop for a moment and consider this situation. Here we have a guy who thinks that aliens from outer space are ruining his career; he obviously had severe problems, and all anybody can think about doing is trying to cover it up! The department head, in all his three-piece-suit administrative glory, is out there frantically pawing through mailboxes to find letters posted by Nash, in order to prevent others from knowing that the guy is going over the edge.

One would think that, for at least medical reasons, these letters would have been preserved for psychiatric review. However, Sylvia Nasar tells us that *none of them have survived*. I find that astonishing. Just what the heck did those letters say?

Some of the people interviewed by Nasar reported that Martin told them that Nash was writing that he had been put in charge of "forming a world government." At another point, Nash wrote in a letter to Adrian Albert, chairman of the math department at the University of Chicago, that he was unable to accept a position there because he was "scheduled to become the Emperor of Antarctica."

The deterioration was swift following the New Year's party. On February 28, John Nash stood before 250 mathematicians to give a lecture. At first what he said seemed like just a series of cryptic, half-formed mathematical remarks, which was not too unusual for Nash. But, halfway through the lecture something happened: "One word didn't fit in with the other. [...] Everybody knew something was wrong. He didn't get stuck. It was his chatter. The math was just lunacy. [...] It was horrible" (Donald Newman, quoted by Nasar, 1998).

The audience, according to another witness, "heaped scorn" on Nash and he was "laughed out of the auditorium."

In his private life, Nash was complaining to his wife that he was "bugged," that something was going on. He stayed up most nights writing his endless letters to heads of state. One night, he painted black spots all over their bedroom walls.

The details of Nash's first involuntary commitment are vague. Nasar and others conjecture that the president of MIT intervened in some way and Nash was "picked up." After his admission to the psychiatric unit at McLean Hospital, an injection of thorazine calmed him down, but did not stop the *flow of ideas* that were coming into his head. He told Arthur Mattuck that he believed that there was "*a conspiracy among military leaders to take over the world, and that he was in charge of the takeover.*"[34] The bizarre and elaborate nature of Nash's psychosis, his beliefs that were simultaneously grandiose and persecutory, and other symptoms, all resulted in a diagnosis of schizophrenia.

In the hospital, Nash quickly learned to stop acting crazy. He wanted out of there, and so he applied himself to learning the rules of the game. Even though he reported that his symptoms had disappeared, his psychiatrists agreed that he was very likely just concealing them, Nash told whoever would listen that he was a "political prisoner."

After much to-do, Nash fled to Europe. In Paris, he was frequently visited by Alexander Grothendieck who, ten years or so later, *founded a survivalist organization, dropped out of academia, and disappeared into the Pyrenees*. Another interesting item from that time consists of the fact that Nash apparently told mathematician Shiing-she Chern that "*four cities in Europe constituted the vertices of a square.*" Which cities they were, and what the implications of this fact are, was apparently not recorded.

---

[34] Oddly enough, considering the global events since 911 and the evidence that Nash's "Game Theory" is an important part of military strategy, what Nash was reported to have said here doesn't seem to be too far from the truth. One wonders what things he wrote in his letters that were destroyed.

Through it all, Nash was talking about numerology, dates, world affairs, and something going on in the Gaza strip. He believed that there were magic numbers, dangerous numbers, and that it was his job to save the world. He lived in constant fear of annihilation: Armageddon, the Day of Judgment, etc. The date May 29 was ominous to him. Eventually, he was hospitalized again and subjected to "insulin therapy." As Nasar reports, "good firsthand accounts of this therapy are difficult to obtain because it destroys large blocks of recent memory." Why are we not surprised?

Nash shuffled in and out of hospitals, back and forth across the Atlantic, fluctuating between various degrees of madness for years. Overall, his condition and symptoms certainly fit his diagnosis: paranoid schizophrenia. And, as is usual with schizophrenics, the delusions were *disconnected bits and pieces of the individual's actual reality.*

My question is: Knowing the "brute force" of Nash's mind, did he actually *do* what he claimed he wanted to do — to "find a different and more satisfying under-picture of a non-observable reality"? Did he penetrate the veil; and did he, in the act, encounter something so dark, so dreadful, that it overloaded his all-too human circuits and triggered his descent into schizophrenia?

What do John Nash and Ira Einhorn have in common besides brilliant minds and a fascination with the game "Go?"

Well, they both had "dominating mothers" who constantly sought to push them forward in the world through assiduous training of their minds, pride in their accomplishments, and a serious lack of allowing them to experience the consequences of their behavior — *if* their mothers were even aware of their "glitches." It actually seems to be so that very good psychopaths learn *very* early to conceal their true nature.

But coming back to the issue: psychopathy and schizophrenia. In all the research I have done on the subjects, there are two things that keep coming up in both conditions: the lack of emotional "connection," and *unusual word usage.* Psychopaths, however, seem to seek to "engage" others for the purpose of getting what they want, to materially fulfill their fantasies. Schizophrenics, on the other hand, withdraw from others into their fantasies, as though the fantasy was more real than the outside world. To adapt the punch line of an old joke: Psychopaths build castles in the air and try to sell them to others; schizophrenics build castles in the air and move into them.

Over and over again, in reading cases of both disorders, we find that "disharmony between the content of patient's words and his emotional expression was striking." However, in the case of the schizophrenic, it has gone to an extreme, in that there is no longer any attempt to fake anything for the sake of deceiving other people. For example, one patient giggled constantly while describing, in *sympathetic words*, an acute illness suffered by his mother. One schizophrenic talked about his child's death with a broad smile on his face. Another patient reacted with rage to a simple question about how he slept at night. Clinically, the degree of *emotional inappropriateness* is often used to indicate the severity of the schizophrenic's condition. But, far from being able to consciously "pretend" appropriate emotions, as the psychopath does for the purposes of deception, the schizophrenic *exhibits a huge discrepancy between what he says and the emotional tone associated with his verbal communications*. The emotions attached to what the schizophrenic talks about are inappropriate and arbitrary, and rarely — if ever — concealed.[35]

Nevertheless, it is clear that both schizophrenics and psychopaths operate largely based on fantasy or delusion.

Does this suggest that psychopathy is a variation of schizophrenia that is outwardly directed in some sense, and which manifests certain coping mechanisms in order to obtain "satisfaction"? Are schizophrenics individuals who have somehow shifted into a mode of being wherein outside stimulation or sources of satisfaction are not only no longer needed, but perceived as completely undesirable? If we think of psychopaths in terms of upright "predators," is it appropriate to think of schizophrenics in terms of upright "prey," even if the predator is in their own mind?

The curious thing about the two cases, Nash and Einhorn, is that both of them exhibited very similar "independence" and "antisocial" behaviors when they were growing up. There was similar aggression and resistance to authority. However, there are most certainly schizophrenics who have been described by their families as very "together" and outgoing, dutiful and giving, before the onset of their symptoms. Many, if not most of them, are shy and introverted as children — seemingly "too sensitive." But that is not always true, and it certainly wasn't true in the case of John Nash.

Mealey has proposed two different aetiologies for sociopathy, but in her framework those displaying chronic antisocial *behaviour* are placed in the same functional category. This implies that they have similar or identical psychological mechanisms. On the other hand, Blair [Blair, R.J.R. (1995), "A cognitive developmental approach to morality: investigating the psychopath," *Cognition*, 57: 1-29] concentrates on the

---

[35] "The Private World of a Schizophrenic,"
http://www.theallengroup.com/members/Schizo_2.html

mechanisms subserving psychopathic behaviour, but concludes that psychopaths have a dysfunctional psychological/neurological mechanism and are disordered in comparison to other members of society....[36]

In one significant study it was found that the *Psychopathy Checklist could not distinguish between psychopathic and schizophrenic offenders* in 50 consecutive male admissions to an English Special Hospital. [Howard, R.C. (1990), "Psychopathy Checklist scores in mentally abnormal offenders: A re-examination," *Personality & Individual Differences*, 11: 1087-1091] *This may indicate that some schizophrenics with a history of antisocial behaviour are suffering from what could be called state-dependent psychopathy....*

What is most outstanding about psychopaths is that they appear extremely at ease with themselves. They can be articulate, are often highly intelligent, and are regularly described as "charming," and "convincing." Psychopathy is not associated with low birth-weight, obstetric complications, poor parenting, poverty, early psychological trauma or adverse experiences, and indeed Robert Hare remarks "I can find no convincing evidence that psychopathy is the direct result of early social or environmental factors." [Hare, R.D. (1993), *Without conscience: The disturbing world of the psychopaths among us*, New York, NY: Simon and Schuster]

No sound evidence of neuroanatomical correlates for psychopathic behavior has been found, though an interesting (and highly significant) *negative correlation* has been found in 18 psychopaths between the degree of psychopathy as assessed by the *Checklist* and *the size of the posterior half of the hippocampi bilaterally.* [Laakso, M.P., Vaurio, O., Koivisto, E., Savolainen, L., Eronen, M., Aronen, H.J., Hakola, P., Repo, E., Soininen, H., & Tiihonen, J. (2001), "Psychopathy and the posterior hippocampus," *Behavioural Brain Research*, 118: 187-93] Lesions of the dorsal hippocampus have been found to impair acquisition of conditioned fear, a notable feature of psychopathy, but *it is not clear whether this neuroanatomical feature is the cause of, or is caused by, psychopathy.* A study of 69 male psychopaths identified by the revised edition of Hare's *Psychopathy Checklist* found no support for the hypothesis that psychopaths are characterized by verbal or left hemisphere dysfunction. [Smith, S.S., Arnett, P.A., & Newman, J.P. (1992), "Neuropsychological differentiation of psychopathic and non-psychopathic criminal offenders," *Personality & Individual Differences*, 13: 1233-1243]

One particularly striking feature of psychopathy is that extremely violent and antisocial behaviour appears at a very early age, often including casual and thoughtless lying, petty theft, a pattern of killing animals, early experimentation with sex, and stealing. [Hare, op. cit.] In a study of 653 serious offenders by Harris, Rice, and Quinsey, childhood problem behaviors provided convergent evidence for the existence of

---

[36] Mealey, L. (1995). "The sociobiology of sociopathy: an integrated evolutionary model." *Behavioral & Brain Sciences.* 18: 523-599.

psychopathy as a discrete class, but "adult criminal history variables were continuously distributed and were insufficient in themselves to detect the taxon." [Harris, G.T., Rice, M.E., & Quinsey, V.L. (1994), "Psychopathy as a taxon: evidence that psychopaths are a discrete class," *Journal of Consulting and Clinical Psychology*, 62: 387-97]

In a recent study psychopathic male offenders were found to score *lower* than non-psychopathic offenders on obstetrical problems and fluctuating asymmetry, and in fact the offenders meeting the most stringent criteria for psychopathy had the lowest asymmetry scores amongst offenders. [Lalumière, M.L., Harris, G.T., & Rice, M.E. (2001), "Psychopathy and developmental instability," *Evolution and Human Behavior*, 22: 75-92] As the authors note this study *provides no support for the idea that psychopathy results from developmental instability* of some kind, but does give partial support for life-history strategy models.

An evolutionary game-theoretic explanation for the low but stable prevalence of psychopathy has been modeled successfully [Colman, A.M., & Wilson, J.C. (1997), "Antisocial personality disorder: An evolutionary Game Theory analysis," *Legal & Criminological Psychology*, 2: 23-34], and though this provides some tentative support for Mealey's suggestion that psychopathy is a frequency-dependent strategy, cross-cultural work using reliable measures will be needed to establish whether there is a stable proportion of sociopaths in traditional societies. [Archer, J. (1995), "Testing Mealey's model: The need to demonstrate an ESS and to establish the role of testosterone," *Behavioural & Brain Sciences*, 18: 541-542] Given the paucity of evidence in favour of developmental instability and brain damage in psychopaths the suggestion that psychopathy is an adaptation is worthy of further exploration. *Particular attention should also be paid to the probability that child psychopaths are mislabeled as suffering from Attention Deficit Hyperactivity Disorder, Conduct Disorder, or Oppositional Defiant Disorder.* [American Psychiatric Association (1994), *Diagnostic and Statistical Manual of Mental Disorders* (DSM IV) (4th ed.), Washington, DC: American Psychiatric Association] According to Hare "none of these diagnostic categories quite hits the mark with young psychopaths. Conduct disorder comes closest, but it fails to capture the emotional, cognitive, and interpersonal personality traits... that are so important in the diagnosis of psychopathy." [Hare, op. cit.]

(Pitchford, 2001; this author's emphases)

As noted above, we have an interesting problem before us: There seems to be some extraordinary correlation between psychopathy and schizophrenia that is, as yet, quite mysterious to researchers. Ian Pitchford has proposed that "some schizophrenics with a history of antisocial behaviour are suffering from what could be called state-dependent psychopathy."

Now notice: This is not saying that schizophrenics are psychopaths. I am personally aware of several people suffering from the horrors of schizophrenia who have never, ever, manifested any psychopathic characteristics. On the contrary, they were gentle, loving, shy and giving until the disease manifested. And, even after, such natures still manifest through the fog of the delusions in numerous ways. No, indeed, we are talking about something else altogether.

As I have noted earlier, when I read the biography of John Nash, it was with a sickening sense of horror that I realized I was reading a vivid description of Frank Scott's life, as he had recounted himself — only without the "pity poor me" spin he had put on it. He had spent so much time lambasting his parents for abusing him, that I was quite shocked to meet them and to immediately sense that there was no way possible that these people had ever abused anyone. At a later point in time, I closely questioned his sister about these things and she assured me that Frank had never been abused. She did acknowledge that he had repeatedly claimed that he had suffered at their hands, and even at her own hands, and that perhaps, since he was so "sensitive," he may have been handled too "roughly" for his delicate sensibilities. The best psychopaths are able to convince another person that they have done something bad, even when they are the one who did it, and they know it wasn't bad — or certainly not as bad as portrayed! As it happened, Frank's sister's versions of events he had recounted as being horrible and violent were just simply not convincing enough to support his claims. It should also be noted that, because she was not "in agreement" with him over many issues, Frank viewed his sister as morally deficient. He also spent a great deal of time interpreting her behavior to his parents as such, even though it was obvious that she was simply an individual who had ideas about living her life that were different from theirs. Frank assiduously "fed" this difference of opinions, ultimately achieving his goal of convincing his parents that his sister was, if not morally deficient, at least psychologically impaired to the point of being incompetent. Heck, until I met her, I was convinced, too! Even after that, for a long period of time, Frank managed to interpret everything she had said and done in my presence in the worst light, so that I was left with nothing but confusion about her.

In any event, as we have noted, the "tests" referred to by the Cassiopaeans on a number of occasions, were obviously related to the ability to "see" something that was not apparent to ordinary vision. Interestingly, this point is emphasized by the fact that the essence of the many learning exercises I was being led through was identified in the course of a discussion about Ted Bundy — a classic psychopath!

01-09-96

Q: (L) I don't quite know how to ask this. It has become increasingly obvious to me that there is some sort of connection where [an unsolved murder of a local 13-year-old girl, with absolutely *no* clues] was concerned, some synchronous connections between that murder and my so-called "awakening," if you want to call it that. And I also noticed a connection between the life pattern, or change in life pattern, of Ted Bundy and certain UFO sightings and cattle mutilations that were in his area of the country. Now, we have another girl who has come up missing at the same time P and I were discussing this case, and this new case has a lot of things that seem to be common to that case. I see that there is an issue here that I would like to get to the bottom. [...] Did my involvement with that case [I was asked by law enforcement official to try to come up with some clues or hints through astrology and psychic impressions] have anything to do with opening the door of my mind to other phenomena, particularly UFOs and aliens?

A: Possible.

Q: (L) You can't give me a clear answer on that?

A: Learn!

Q: (L) Okay. I had dreams about it. The work that I did on the case astrologically, the dreams I had about it, as well as certain impressions I received, was that an opening of my instinctual awareness in some way?

A: Maybe. [...]

Q: (L) Was there some connection between murder and "alien" activity?

A: There is always this connection in one way or another, at one plane convergence or another.

Q: (L) Was the murder of this child a "mini-plane convergence"?

A: What did we just say?

Q: (L) It seemed to me that was what you said, and I was trying to clarify it — is that, in fact, a plane convergence, where one person's plane of reality converges with another person's plane of reality, and one or the other gets annihilated?

A: 4th, 5th and 3rd density is involved.

Q: (L) Is this true with all murders?

A: Discover, and yes.

Q: (L) Was my interaction into that reality a sort of entering into a point of plane convergence?

A: Flirting with the edges.

Q: (L) So, when a person is working on a murder investigation, or thinking about it, or applying thoughts, talents, instincts or whatever to the solving of this kind of puzzle, they are interacting with a plane convergence?

A: This represents one manifestation of the always present desire to return "home" to 5th density.

Q: (L) Okay. Well. Now, I want to get to the 64,000-dollar question. In the JO case, was my conclusion correct?

A: "Correctness" takes many forms and *provides a window to many conventions.*

Q: (L) (F) What does that mean? (L) I don't know.

A: Learn.

Q: (L) Was the man who killed her known to her?

A: We recall advising a cautious approach, in order to insure that your lessons are learned not only accurately, but painlessly as well.

Q: (L) Could you suggest, just to get me on track here, a form of question that would be a "cautious" question? Then I can frame subsequent questions on that model.

A: The issue here is not how to "frame" a question in such a way as to lure us into answering in the way you desire, but for you to learn most effectively. Do not have prejudice that there is only one thing to be learned from each response. "You never know what there is to be learned when you inquire with innocence and freedom from supposition."

Q: (L) I just played the tape back and it is all muddy. Could you tell us why we are having this problem with the tape?

A: Telekinetic wave transfer.

Q: (L) What is this telekinetic wave transferring?

A: Evolving energy.

Q: (L) Given off by us?

A: Both to and from.

Q: (L) From us to you?

A: *You and others, not us.* [The only other person present was Frank.]

Q: (L) Who are these others?

A: 4th density eavesdroppers, P's involvement should "heat things up."

Q: (L) Is P's involvement going to be beneficial to this work?

A: Yes, but also expect anomalies.

Q: (L) That is interesting. Are you going to tell me who killed the child? I am willing to give up my conclusion if necessary.

A: Learn. Review our previous response. [...]

Q: (L) Okay. Learn. Was there something about Ted Bundy, and the fact that his life seemed to disintegrate at the same time a lot of UFOs were sighted?

A: Yes.

Q: (L) Was Ted Bundy abducted?

A: Yes.

Q: (L) Was Ted Bundy programmed to do what he did?

A: Yes.

Q: (L) What was the purpose behind that programming?

A: We must withhold answer for the present.

Q: (L) Okay. Bundy described his murdering urges as a "pressure building inside" him that he couldn't overcome, and it seemed to cause him to stop being "human," as we think of it. That seems to me to be an example of an implant being able to overcome a person's social behavior, or controls over antisocial tendencies. Is this also what happened to the person who killed JO?

A: Maybe.

Q: (L) Is there a connection between the newly missing girl, CB, and JO?

A: You are doing well in your probing of the knowledge within on this issue, we suggest continuance, after all, learning is fun!

Q: (L) So, it seems to me that there was a connection between the appearance of CB and JO. Could it be that the individual who killed one or both of them was programmed to respond to this particular type facial characteristic? Could that be part of the programming?

A: End subject.

Q: (L) What do you mean?

A: We have helped you all that is necessary for now on this matter. It is beneficial for you to continue on your own for growth.

Q: (L) Can I ask just one or two more *little* questions in a different direction? I mean, this is like walking away and leaving me in the dark!

A: No it is not!

Q: (L) I would like to be able to solve this because the families are in pain and have asked for help.

A: Why don't you trust your incredible abilities? If we answer for you now, you will be helpless *when it becomes necessary for you to perform this function on a regular basis*, as it will be!!!!

Q: (L) Well, frankly, I don't want to be involved in any more murder investigations. It is too upsetting. Am I supposed to *do* this sort of thing regularly?!

A: Not same arena.

Q: (L) Well, then how do you mean "perform this function"?

A: No, seeing the unseen.

And, regarding the issue of schizophrenia:

10-25-94

Q: (L) Are Lizards responsible for paranoid schizophrenia?

A: Some.

Q: (L) In a general sense, in the majority of cases, what is the cause of paranoia or schizophrenia?

A: Lizard manipulation of energies.

Q: (L) Why?

A: To feed off the negative results.

Q: (L) So it isn't necessarily attachments?

A: No.

Q: (L) Do Lizards use attachments of dark energies to effect their purposes?

A: Yes.

Q: (L) In a lot of cases of paranoid schizophrenia are attachments used?

A: Yes.

Q: (L) Are they perpetuating schizophrenia through genetics?

A: Can. Or mental and emotional. Environmental life experiences.

Q: (L) Why does it not usually show up until adolescence? Is this because adolescents are being abducted and having implants put in?

A: Not necessarily.

Regarding the next excerpt, there were several sessions at this point in time that were held without Frank's presence. They have never been included in the general transcripts because they were primarily about Frank himself, and we thought it was a kindness to protect him from the shock of certain knowledge, all the while encouraging positive interactions in an effort to help him. In the session just prior to this one, we were told that there was danger around Frank. We were not entirely sure if the remarks meant that Frank was *in* danger because of his being unconsciously manipulated by other forces, or if he was a danger *to us*. This was the event that motivated me, once again, to try to save Frank. In this excerpt below, he had just rejoined the group and I wanted to inquire about this information with him present, so that if there was anything he needed to do to protect himself he would be made aware.

04-15-95

Q: (L) The other night when we were working without Frank, we got some information that indicated that Frank was in danger via the government. Is that true, or was that true?

A: Partly.

Q: (L) What is the source of this danger?

A: Source?

Q: (L) I mean, like, the IRS, the FBI, the CIA, or what?

A: Not initialed as such.

Q: (L) Is this physical danger or just harassment danger?

A: Mind attack for purpose of self-destruction.

Q: (L) Is there anything that can be done to shield against this kind of attack?

A: Yes.

Q: (L) What can be done for shielding?

A: Knowledge input on a continuous basis.

Q: (L) And what form should this knowledge take? Does this mean channeled information, books, videos, what?

A: All and other.

Q: (L) A specific other?

A: Networking of information. Now, warning!!! All others will very soon experience great increase of same type of attack, two of you have had episodes in past from same source for similar reasons, but now your association puts you in different category!! Remember all channels and those of similar make-up are identified, tracked, and "dealt with."

Q: (T) Which two have experienced similar types of attack?

A: Up to you to identify for learning. [...] Suicidal thoughts?

Q: (L) Have you had suicidal thoughts? (J) No. (T) Not me. (F) I have had them constantly. (T) Laura, did you? (L) I was pretty damn low. I wasn't contemplating suicide, I was just thinking how nice it would be if we could just turn out the lights and end the illusion. (T) Okay, so we have identified the two, you and Frank. (L) So, in other words, Jan, it is going to get worse. [...] (T) So, we have the knowledge and all we have to do to prevent the attacks from being nasty?

A: You do not have all the awareness you need! Not by any means!

Q: (J) Is one of the reasons why this whole thing between Frank and Laura happened to show us that we could establish, albeit a very weak and very jumbled, connection with the channel without Frank's presence. A sort of verification of the channel's integrity. Was that one of the byproducts of this, or one of the purposes of it?

A: Byproduct is good way of putting it. Remember, all there is is lessons.

Q: (L) The attack was more internal in terms of doubt, not only of the channel and the information, but of the very foundations of existence. I mean, the realization that we may not be at the top of the food chain was shattering. (T) That snowballed on its own once the initial conflict was established. (J) Maybe the way to look at it is: Yes, we went through all this crap, and you and Frank went through all this anguish, but maybe one good thing that came out of it, and maybe wasn't intended, was the fact that, yes, we were able to see that there is a channel separate and distinct from all of us. It is not dependent on any one of us to be present. Yes, we do need to have all of us together for optimum contact....

A: All are able to channel, but practice is required to establish the same extent of grooving, but be aware of ramifications!

Q: (L) What ramifications?

A: Observe "Frank."

Q: (T) We are observing you. (J) Yeah. And? (F) I think what they mean is, when you can channel as I can, because I channel almost continuously, this has a good side and a bad side. Now, the good side you know. The bad side you don't know. The bad side is very hard to live with. I cannot even describe the state of my mind.

It is notable that Frank indicates that his "constant channeling" was so negative, that he thought about suicide constantly. One has to wonder about what he called the "bad side" of channeling that was so "very hard to live with." It is also notable that the Cassiopaeans indicated that this state that Frank was calling "constant channeling" was actually "mind attack," and that the cure was "knowledge input" on a continuous basis, which obviously was *not* happening in his so-called "constant channeling." This implies that, when he was not working with the group, he was not networking to the "knowledge base" which the Cassiopaeans have numerous times described as "light." In short, the many things that can be inferred from the above series of remarks merely confirms the ideas regarding Frank Scott already presented in these pages. It was this test, this seeing through the enormously successful deception perpetrated by Frank, that I was being challenged to "pass."

We have already had a look at the fact that interactions with a psychopath are, as the Cassiopaeans describe it, a "dance" that can be extremely dangerous. It was this "dance" that I was being challenged to discern by "seeing the unseen." One could say that being under the control of a psychopath is rather like dancing with the devil. What most people don't realize is exactly how serious this is in the hyperdimensional sense. To allow oneself to be conned or used by a psychopath is to effectively become part of his "hierarchy" of feeding. To believe the lies of the psychopath is to submit to his "bidding" (he bids you to believe a lie, and you acquiesce), and thus, to relinquish your free will.

In strictly material terms, this doesn't seem to be much of an issue, right? After all, somebody lies to us and who really cares? Is it going to hurt us to just let them lie? Is it going to hurt us to just go along with them for the sake of peace, even if we know or suspect they are lying? After all, checking the facts and facing the psychopath with truth, and telling them "no," is generally very unpleasant. Remember, the game is set up so that we pay a lot for being ethical in dealing with the psychopath. In material terms, it really doesn't seem to be worth it because we suffer all kinds of attack — verbal, psychological, and even physical abuse — so it's just easier to let sleeping dogs lie, right? This is where we again enter the realm of hyperdimensional realities.

It may not seem to mean much in the material world whether or not we let things slide, accommodate a psychopath or two, or engage with them thinking that we are well able to "draw the line," or that when things get "out of hand," we can just walk off the dance floor. We may wish to think that making decisions about our associations with others based on only probable hyperdimensional realities is an "iffy business." After all, we cannot see any of the things that the Cassiopaeans talk about when they discuss 4th density realities. We cannot see "emotional feeding" that is "draining" us or depriving us of the ability to raise our awareness. We cannot see any "tenuous filamentary connections" between ourselves and others that are part of a "hierarchy of control." At best, we can only really penetrate to the level of the psychological reality, observed behavior that is discordant or self-destructive. We are thoroughly programmed to help by giving until it hurts, or trying to fix or make nice.

All of these things, all of these accommodations of psychopathy, on just a practical level, can be seen to "select for psychopathy" in terms of the gene pool. But on the hyperdimensional level, considering the great amount of evidence we have that there is something very mysterious going on that has to do with "controlling the minds of humanity," and covering up something that may affect every single human being on this planet, we find that the issue is crucial. Refusing to accommodate the manipulations and maneuvers of the psychopath may, indeed, be critical to the positive transformation of our planet. In fact, this matter has already been extensively discussed by Michael Topper in his "Precis on The Good and The Evil," published in the now defunct *Thunderbird Journal*. Most interestingly, he discusses it in terms of economy! We thus see the Hermetic connection: As above, so below, and vice versa. We can surmise the agenda above, by observing the dynamics below! (The following excerpts from Topper's article have been edited for clarity and brevity, though it's hard to edit Topper since his word usage is so dense.)

In the case of both Positive and Negative beings of 4th density, the negotiable currency of their transactions is a bio-psychic energy; the mode of both is accumulation, in the sense of storing and putting such bio-psychic energy-capital to work in powering or transforming the centers toward deeper integration and functional unity.

Due to the character of our traditional spiritual and esoteric teachings, we may have a conceptual difficulty understanding how a being of a higher density manages to acquire its status without benefit of any basic heart-development.

Positive beings at 4th density have achieved the necessary intensity and developmental alignment through conscious decisions based on recognition of the abstract propriety of identifiable divine law. [...] Such beings display a recognizably "scientific" approach to spiritual considerations; they openly regard Divine Light as a mensurable magnitude. [...] Their apparent "coldness" or objectivity is only apparent. They register the distress of others and modify their approach accordingly. [...] One need only compare the behavior of truly Negative beings to appreciate the difference. In the famous account of Whitley Strieber [there is] an adequate example. [...]

In Strieber's account, we witness the astonishing effort to transmute those horrific experiences into a positive outline. Thus Strieber, with almost excruciating transparency invokes the standard "humanistic" saw to the effect that dichotomies of good-and-evil are too simplistic and medieval, truth always being some "gray" blend of opposites; in this way he shields from himself the obvious implications of his ongoing ordeal.

But more importantly, he demonstrates to perfection the procedure of how one "falls into the hands" of the Negative beings and, by the denial mechanism of 3rd density psychology, creates the belief that "good" things, developmental things, positively proceed from such ordeals. [...]

His conclusions, his distillates of what he's learned, insist almost schizophrenically that these entities must in some way have the "good of mankind" at heart, but that through the apparent terrorism of their utterly unworldly appearance and vile behavior they function something on the order of "cosmic zen masters," taking a stick to our stubborn skulls. [...] As "proof" of the actually liberating work they're performing, Strieber invokes the fact that owing to his jarring experiences he's "come loose" and is able to sample in waking consciousness the phenomenon of astral travel.

Strieber's inventory of "positive side effects" on the whole describe a definitive list of what would be characterized as distinct inroads in the Negative program of conquest and ultimate Soul-capture. Like diabolical chessmen, Strieber inadvertently shows that the "space beings" have maneuvered and bullied his thoroughly beleaguered psyche into actively choosing the hypothesis with which they've implicitly enveloped him....

In further "defending" his tormentors and interpreting their tactics as a strict but ultimately benevolent discipline, Strieber helpfully displays for us one of the common vulnerabilities on which the Negative tactics count, a kind of hook upon which the Soul is sure to be snagged: the persistent intellectual pride which refuses to be counseled when the counsel seems to touch too close to truth; for any suggestion that his entities are plain evil — that he might be being deceived — seems to cause him to clutch his experiences the more covetously, and guard his interpretation jealously from any who might have a revealing word. He proclaims over and again that no one can explain his experiences to him since they're uniquely his, that anyone with another interpretation *ipso facto* has an ax to grind; and finally, his intellectual superiority makes him uniquely qualified to pioneer this field which he acknowledges sharing with other "abductees," inferentially not so well qualified. [In other words, he invokes the ever popular "it's *my* truth!"]

It is this type of rationalization and self-protective recoil upon which the Negative design counts. It is these psychological properties of 3rd density consciousness which serve all too predictably to convert scenarios of coercion into full volitional acceptance [Stockholm syndrome]. [Strieber] accepts and defends in full Will, like a snapping terrier protecting its bone against all comers; and that is too bad, because by his own account and according to his public history he is a man of gentle instinct and kind, overtly benevolent traits. [...]

The natural question to ask is how, considering factors such as "karma" and psychic "laws" of like attracting like, etc., that an apparently positively-inclined personality such as Strieber should be caught up in the net of Negativity which he details? Isn't his tendency toward "goodness" enough? Is there some unknown element involved in all this which accounts for the seeming collapse of protection that ought to surround a "good man" ? [...]

In *Transformation* Strieber recounts the otherworldly interdiction whereby a "voice" bade him refrain forever from sweets, his one true vice. Addicted as he was, Strieber couldn't stop, even though the "beings" engineered circumstances so as to bombard him with dire implications. As a result, one evening he is visited by a malevolent presence which he himself — as always — describes best, i.e. as "monstrously ugly, so filthy and dark and sinister. Of course they were demons. They had to be." Again, "the sense of being infested was powerful and awful. It was as if the whole house were full of filthy, stinking insects the size of tigers." The entity, rising up beside his bed like a "huge, predatory spider," places something at his forehead and with an electric tingle he is "transported" to a dungeon-like place where his attention is fixed upon a scene of excruciating torture. The victim, a normal looking though quite naked man, is being whipped to shreds amidst agonized screams by a cowled figure. His "entity" explains to him that "he failed to get you to obey him and now he must bear the consequences." This disclosure is followed by a very interesting and significant "assurance" that "it isn't real, Whitty, it isn't real." [...]

The purpose of soothing Strieber with such assurance as to the ultimate unreality of the convincing scene experienced, should be familiar to anyone who's heard of the torture tactics employed in any good Banana Republic (i.e. those in which the victim is subjected to excruciating pain on the one hand while being simultaneously stroked and reassured on the other, often by the same party). The object is to elicit the full cooperation of the victim under duress, by making him instinctively gravitate toward the implicit salvation extended through the "motherly" touch demonstrated in that schizoid Grasp. [...]

Indeed, Strieber proves himself the compliant guinea pig; even having been told that it's all a thought form, his compassion for the unsuccessful "bidder" persists so that finally he collapses into repentant love for the very roaches that bedevil him. "Again, though, I felt love. Despite all the ugliness and the terrible things that had been done, I found myself longing for them, missing them! How was this possible?" Again, "I regretted the contempt I had shown for its [the other reality's] needs and its laws and felt a desperate desire to make amends. [...] I had felt a pain greater than the pain of punishment. It was the pain of their love.... I had the sense that they had on my behalf turned away from perfect love, and that they had done this to help me. [...] I suspect that the ugliness I had seen last night was not them, but me. I was so ashamed of myself that I almost retched."

Giving Love to such a being is a yielding to the Negative requirements.

Should there remain any reluctance to grasp this point, or some desire to conserve the liberal-humanistic proposal to which Strieber often turns (i.e. to call such things truly Negative or Evil is "simplistic," you know) we find a passage in the *Ra Material* that anticipates Strieber's account by years and furnishes a framework before the fact, which not only fits the Strieber-entities' behaviors like a key in a lock, but gives us a needed perspective of evaluation.

On page 21 of *Volume III, The Law of One*, the Ra entity characterizes a prototypical tactic of the [4th density STS], that of "bidding." "Bidding" is described in such a way to make it clear that Strieber's experience represents a concrete instance of the phenomenon.

"Bidding" is *a contest of will, rendering the consciousness that obeys into enslavement through its own free will*. It is a command of obedience, precisely such as that issued without explanation against Strieber's lust for sweets. Its sole purpose is to *bend the subject into accepting* the command, the actual content of the order being largely beside the point. [...] To possess a legion of servants in this way is an actual nourishment to the centers and systems of 4th density; a kind of "food-chain pyramid." [...]

Thus we find the Strieber entity virtually paraphrasing the earlier Ra recitation of the *modus operandi* that identifies the Negative beings — *the failure to exact obedience bears punishable consequence*. It is a continuing illustration of the way in which the Negative polarity extorts the desired obedience — and thus soul capture — *through manipulation of Love*. [This author's emphases]

The higher-density Positive entities are Light beings. The higher-density Negative entities are "Light eaters." Love is Light is knowledge. When they induce belief *against what is objectively true*, they have "eaten" the Light-knowledge of the person who has chosen blind belief over fact! When you believe a lie, you have allowed the eating of your energy of awareness! When you do not take the time and trouble to check things out for yourself, to do the research, to compare, to network, to get a consensus, you have given away your power. You have failed in the creative act of learning.

Such beings are associated with darkness because the Light-knowledge is drawn into the cavernous "black hole" of their congenital emptiness. [...] All the massive, cosmic project they are engaged in, in full consciousness and on the grand scale, is ultimately a means of *"cornering the market" on energy*, monopolizing all the known fields of Light or Light potential. The expanding order they attempt to impose, the totalitarian control over increasingly large numbers they attempt to exert, is the fantastical and internally self-contradictory project of *coercing everything in creation to work for them*, to cultivate and keep the fields of their energy-reserves and to furnish self-replenishing "herds" of emotional source-nutriment which can be converted into useful energy or *Light-capital*. Since the Negative beings can't generate an important Light-energy source themselves, they use the reserves of the beings effectually harnessed in thrall to them. [...]

There is an immediate psychic bond produced by belief. There is an instantaneous linkage and interpenetration with the individual who has chosen to believe a lie. The higher-dimensional beings have subtle, vertical filamental axes fixed on human beings. Those subtle nerve-networks process radiant-energy values, drawn in through the etheric "chakras" of the higher-dimensional systems, represented by the pineal/pituitary glands.

The network of the STS hierarchy extends in myriad psychic webs of specialized powers, forces and functions like a voracious net flung across the heaven of stars, the sum energy comprising the group consciousness of that net redounds to the basic benefit of the Being at the Apex of the control pyramid. This apex is composed of the most persistently Negative being — the one who has stuck it out against all evidence of progressively-diminishing returns. This being can be described as the Desolate One, a being who most directly embodies and promotes the ultimate Negative objective.

The consciousness of that being is literally fed and magnified by the number and relative strength of the subordinate souls who have been voluntarily subsumed to the network. The greater the development of the psychic potentials of the individual who has been co-opted, the more "energy" he contributes to the whole system. The more psychic energy available to the "Commander" of the Negative soul hierarchy, the greater his effective power to co-opt even more potent and more difficult-to-capture souls.

The "contributions of consciousness" consist effectively of the energy a soul would otherwise utilize to encompass objective knowledge. Each time they choose a lie over the effort required to dig down to the truth, or the effort required to adjust their own psyche to adapt to Truth, that "Love" energy is effectively transferred to the individual who is producing the lie in which they are believing without effort on their part to ascertain its truth for themselves.

Since the flow of contributions is a vertically-hierarchic flow, it may be seen that development of the mechanical psychic "food pump" — the opening and development of the vital-psychic powers of the field troops — contributes energy-sums upwardly to the "cortical" station occupied by the Commander, and serve to literally amplify the Intelligence, the effective Presence of Wisdom (negatively influenced in this case). *The use of chemicals to achieve higher states of awareness, especially those related to the Pineal gland — the "outflow valve," are emphasized in the teachings of the Negative hierarchy.*

The subordinates of the Negative hierarchy are all connected like tiers of an immense structure, functioning as regimented extensions and mind/body "parts" — organs and processes serving the Negative agenda from their respective levels — of the Overarching Apex, the "Eye of the Pyramid." The beings at the different tiers do not perceive the object of the Ultimate Objective because it is a characteristic of the Negative hierarchy to deliberately mask and distort that which is higher and more comprehensive from that which is lower and more "specialized."

In the Positive realms, it is more intrinsically possible for the lower levels to perceive the objective and functions of the higher levels with a minimum of distortion because of the characteristic of the Positive realms to share and exchange rather than producing a one way channel of energies to the top, so to say.

As far as the beings of the lower levels of the STS hierarchy are aware, their objective is to crystallize, *under artificial pressure of economics*, a global power elite structure of indigenous and alien beings that will completely dominate the earth and its inhabitants, thus making it a captured negative resource to "feed" the hierarchy.

This is what the 4th density STS beings perceive. This is their goal. This is all they know: ruthlessness and domination, power and control. The *real* objective is, of course, concealed from them. The even higher density tiers of the Negative hierarchy deliberately distort and mask the character of their objectives from the lower planes of "subordinate functionaries" that are viewed by them as a regimented machinery of beings serving as expendable cogs in an Engine of Conquest.

At the highest levels, the objective is to convert all energies and specialized powers of the Negative troops into potable information patterns of a far-flung Matrix contributing to the knowledge amplification and awareness of the pyramidal generals, commanders, and overlords. The fruits of those energies and powers of the "pawns" in terms of actual ground gained, real elements subdued, contributes to the progressive compounding of highly integral control for the very topmost echelon of the Negative hierarchy, enabling them to even further expand the field that comes under their regulatory jurisdiction, thereby ensuring progressively more voluminous "farms" of energy-nutriment on which to vampirically suckle.

There is a consequence to this one sided intake of Light energy, and that consequence is only progressively manifested in the higher densities.

At 3rd density, the process of perception is a continuous two-way circulation of Light values transmitting noetic patterns of the environment through the locus corresponding to the pineal-pituitary glands. Coded Light values charged with the psychic imprint of Nature come in and are "mixed" with the radiant energy substances taken in through food and other sensory impressions. These Light values generally are taken in via the sensory system and are transmitted out via the Pineal gland with the "charge" of the "identity impressions" of the individual. As a rule, there is a lot of waste because of the comparatively low body/mind integration.

In order to move to the higher order of circulating and incorporating Light with greater efficiency, purer impressions are required. "Purer" means more objective. The less "twist" or "spin" put on the Light impression by the receiving entity, the purer it is. This purity or objectivity of impressions is optimized in correspondence to a deep, unitive apprehension of consciousness.

This term "unitive" does not mean that the "other" is perceived exactly *as* self, but rather that all others are perceived as Light-energy units to be conserved, aligned, balanced, as part of a whole structure, rather than something to be consumed in a one-way discharge to a single typology. This "unitive consciousness" promotes the cognitive field of the Whole-being, as being wholly of value through all patterns and forms. It perceives the sum of the different identities as possessing an *axis* of Conscious Identity, expressed by endless forms of existence/creation. It is thus an ecstatic, celebration of the congruence of Love without any desire or need to "possess" or "eat" or force a "one way flow" of being. The Positive forces, the STO beings at higher levels are actively working to sustain the Creative fields toward achievement of the Positive Logoic purpose which is the realization of Absolute Consciousness under all conditions and in all forms.

However, the Negative hierarchy is oriented toward the consuming of radiant Light energy in a one way flow. The progressive power that devolves from the "capture" and incorporation of radiant Light sources serves to feed and enhance an exclusive subjectivity of consciousness since the effort is toward the subordination of all things to the magnified narcissism belonging to devout ego consciousness.

At the higher levels, the absorptive framework of self-enhanced ego consciousness takes on a severe functional contraction and effective withdrawal from interest/involvement in the created fields of being, maintaining a minimalist interaction with only its closest contacts in the pyramid.

Thus, the Negative being of higher densities takes on the configuration of a forebodingly lonely presence, lurking in caves and desolate grottos of the astrophysical realms. It becomes a fiercely mental entity of 5th density power-knowledge, possessing the proverbial basilisk gaze and only turning the stream of its attention "away" from that intensified/contractile self-absorption toward the created worlds in token deference of the need to canalize the funneling food source — sucking vitality from the extravagances and pastimes comprising the follies of the created worlds, imbibing the "Light units" to insure the uninterrupted power that it needs, the inconceivable "wattage" required, to maintain that monumental self-absorption and narcissistic self-luminance of the Negative Ego-Postulate — the Anti-Logos, the Selfness of Consciousness.

So it is that the Anti-Logos cannot simply withdraw from the worlds of creation — it must absorb them into itself — it feels the necessity of undoing creation — it *needs* that energy to fuel its infinite self-contemplation.

This is the Ultimate Objective of the Being at the Apex of the pyramidal food chain. And this is why its agenda is masked in the lower levels of the hierarchy. Until such lower level Negative beings have consumed sufficient energy — a sort of critical mass — to trigger the implosion of such extraordinary self-concentration, they are only interested in destroying that which resists their domination and preserving a vital minimum of captured resources so as to possess an ongoing supply of nutriment. They wish to control, or freeze the rate of planetary destruction and disintegration so as to technically conserve the intelligent life-form in a tractable state so as to render it a good "servant." [This author's emphases]

This is where Game Theory is so important. It is also a matter of economics, as the reader will have noted. The 4th density STS beings and their human cohorts perceive that there is no advantage in total destruction, because there can be no profit. They seek to establish an optimum ratio that is just a delicate balance between diminishing returns and unbridled death — Alternative 3 and the Sixth Extinction.

The objective — the ultimate total destruction that is desired by the Being at the Apex — the total consumption of even its minions — is masked from the higher density Negative beings below it. The lower levels may "feel" or "sense" the looming black hole of absolute annihilation as they mount through the hierarchy, but the immediate pleasures of their feeding-frenzies keep them occupied, reinforcing the discouragement of the idea of looking any deeper into the Heart of Darkness, which has no qualms about making food out of even *them*!

Because the STO contingent honors and conserves the realization of consciousness in *all forms and under all conditions*, it has greater resources to sustain creation. It has access to an even more powerful allegiance of unified conscious resolve and collective intent, so that the Negative hierarchy is met with resistance at every turn. This STO resistance is a great deal more in harmony and alignment with the Logoic pattern, so that it possesses the sum value of the Creative as its enforcement.

The Negative hierarchy, however, because of its intrinsic nature, must "borrow" the creative extensions and waking tools-of-being as much as anything or anyone else, while at the same time being antagonistic toward them! In short, its intrinsic nature is that of self-loathing, from which it must continually flee, requiring more and more energy to protect itself from its own "truth."

However, all of this is the Theological Reality of the hyperdimensional game of chess being played on our planet. When the Negative hierarchy succeeds in conquering a planet or even a galaxy — as it occasionally does — it only accomplishes this by a long cerebral tournament of moves and counter-moves, plays of the most subtle and surreptitious type in which the idea is always to draw on the given Positive elements and attributes of the game board and progressively co-opt them, slyly compromising their positive effectiveness and gradually integrating their characteristic moves to deviant patterns, secretly optimizing the Negative potential — either by neutralizing their Positive effectiveness, or actually "taking them over" by progressive, imperceptible distortions of the straight-and-true alignment until they add their own dimension of deliberate of conscious negativity to the overall strategic setup.

The Negative forces can demonstrate remarkably far-sighted restraint. If one group has been effectively captured and could be completely annihilated, instead, the game will be preserved with the conquerors holding their positions intact, poised to parlay their gains into even greater Negative glory of "galactic conquest." This is just superior strategy, trying to include as much as can be included at once so that a comparatively larger portion of the multidimensional cosmos can be wiped out in the twinkling of an eye.

What we find, at the end, is that the STS consciousness *must* subtract Love from the equation. It is only through a lack of Love that an individual is moved to suppress or control or convert another person. Love expresses the eternal condition that, within the unqualified identity of the Absolute, *all* qualifications are allowed by largesse of Unlimited Potential. Love is the Power of the Absolute that makes it Absolute and unqualified, by permitting even the apparent antithesis of the STS hierarchy.

Reality in its fullness, by its very nature, surpasses the STS hierarchy in an ultimate sense. The STO perspective pairs Love with Infinite Potential — *the endlessly proliferating properties of Creation in all its forms and varieties. The STO perspective celebrates differences in form, while acknowledging identity in consciousness.* It is for this reason that the STS hierarchy is "bitter" toward Love — because Love *allows and accepts that attention can be shared, and STS craves exclusivity.*

Love is that which moves out into the multiplying streams of Creative potential, all the while retaining that awareness of the dawn state of infinite possibility — non-anticipation. Love is that which knows that no matter how many times the Logos is subdivided or multiplied, it remains intact. Love is consciousness that knows it may be provisionally focused on chosen delimiting attributes, but it also knows that it is never diminished.

But the STS hierarchy sees this Creative Expression as a "flirtatious wanton" whoring after other "gods." It sees its job as "arresting" the indiscriminate proliferation and freedom-granting bounty of Love. It wishes to "capture" the Mother, to keep her under lock-and-key, to utilize her power by appropriating her means, mimicking her actions and functions while strategically altering them and incorporating them into a restricted simulacrum suited to its own ends.

STS forces, remember, have no power of Creativity; they cannot generate work of their own. They need the power of the Mother to do that. That is why they capture women with power, keep them half-alive so as to maintain a minimal continuity of creative interaction and the suitable production of form.

It is through tailored regulation of Love's forms that STS intelligence derives the means of coercing soul-energy into converting abstract-conscious capital to specific psychic and emotional coinage. It is through the cumulative psychic and emotional energy that the STS forces hope to obtain the energy-keys to timelocks and spatial corridors of even richer and still-virgin terrains, portions of the cosmos intact with creative life, ripe for plunder and privileged profit.

> The absolute trust my mother's strong relationship imposed on my psyche. Do I wish to master a woman sufficiently so that she will take care of me as my mother did?
>
> I'm slowly beginning to realize the enormity of the problem which my development has created in respect to women. The interaction with Rita is just an example of how difficult, even at that age and with such a magnificent partner, any final linking is to be. Judy provided in her striking beauty a repository for always wandering projections, and the strength of our deathlike struggle is a good indication of how impossible my quest is to be. I refuse to admit the inevitable — that I can live without a woman (my mother). Until I accept this my productivity will be intense, like my countless infatuations, but sporadic. *I'm faced with a hell that is somewhat relieved by my incredible energy which is so capable of constantly creating that joy which is deeper than sorrow.* (Ira Einhorn, quoted by Levy, 1988; this author's emphasis)

We see that the ultimate aim of the psychopath, as living representatives of the STS hierarchy, is to *master* creative energy. To assimilate it to the self, to deprive others of it by inducing them to believe lies. Because, when you believe the lie of the psychopath, you have given him control of your free will — the essence of creativity.

The planetary entity is the focal point of a specific density of mind/body interaction. At certain cosmic moments, or "crossroads," such a planetary entity may be scheduled to polarize into a higher density. The Negative hierarchy sees this as a "ripe moment" to induce that polarization to take place negatively, so that the planetary entity will participate wholly in the Negative 4th density reality rather than the Positive reality. Negatively polarized beings require a negatively-polarized planetary base from which to function, just as higher-density Positive beings need positively-polarized planetary bases.

The Hermetic maxim again: economics of Light energy above, and economics of control of minds and will below. They want to use humanity's own creative energy to "lock" our planet under their domination.

What we see now in terms of the diminishing resources of our planet is that the intensified UV bombardment of our atmosphere is *not* an "unfortunate but inevitable byproduct of industrialization", *it is part of the deliberate, covert effort of the Negative hierarchy to prepare the biochemical and electrical composition of this planet for Negative polarization.*

There *are* such things as "evil planets" and dark stars. The real question at this time is: *Is Mother Earth about to become one?*

Appendix A
Meteorites, Asteroids and Comets: Damages, Disasters, Injuries, Deaths and Very Close Calls

Astronomy books and papers far too numerous to cite offer the assurance that "no one has ever been killed by a meteorite." (John S. Lewis, University of Arizona)

Over the past few years, while SOTT.net has been tracking the increasing flux of fireballs and meteorites entering the earth's atmosphere, we have been, by turns, amused and horrified at the ignorant reactions and declarations that issue from academia and the media regarding these incursions. A few years ago, we read that "this is a 'once in a hundred years' event!" Not long after it was a "once in a lifetime" event. Still later, after many more incidents it became a "once in a decade" event. More recently, it has been admitted in some quarters that meteorites hit the ground (as opposed to safely burning up in the atmosphere) several times a year!

We have discovered the fact that the governments of our planet are well aware that there are atmospheric explosions from such bodies numerous times a year. We have also learned that the frequent reports of unusual booms and shaking of the ground is often due to such overhead explosions. Yet the media steadfastly refuses to honestly address this issue, offering instead a plethora of articles presenting opposing academic arguments designed to put the populace back to sleep, reassuring them that there is nothing to worry about. We most certainly can see that the issue of meteorite, cometary and asteroid impacts on our planet, and their true potential danger to each and every one of us, must be added to this list of unfunded research. This is a very bad and dangerous state of affairs. As Victor Clube wrote in his letter to SOTT.net:

First, I should say your references to the (cosmically complacent) paleoclimate community and to my otherwise unread narrative report to the USAF European office strike a very considerable chord with me. After all, neither Ms. Victoria Cox nor your good self can be aware how very much Bill and I had reason to appreciate the timely injection of USAF funds at a time when *the line of research we championed appeared to be successfully closed down by the UK scientific establishment. Thus we were both in turn obliged to relinquish our career posts at the Royal Observatory, Edinburgh on account of this line of research* — which gave rise to our reincarnation at a more tolerant haven, namely my alma mater (Oxford).

Also, whilst I broadly accept your commentary regarding the role of "national elites" in the face of near-Earth threats, I am quite certain the elites in practice currently know VERY "much LESS than they let on" and that the situation for humanity is dire. Any comfort you may draw from the opposite opinion seems to me to be entirely misplaced. Thus although the globally modest efforts to assess the NEO threat with telescopes by a few semi-enlightened national administrations (e.g. USA) or by a few private enterprises (e.g. Gates) are certainly to be commended, I look upon this aspect of the NEO threat as basically intermittent and therefore more or less symbolic, so far as generally more urgent and still largely undetected low-mass NEO flux (which is demonstrably climatological in its effect) is concerned. This particular threat (evidently responsible for our planet's evolving glacial/interglacial condition during the past three million years) is of course FUNDAMENTALLY ignored by the current Body Scientific and hence by most of humanity as well. [This author's emphases]

I have prepared a list, by no means exhaustive, of all the incidents I have been able to uncover of meteorite, asteroid, or cometary impacts that have caused death and destruction, property damage, or were near misses. Major parts of this list are extracted from the work of John S. Lewis, Professor of Planetary Sciences at the Lunar and Planetary Laboratory, Co-director of the NASA/University of Arizona Space Engineering Research Center, and Commissioner of the Arizona State Space Commission; in specific, his books entitled *Rain of Iron and Ice* and *Comet and Asteroid Impact Hazards on a Populated Earth*. In this latter volume, he writes:

The most intensively studied impact phenomenon, impact cratering, is of limited importance, due to the rarity and large mean time between events for crater-forming impacts. *Almost all events causing property damage and lethality are due to bodies less than 100 meters in diameter, almost all of which, except for the very largest and strongest, are fated to explode in the atmosphere. […]* [W]e are forced to conclude that the complex behavior of smaller bodies is closely relevant to *the threat actually experienced by contemporary civilization.* [This author's emphases]

Based on the data he collected, Lewis noted that:

[O]n the century time scale, *firestorm ignition and direct blast damage by rare, strong, deeply penetrating bodies are the most common threats to human life,* with average fatality rates of about 250 people per year. [...] On a 1000-year scale, the most severe single event, which is usually a 10 to 100-megaton Tunguska-type airburst, accounts for most of the total fatalities. On longer time scales, regional *impact-triggered tsunamis* become the most dangerous events. [...] The exact impactor threshold size for global effects remains poorly determined. [...] Perhaps most interesting is the implication that the large majority of lethal events (not of the number of fatalities) are *caused by bodies that are so small, so faint, and so numerous that the cost of the effort required to find, track, predict, and intercept them exceeds the cost of the damage incurred* by ignoring them. [This author's emphases]

Unfortunately, Professor Lewis did not have at hand the information presented by Mike Baillie in his book *New Light on the Black Death*, nor did he consider the global events of 12,000 years ago revealed by the work of maverick scientists, Firestone, West and Warwick-Smith. If he had added the estimated numbers of fatalities from those events into his calculations, he might not have decided that the small, faint, and numerous bodies were so easily ignored. I think that if *all* the data were plugged in, the average deaths per year would be a lot higher than 250. Regarding impacts from history, Lewis writes in *Comet and Asteroid Impact Hazards on a Populated Earth*:

Many ancient sources from many cultures treat comets as literal, physical harbingers of doom. Such phenomena as the burning of cities and the overthrow of buildings and walls by aerial events are mentioned many times in Latin, Greek, Hebrew, and Chinese records, but there is no evidence of physical understanding of the nature of the bombarding objects or their effects until quite recently. [...] There is indeed a language problem in understanding the ancient reports, but it is largely a matter of the lack of an appropriate technical vocabulary in the older writings. [...] *In certain locations and periods, especially in medieval Europe, all unusual heavenly events were interpreted as signs sent by God. Therefore, the surviving accounts are strongly biased toward explaining the moral purpose of these events, not their physical nature.* Such fundamental information as exact date and time, exact location, place of appearance of the phenomenon in the sky, its duration and physical extent, luminosity, precise nature of the damage done, and the like, were generally regarded as unimportant, and therefore rarely recorded for posterity. [...] Even in 20th-century newspapers, bolide explosions may be described (and indexed) as "mysterious explosions," aerial blasts, aerolites, aeroliths, bolides, earthquakes, fireballs, meteorites, meteors, shocks, thunder, and so on. [...] Reports of meteorite falls, often with consequent damage, extend back to the fall of a "thunderstone" in Crete in 1478 BC, described by Malchus in the *Chronicle of Paros*. The earliest Biblical source is the account of a lethal fall of stones in... Joshua 10:11. [...] Other ancient reports in the West are found in the writings of Pausanius, Plutarch, Livy, Pindar, Valerius Maximus, Caesar, and many others. The report of a great fall of black dust at Constantinople in 472 BC, perhaps the result of a high-altitude airburst, is documented by Procopius, Ammianus Marcellinus, Theophanes, and others.

Colonel S. P. Worden has called to my attention the following passage in *The History of the Franks*, written by Bishop Gregory of Tours: "580 AD in Louraine, one morning before the dawning of the day, a great light was seen crossing the heavens, falling toward the east. A sound like that of a tree crashing down was heard over all the countryside, but it could surely not have been any tree, since it was heard more than fifty miles away... the city of Bordeaux was badly shaken by an earthquake... *a supernatural fire burned down villages* about Bordeaux. It took hold so rapidly that houses and even threshing-floors with all their grain were burned to ashes. Since there was absolutely *no other visible cause of the fire*, it must have happened by divine will. The city of Orleans also burned with so great a fire that even the rich lost almost everything."

Astronomers who have sought documentary evidence of ancient astronomical phenomena (eclipses, comets, fireballs, etc.) have found that East Asian records are far superior to European records for many centuries. Kevin Yau has searched Chinese records and found many reports of deaths and injuries (Yau *et al.*, 1994). The Chinese records of lethal impact events include the death of ten victims from a meteorite fall in 616 AD, an "iron rain" in the O-chia district in the 14th century that killed people and animals, several soldiers injured by the fall of a "large star" in Ho-t'ao in 1369, and many others. The most startling is a report of an event in early 1490 in Ch'ing-yang, Shansi, in which many people were killed when stones "fell like rain." Of the three known surviving reports of this event, one says that "over 10,000 people" were killed, and one says that "several tens of thousands" were killed.

On September 14, 1511, a meteorite fall in Cremona, Lombardy, Italy, reportedly killed a monk, several birds, and a sheep. In the 17th century we find reports of a monk in Milano, Italy, who was struck by a meteorite that severed his femoral artery, causing him to bleed to death, and of two sailors killed on shipboard by a meteorite fall in the Indian Ocean.

In addition to these shipboard fatalities, there have been several striking accounts of near disasters involving impacts very close to ships. Near midnight of February 24, 1885, at a latitude of 37 degrees N and a longitude of 170 degrees 15 minutes E in the North Pacific, the crew of the barque Innerwich, en route from Japan to Vancouver, saw the sky turn fiery red: "A large mass of fire appeared over the vessel, completely blinding the spectators; and, as it fell into the sea some 50 yards to leeward, it caused a hissing sound, which was heard above the blast, and made the vessel quiver from stem to stem. Hardly had this disappeared, when a lowering mass of white foam was seen rapidly approaching the vessel. The noise from the advancing volume of water is described as deafening. The barque was struck flat aback; but, before there was time to touch a brace, the sails had filled again, and the roaring white sea had passed ahead."

A strikingly similar event occurred only two years later on the opposite side of the world. Captain C.D. Swart of the Dutch barque J.P.A. reported in the *American Journal of Meteorology* 4 (1887) that, when sailing at 37 degrees 39 minutes N and 57 degrees W, at about 5pm on March 19, 1887, during a severe storm in which it was "as dark as night above," two brilliant fireballs appeared as in a sea of fire. One bolide "fell into the water very close alongside the vessel with a roar, and caused the sea to make tremendous breakers which swept over the vessel. *A suffocating atmosphere and perspiration ran down every person's face on board and caused everyone to gasp for fresh air. Immediately after this, solid lumps of ice fell on deck, and everything on deck and in the rigging became iced, notwithstanding that the thermometer registered 19 degrees C.*" [...] Next, according to the *Times*, on September 13, 1930, a fireball plunged into the sea near Eureka, California, barely missing the tug Humboldt, which was towing the Norwegian motorship *Childar* out to sea. It requires little imagination to appreciate that such an event, if it were to strike a ship, should easily cause fatalities, or even the loss of the vessel with all hands. (Lewis, 1999) [This author's emphases]

Now, that just gives you a taste of what is to come.

THE LIST: Damages, Disasters, Injuries, Deaths, and Very Close Calls

10000–11000 BCE (10700) – The earliest disaster we know of from our historical or mythic records is, of course, the legendary Deluge of Atlantis. The description of the end of Atlantis given by Plato in the *Timaeus* and *Critias* dialogues bears striking resemblance to what many scientists are now agreed would be the inevitable result of an oceanic impact by a disintegrating comet or large asteroid. The resultant "tsunami," or tidal waves, would easily reach 2,000-feet high as they approached land, wiping out any and all coastal settlements. The deluge traditions, of which there are literally hundreds worldwide, appear in this light to be variations on Plato's account, and could even be actual observation-based tales, eye-witness accounts of the same, or similar, events. This is very likely the event discussed by Firestone, West and Warwick-Smith in *The Cycle of Cosmic Catastrophes: How a Stone-Age Comet Changed the Course of World Culture*. As I have discussed in my book, *The Secret History of the World*, the North and South American continents in the Western Hemisphere fit all the descriptions of "Atlantis," and it is very likely that the event that led to the extinction of about 30 species of large mammals about 12,000 years ago was the source of the legends of Atlantis and probably the legends of a global deluge, such as Noah's Flood. Let's look at some descriptions of what such an event can do.

Back in the 1940s, Dr. Frank C. Hibben, Professor of Archeology at the University of New Mexico, led an expedition to Alaska to look for human remains. He didn't find human remains; he found miles and miles of icy muck just packed with mammoths, mastodons, and several kinds of bison, horses, wolves, bears and lions. Just north of Fairbanks, Alaska, the members of the expedition watched in horror as bulldozers pushed the half-melted muck into sluice boxes for the extraction of gold. Animal tusks and bones rolled up in front of the blades "like shavings before a giant plane". The carcasses were found in all attitudes of death, most of them "pulled apart by some unexplainable prehistoric catastrophic disturbance." (Hibben, Frank, *The Lost Americans*. New York: Thomas & Crowell Co. 1946)

The killing fields stretched for literally hundreds of miles in every direction. [*ibid*] There were trees and animals, layers of peat and moss, twisted and tangled and mangled together as though some Cosmic mixmaster sucked them all in circa 12,000 years ago, and then froze them instantly into a solid mass. (Sanderson, Ivan T., "Riddle of the Frozen Giants", *Saturday Evening Post*, No. 39, January 16, 1960.)

Just north of Siberia entire islands are formed of the bones of Pleistocene animals swept northward from the continent into the freezing Arctic Ocean. One estimate suggests that some ten-million animals may be buried along the rivers of northern Siberia. Thousands upon thousands of tusks created a massive ivory trade for the master carvers of China, all from the frozen mammoths and mastodons of Siberia. The famous Beresovka mammoth first drew attention to the preserving properties of being quick-frozen when buttercups were found in its mouth.

What kind of terrible event overtook these millions of creatures in a single day? The evidence suggests an enormous tsunami raging across the land, tumbling animals and vegetation together, to be finally quick-frozen for the next 12,000 years. But the extinction was not limited to the Arctic, even if the freezing at colder locations preserved the evidence of Nature's rage.

Paleontologist George G. Simpson considers the extinction of the Pleistocene horse in North America to be one of the most mysterious episodes in zoological history, confessing "no one knows the answer." He is also honest enough to admit that there is the larger problem of the extinction of many other species in America at the same time. (Simpson, George G., *Horses*. New York: Oxford University Press, 1961) The horse, giant tortoises living in the Caribbean, the giant sloth, the saber-toothed tiger, the glyptodont and toxodon, these were all tropical animals. These creatures didn't die because of the "gradual onset" of an ice age; "unless one is willing to postulate freezing temperatures across the equator, such an explanation clearly begs the question." (Martin, P. S. & Guilday, J. E., "Bestiary for Pleistocene Biologists", *Pleistocene Extinction*, Yale University, 1967)

Massive piles of mastodon and saber-toothed tiger bones were discovered in Florida. (Valentine, quoted by Berlitz, Charles, *The Mystery of Atlantis*, New York, 1969) Mastodons, toxodons, giant sloths and other animals were found in Venezuela quick-frozen in mountain glaciers. Woolly rhinoceros, giant armadillos, giant beavers, giant jaguars, ground sloths, antelopes and scores of other entire species were all totally wiped out at the same time, at the end of the Pleistocene, approximately 12,000 years ago.

This event was global. The mammoths of Siberia became extinct at the same time as the giant rhinoceros of Europe, the mastodons of Alaska, the bison of Siberia, the Asian elephants and the American camels. It is obvious that the cause of these extinctions must be common to both hemispheres, and that it was not gradual. A "uniformitarian glaciation" would not have caused extinctions because the various animals would have simply migrated to better pasture. What is seen is a surprising event of uncontrolled violence. (Leonard, R. Cedric, Appendix A in "A Geological Study of the Mid-Atlantic Ridge", Special Paper No. 1, Bethany: Cowen Publishing, 1979) *In other words, 12,000 years ago, something terrible happened — so terrible that life on earth was nearly wiped out in a single day.*

Harold P. Lippman admits that the magnitude of fossils and tusks encased in the Siberian permafrost present an "insuperable difficulty" to the theory of uniformitarianism, since no gradual process can result in the preservation of tens of thousands of tusks and whole individuals, "even if they died in winter." (Lippman, Harold E., "Frozen Mammoths", *Physical Geology*, New York 1969) Especially when many of these individuals have undigested grasses and leaves in their belly. Pleistocene geologist William R. Farrand of the Lamont-Doherty Geological Observatory, who is opposed to catastrophism in any form, states: "Sudden death is indicated by the robust condition of the animals and their full stomachs... the animals were robust and healthy when they died." (Farrand, William R., "Frozen Mammoths and Modern Geology", *Science*, Vol.133, No. 3455, March 17, 1961) Unfortunately, in spite of this admission, this poor guy seems to have been incapable of facing the reality of worldwide catastrophe represented by the millions of bones deposited all over this planet right at the end of the Pleistocene. Hibben sums up the situation in a single statement: *"The Pleistocene period ended in death. This was no ordinary extinction of a vague geological period, which fizzled to an uncertain end. This death was catastrophic and all inclusive."* (Hibben, *op. cit.*) [Excerpted from Laura Knight-Jadczyk, *The Secret History of The World*]

Firestone, West and Warwick-Smith write:

> Until recently, the astronomical mainstream was highly critical of Clube and Napier's giant-comet hypothesis. However, the crash of comet Shoemaker-Levy 9 on Jupiter in 1994 has led to a change in attitudes. The comet, watched by the world's observatories, was seen split into 20 pieces and slammed into different parts of the planet over a period of several days. A similar impact on Earth, it hardly needs saying, would have been devastating.

The Carolina Bays date to this time. The Carolina bays are mysterious land features often filled with bay trees and other wetland vegetation. Because of their oval shape and consistent orientation, they are considered by some authorities to be the result of a vast meteor shower that occurred approximately 12,000 years ago. What is most astonishing is the number of them. There are over 500,000 of these shallow basins dotting the coastal plain from Georgia to Delaware. That is a frightening figure.

Let me repeat: *There are over 500,000 of these shallow basins.*

> Unlike virtually any other bodies of water or changes in elevation, these topographical features follow a reliable and unmistakable pattern. Carolina Bays are circular, typically stretched, elliptical depressions in the ground, oriented along their long axis from the Northwest to the Southeast. [T]hey are further characterized by an elevated rim of fine sand surrounding the perimeter. [...]
>
> Robert Kobres, an independent researcher in Athens, Georgia, has studied Carolina Bays for nearly 20 years in conjunction with his larger interest in impact threats from space. His recent self-published investigations have profound consequences for Carolina Bay study and demand research by academia as serious, relevant and previously unexamined new information. The essence of Kobres' theory is that *the search for "debris," and the comparison of Bays with "traditional" impact craters, falsely and naively assumes that circular craters with extraterrestrial material in them are the only terrestrial evidence of past encounters with objects entering earth's atmosphere.*
>
> Kobres goes a logical step further by assuming that forces associated with incoming bodies, principally intense heat, should also

leave visible signatures on the earth. And, finally, that *physics does not demand that a "collision" of the bodies need necessarily occur to produce enormous change on Earth.* To verify that such encounters are possible outside of the physics lab, we need look no further than the so-called "Tunguska Event."

At the epicenter of the explosion lay not a large crater with a "rock" in it, as might be expected, but nothing more than a number of "neat oval bogs." The Tunguska literature generally mentions the bogs only in passing, since the researchers examining the site failed to locate any evidence of a meteorite and went on to examine other aspects of the explosion. (*The Secret History of The World*)

Now, how many human deaths ought we to assign to this event? As Firestone *et al* discuss, it was global in effect and the evidence of a sharply reduced population of not only animals, but humans, is there in the geological record. But what was the total human population? What kind of numbers can we plug into Lewis' calculations? Frankly, we don't know. Undoubtedly, multiplied millions of human beings perished at that time, along with the extinction of many animal species. One thing that seems certain is that if these numbers were included in Lewis' assessment, it would make a significant change in the "average number of deaths per year." Though, of course, this was a very big event, and those don't happen every year, or even every century. They happen on a scale of thousands of years and there hasn't been one like that for 12,000 years.

9100 BCE – Extinction of the woolly mammoths.

7500 BCE – Brings ice age to an end.

5900 BCE – Metals found smelted "naturally," which gives rise to humans smelting metals.

4300 BCE – Metals smelted naturally; beginning of Homeric "religions." Possible time for event on which part of the Exodus story is based.

3195 BCE – Eco-disaster as shown in tree rings. What evidence is there then that something unusual happened around 3100 BCE, other than the Mayan year-zero, supposedly relating to 3114 BCE? We have the construction of Newgrange in Ireland; floods in the paleoclimatic data; the construction of Stonehenge number one; the unification of Egypt; methane peak fires; cold time, according to bristlecone pines; and the erection of the coastal menhirs in Brittany. Although any one of these in itself would not be unusual, the timing of them within a frame of only 100 years is what makes us suspect that something unusual was going on. The next 1,000 years or so were a very restless time globally:

The postulated bombardments and dust-veils at around 3195 BCE, another narrowest tree-ring date, would have wreaked havoc on both the local and global climate, and any and all cultures affected would have taken many decades, maybe even centuries, to recover. The sheer terror that "multiple-Tunguska-class fireballs" would have instilled into the peoples of those times would have understandably motivated them towards building some form of observatories to help predict future meteor showers/storms as a matter of perceived urgency. . (*The Secret History of The World*)

Stonehenge may very well have been built to help in the watch for comets. And, yet again, we have no numbers of human fatalities to plug into the calculations, but they must have been enormous.

3123 BCE, June 29 – Germany: "The clay tablet that tells how an asteroid destroyed Sodom 5,000 years ago".[37]

A clay tablet that has baffled scientists for more than a century has been identified as a witness's account of an asteroid that destroyed the Biblical cities of Sodom and Gomorrah 5,000 years ago.

Researchers believe that the tablet's symbols give a detailed account of how a mile-long asteroid hit the region, causing thousands of deaths and devastating more than one million sq km (386,000 sq miles).

The impact, equivalent to more than 1,000 tons of TNT exploding, would have created one of the world's biggest-ever landslides.

The Old Testament story describes how God destroyed the "wicked sinners" of Sodom with fire and brimstone but allowed Lot, the city's one good man, to flee with his family.

The theory is the work of two rocket scientists — Alan Bond and Mark Hempsell — who have spent the past eight years piecing together the archaeological puzzle.

At its heart is a clay tablet called the Planisphere, discovered by the Victorian archaeologist Henry Layard in the remains of the library of the Royal Palace at Nineveh.

Using computers to recreate the night sky thousands of years ago, they have pinpointed the sighting described on the tablet — a 700 BC copy of notes of the night sky as seen by a Sumerian astrologer in one of the world's earliest-known civilisations — to shortly before dawn on June 29th in the year 3123 BC.

Half the tablet records planet positions and clouds, while the other half describes the movement of an object looking like a "stone bowl" travelling quickly across the sky.

37

http://www.dailymail.co.uk/pages/live/articles/news/news.html?in_article_id=551010&in_page_id=1770

The description matches a type of asteroid known as an Aten type, which orbits the Sun close to the Earth. Its trajectory would have put it on a collision course with the Otz Valley. [In Germany, in other words. In short, the story wasn't about Abraham and Lot in Palestine!]

"It came in at a very low angle — around six degrees — and then clipped a mountain called Gaskogel around 11 km from Köfels," said Mr. Hempsell.

"This caused it to explode — and as it travelled down the valley it became a fireball.

"When it hit Köfels it created enormous pressures which pulverised the rock and caused the landslide. But because it wasn't solid, *there was no crater*."

The explosion would have created a mushroom cloud, while a plume of smoke would have been seen for hundreds of miles.

Mr. Hempsell said another part of the tablet, which is 18 cm across and shaped like a bowl, describes a plume of smoke around dawn the following morning.

"You need to know the context before you can translate it," said Mr. Hempsell, of Bristol University.

Geologists have dated the landslide to around 9,000 years ago, far earlier than the Sumerian record. However, Mr. Hempsell, who has published a book on the theory, believes *contaminated samples from the asteroid may have confused previous dating attempts.*

Academics were also quick to disagree with the findings, which were published in *A Sumerian Observation of the Köfels's Impact Event*.

John Taylor, a retired expert in Near Eastern archaeology at the British Museum, said there was no evidence that the ancient Sumerians were able to make such accurate astronomical records, while our knowledge of Sumerian language was incomplete.

"I remain unconvinced by these results," he added. [This author's emphases]

2345 BCE – Eco-disaster focused in the Levant as shown in tree-rings. End of Egyptian Old Dynasty?

The French archaeologist, Marie-Agnes Courty, presented a paper at the Society for Inter-Disciplinary Studies July 1997 conference at Cambridge University, in which she first detailed the findings of excavations at a site in northern Syria, at Tell Leilan. This was the first time ever that an archaeological excavation had been initiated where the main purpose was to examine the stratigraphical record of the area with a view to searching for evidence of "scorched earth" due to a suspected episode of extra-terrestrial "fireball bombardment."

She and her team found much evidence of microscopic glass spherules typical of melted sand and rock which is caused by the intense heat resulting from an asteroid impact or air-burst. She recommended further excavations there and at other sites. It would make sense that attention should be focussed on sites once occupied at dates where the tree-ring chronologies show evidence of abrupt climate changes — as at Tell Leilan in northern Syria, where the "burn event" has now been dated by Courty as immediately prior to 2345 BC, a "narrowest tree-ring" date.

Another with no human fatality numbers included in the calculations.

> Scientists have found the first evidence that a devastating meteor impact in the Middle East might have triggered the mysterious collapse of civilisations more than 4,000 years ago.
>
> Studies of satellite images of southern Iraq have revealed a two-mile-wide circular depression which scientists say bears all the hallmarks of an impact crater. If confirmed, it would point to the Middle East being struck by a meteor with the violence equivalent to hundreds of nuclear bombs. Today's crater lies on what would have been shallow sea 4,000 years ago, and any impact would have caused devastating fires and flooding. The catastrophic effect of these could explain the mystery of why so many early cultures went into sudden decline around 2300 BC. The crater's faint outline was found by Dr Sharad Master, a geologist at the University of Witwatersrand, Johannesburg, on satellite images of the Al 'Amarah region, about ten miles north-west of the confluence of the Tigris and Euphrates and home of the Marsh Arabs. (Robert Matthews, Science Correspondent, *The Telegraph*, 11-4-1)

1628 BCE – "The Exodus": Biblical scholars have been debating the date of the so-called Exodus for hundreds of years. The most recent researches have indicated that there was no "exodus" as depicted in the Bible, it was all made up by post-exilic priests to create a "history" justifying their elite status and privileges. More than that, based on historical knowledge of how things were done in those times, they probably were not even related to any of the people "carried away to Babylon" in the first place. And so, it seems logical to speculate that the background information contained in the Exodus story — and other related stories in the Bible, such as the collapse of Jericho and the destruction of Sodom and Gomorrah — were legendary stories of events that occurred around the time of the eruption of Thera, which has been fairly securely fixed around 1600 BCE, plus or minus fifty years. Mike Baillie reports that whatever happened during this period of history that includes this monstrous eruption, it was global in effect as is shown in the tree-ring chronologies. In other words, more was going on than just a volcanic eruption. Again, no numbers of fatalities to plug into the calculations, though there are many ancient reports of plague and mass death, and Egyptian records report many strange sky, weather and plague phenomena.

1159 BCE – Collapse of Shang and Mycenean cultures. Collapse of the Bronze Age in the Mediterranean region. Possible origin of more "Exodus" stories amalgamated with older tales of similar events. Wikipedia tells us:

> The Bronze Age collapse is the name given by those historians who see the transition from the Late Bronze Age to the Early Iron Age, as violent, sudden and culturally disruptive, expressed by the collapse of palace economies of the Aegean and Anatolia, replaced after a hiatus by the isolated village cultures of the Dark Age period of history of the Ancient Middle East.

Mike Baillie points out that a series of impacts/overhead explosions would more adequately explain the longstanding problem of the end of the Bronze Age in the Eastern Mediterranean in the 12th century BCE. At that time, many — uncountable — major sites were destroyed and totally burned, and it has all been blamed on those supernatural "Sea Peoples." If that was the case, if it was invasion and conquest, there ought to at least be some evidence for that, like dead warriors or signs of warfare; but for the most part, that is not the case. There were almost no bodies found, and no precious objects except those that were hidden away as though someone expected to return for them, or didn't have time to retrieve them. The people who fled (extra-terrestrial events often have precursor activities and warnings because a comet can often be observed approaching for some time) were probably also killed in the act of fleeing, and the result was total abandonment and total destruction of the cities in question.

John Lewis did not include this in his calculations either.

207 BCE – "Scientists Say Comet Smashed into Southern Germany in 200 BCE" [38]

> A comet or asteroid smashed into modern-day Germany some 2,200 years ago, unleashing energy equivalent to thousands of atomic bombs, scientists reported on Friday.
>
> The 1.1-kilometre (0.7-mile) diameter rock whacked into southeastern Bavaria, leaving an "exceptional field" of meteorites and impact craters that stretch from the town of Altoetting to an area around Lake Chiemsee, the scientists said in an article in the latest issue of US magazine *Astronomy*.

---

[38] http://www.spacedaily.com/news/comet-04l.html

Colliding with the Earth's atmosphere at more than 43,000 km per hour, the space rock probably broke up at an altitude of 70 km, they believe.

The biggest chunk smashed into the ground with a force equivalent to 106 million tonnes of TNT, or 8,500 Hiroshima bombs.

"The forest beneath the blast would have ignited suddenly, burning until the impact's blast wave shut down the conflagration," the investigators said.

"Dust may have been blown into the stratosphere, where it would have been transported around the globe easily…. The region must have been devastated for decades."

The biggest crater is now a circular lake called Tuettensee, measuring 370 metres (1,200 feet) across. Scores of smaller craters and other meteorite impacts can be spotted in an elliptical field, inflicted by other debris.

The study was carried out by the Chiemgau Impact Research Team, whose five members included a mineralogist, a geologist and an astronomer. […]

Additional evidence comes from local discoveries of Celtic artifacts, which appear to have been scorched on one side.

That helped to establish an approximate date for the impact of between 480 and 30 BCE.

The figure may be fine-tuned to around 200 BCE, thanks to tree-ring evidence from preserved Irish oaks, which show a slowing in growth around 207 BCE.

This may have been caused by a veil of dust kicked up by the impact, which filtered out sunlight.

In addition, Roman authors at about the same time wrote about showers of stones falling from the skies and terrifying the populace.

The object is more likely to have been a comet than an asteroid, given the length of the ellipse and scattered debris, the report says.

44 BCE – Pliny states that "Portentous and protracted eclipses of the sun occurred, such as the one after the murder of Caesar the dictator." Yet there were no solar eclipses visible from anywhere in the Roman Empire from February of 48 BCE through December of 41 BCE, inclusive. There was a spectacular daylight comet in 44 BCE, perhaps the most famous comet in antiquity. A dust veil occluded the sky over Italy in the spring of 44, and has often been attributed to an (unconfirmed) eruption of Mt Etna. There are sulfate deposits in the Greenland ice cores for this year and there is tree ring evidence from North America, where dendrochronology points to a climatic change in the late 40s BCE. What hit and where it hit, has yet to be determined, and whether or not there was death and destruction somewhere on the globe, is unknown.

John S. Lewis does not include this event in his calculations.

60–70 AD – The destruction of Jerusalem.

> The story Josephus tells of the sixties is one of famine, social unrest, institutional deterioration, bitter internal conflicts, class warfare, banditry, insurrections, intrigues, betrayals, bloodshed, and the scattering of Judeans throughout Palestine.... There were wars and rumors of wars for the better part of ten years and *Josephus reports portents, including a brilliant daylight in the middle of the night*! (Burton Mack, *A Myth of Innocence*: *Mark and Christian Origins*, 1988, 2006) [This author's emphases]

We recognize that "brilliant daylight at night" from the Tunguska Event. Josephus gives several portents of the evil to befall Jerusalem and the temple. He described a star resembling a sword, a comet, a light shining in the temple, a cow giving birth to a lamb at the moment it was to be sacrificed in the Jerusalem Temple, *armies fighting in the sky*, and a voice from the Holy of Holies declaring, "We are departing" (Josephus, *Jewish Wars*, 6). [Obviously, the voice was apocryphal.]

Some of these portents are mentioned by other contemporary historians, Tacitus for example. However, in book five of his *Histories*, Tacitus castigated the superstitious Jews for not recognizing and offering expiations for the portents to avert the disasters. He put the destruction of Jerusalem down to the stupidity or willful ignorance of the Jews themselves in not offering the appropriate sacrifices.

> Thus there was a star resembling a sword, which stood over the city [Jerusalem], and a comet, that continued a whole year.... (Josephus, *Jewish Wars*, 6.3)

In short, it very well may be that the eschatological writings in the New Testament — the very formation of the Myth of Jesus — was based on cometary events of the time, including a memory of the "Star in the East." The destruction of the Temple at Jerusalem may very well have been an "Act of God," as reported by Mark in his Gospel.

312 AD – Italy: A team of geologists believes it has found the incoming space rock's impact crater, and dating suggests its formation coincided with the celestial vision said to have converted a future Roman emperor to Christianity. The small circular "Cratere del Sirente" in central Italy is clearly an impact crater, said the geologists, because its shape fits and it is also surrounded by numerous smaller, secondary craters, gouged out by ejected debris, as expected from impact models.

Radiocarbon dating puts the crater's formation at about the right time to have been witnessed by Constantine and there are magnetic anomalies detected around the secondary craters — possibly due to magnetic fragments from the meteorite. It would have struck the Earth with the force of a small nuclear bomb, perhaps a kiloton in yield. It would have looked like a nuclear blast, with a mushroom cloud and shockwaves.

476 AD – I-hsi and Chin-ling, China: "Thundering chariots" fell to ground "like granite"; vegetation was scorched.

526 AD – Great Antioch earthquake.

> ...those caught in the earth beneath the buildings were incinerated, and sparks of fire appeared out of the air and burned everyone they struck like lightning. The surface of the earth boiled and foundations of buildings were struck by thunderbolts thrown up by the earthquakes, and were burned to ashes by fire.... It was a tremendous and incredible marvel with fire belching out rain, rain falling from tremendous furnaces, flames dissolving into showers... as a result Antioch became desolate... in this terror up to 250,000 people perished. (John Malalas, quoted by Jeffreys, E., Jeffreys, M. and Scott, R. in "The Chronicle of John Malalas," *Byzantina Australiensia*, Australian Association, Byzantine Studies 4, 1986.)

536–45 AD – Reduced sunlight, mists or "dry fogs," crop failures, plagues, and famines in China and the Mediterranean. The Roman Praetorian Prefect, Magnus Aurelius Cassiodorus, wrote a letter documenting the conditions:

> All of us are observing, as it were, a blue-coloured sun; we marvel at bodies which cast no mid-day shadow, and at that strength of intensest heat reaching extreme and dull tepidity [...] So we have had a winter without storms, spring without mildness, summer without heat [...] The seasons have changed by failing to change; and what used to be achieved by mingled rains cannot be gained from dryness only.

Procopius of Caesarea, a Byzantine, wrote:

> And it came about during this year that a most dread portent took place. For the sun gave forth its light without brightness, like the moon, during the whole year, and it seemed exceedingly like the sun in eclipse, for the beams it shed were not clear nor such as it is accustomed to shed.

John of Ephesus, cleric and a historian, wrote:

> The sun was dark and its darkness lasted for eighteen months; each day it shone for about four hours; and still this light was only a feeble shadow... the fruits did not ripen and the wine tasted like sour grapes.

In the wake of this inexplicable darkness, crops failed and famine struck. And then, pestilence. But here we mean "pestilence" as Jacme d'Agramaont, a doctor writing in 1348, described it in reference to the "Black Death":

Agramont said nothing concerning the term *epidemia*, but he extensively developed what he meant by *pestilencia*. He gave this latter term a very peculiar etymology, in accordance with a from of knowledge established by Isidore of Seville (570-636) in his *Etymologiae*, which came to be widely accepted throughout Europe during the Middle Ages. He split the term *pestilencia* up into three syllables, each having a particular meaning: *pes = tempest*, "storm, tempest"; *te = temps*, "time"; *lencia = clardat*, "brightness, light"; hence, he concluded, the *pestilencia* was *"the time of tempest caused by light from the stars."* (Jon Arrizabalaga, see Part One) [this author's emphasis]

During the time of Justinian, this "pestilence" ravaged Europe, *reducing the population of the Roman Empire by a third*, killing four-fifths of the citizens of Constantinople, reaching as far East as China and as far Northwest as Great Britain. John of Ephesus documented the progress of this "pestilence" in 541-542 AD in Constantinople, where city officials gave up trying to count the dead after two-hundred and thirty thousand:

The city stank with corpses as there were neither litters nor diggers, and corpses were heaped up in the streets. [...] It might happen that [a person] went out to market to buy necessities and while he was standing and talking or counting his change, suddenly the end would overcome the buyer here and the seller there, the merchandise remaining in the middle with the payment for it, without there being either buyer or seller to pick it up.

This was also the time assigned to the legendary King Arthur, the loss of the Grail, and the manifestation of the Wasteland. Although scholars place the historical King Arthur in the 5th century, the date of his death is given as 539 AD. According to Mike Baillie, *the imagery from the Arthurian legend is in accordance with the appearance of a comet and subsequent famine and plague*: the "Waste Land" of legend. Ireland's St. Patrick stories feature a wasteland as well. And although St. Patrick is credited with ridding Ireland of snakes, we might consider that there never were snakes in Ireland, and that snakes and dragons are images associated with comets.

Until that point in time, the Britons had held control of post-Roman Britain, keeping the Anglo-Saxons isolated and suppressed. After the Romans were gone, the Britons maintained the *status quo*, living in towns, with elected officials, and carrying on trade with the empire. After 536 AD, the year reported as the "death of Arthur", the Britons, the ancient Cymric empire that at one time had stretched from Cornwall in the south to Strathclyde in the north, all but disappeared, and were replaced by Anglo-Saxons. There is much debate among scholars as to whether the Anglo-Saxons killed all of the Britons, or assimilated them. Here we must consider that they were victims of possibly many overhead cometary explosions which wiped out most of the population of Europe, plunging it into the Dark Ages which were, apparently, literally *dark*, atmospherically speaking.

The mystery of the origins of the red dragon symbol, now on the flag of Wales, has perplexed many historians, writers, and romanticists, and the archæological community generally has refrained from commenting on this most unusual emblem, claiming it does not concern them. In the ancient Welsh language it is known as *Draig Goch*, "red dragon"; and in... the *University of Wales Welsh Dictionary* (University of Wales Press, 1967, p. 1082) there are translations for the various uses of the Welsh word *draig*. Amongst them are common uses of the word, which is today taken just to mean a "dragon," but in times past it has also been used to refer to *Mellt Distaw* (sheet lightning) and also *Mellt Didaranau* (lightning unaccompanied by thunder).

But the most interesting common usage of the word in earlier times, according to this authoritative dictionary, is *Maen Mellt*, the word used to refer to a "meteorite." And this makes sense, as the Welsh word *maen* translates as "stone," while the Welsh word *mellt* translates as "lightning" — so literally a "lightning-stone." That the ancient language of the Welsh druids has words still in use today which have in the past been used to describe both a dragon and also a meteorite, is something that greatly helps us to follow the destructive "trail of the dragon" as it was described in early Welsh "riddle-poems." [...]

The exact nature and sequence of events in mid 6th-century AD that gave rise to the period we refer to as the European "Dark Age" is still a matter for speculation amongst historians and archaeologists. Over the past twenty years or so, certain paleo-climatologists have begun comparing notes with archaeologists and astronomers, and interestingly, in the absence of written records, many have begun to look a little more closely at mythology in their efforts to corroborate the findings of their researches. While much of this recent bout of inter-disciplinary brainstorming has focussed on the 6th century AD start of the European Dark Age, earlier dates are also of great interest to those embroiled in this veritable "paradigm shift". [...]

In recent years certain astronomers have increasingly come to appreciate that encoded in the folklore and mythologies of many cultures are the accurate observations of ancient skywatchers. Almost all tell of times when death and mass destruction came from the skies, events that are often portrayed as "celestial battles" between what they variously depicted as "the Gods." And curiously the imagery in these "myths" have many common features, even between the mythologies of cultures widely spaced in time and location. (*The European "Dark Age" and Welsh Oral Tradition*[39])

Out on the Asian steppes, whatever happened in 536 AD caused political upheaval. The horse-based economy of the war-like Avars foundered, and their vassals, the cattle-herding Turks, overthrew them. Driven from the steppes, the Avars joined forces with the Slavs in Hungary on the borders of the Roman Empire.

Gildas, who was writing at approximately 540 AD, says that "*the island of Britain was on fire from sea to sea... until it had burned almost the whole surface of the island and was licking the western ocean with its fierce red tongue.*" In *The Life of St. Teilo* contained in the Llandaf Charters of St. Teilo, who had recently been made Bishop of Llandaf Cathedral in Morganwg, South Wales, it says:

> [H]owever he could not long remain, on account of the pestilence which nearly destroyed the whole nation. It was called the Yellow Pestilence, because it occasioned all persons who were seized by it to be yellow and without blood, and it appeared to men a column of a watery cloud, having one end trailing along the ground, and the other above, proceeding in the air, and passing through the whole country like a shower going through the bottom of valleys. Whatever living creatures it touched with its pestiferous blast, either immediately died, or sickened for death ... and so greatly did the aforesaid destruction rage throughout the nation, that it caused the country to be nearly deserted.

St. Teilo is recorded as having left South Wales for Brittany to escape the Yellow Pestilence, and that it lasted for some eleven years.

In 540, in Yemen, the Great Dam of Marib, dating from around the 7th century BCE, one of the engineering wonders of the ancient world and a central part of the south Arabian civilization, broke and began to collapse. By 550, the dam was a complete loss and thousands of people migrated to another oasis on the Arabian peninsula, Medina. The Arab tribes, traumatized by the environmental disasters around them, began to think of conquest for the sake of survival. In 610, a new leader unified them: Muhammad.

Although a great many historical changes happened in the seventh century, such as the Roman war with Persia, the rise of Islam, rebellion and civil war in the Roman empire, and the advance of the Slavs driven by the Avars, it can be said that the seeds of these changes, the

---

[39] http://www.morien-institute.org/darkages.html

destruction of the old that made way for the new, can be traced to the environmental catastrophe of 536.

John Lewis does not include any estimates of the death and destruction occurring at that time in his "average number of annual deaths by comets."

580 AD – France: Great fireball and blast; Orleans and nearby towns burned.

588 AD – China: "Red-colored object" fell with "noise like thunder" into furnace; exploded; burned several houses.

616 AD – China: Ten deaths reported in China from a meteorite shower; seige towers destroyed.

679 AD – Coldingham, England: Monastery destroyed by "fire from heaven" as reported in the *Anglo-Saxon Chronicle*.

764 AD – Nara, Japan: Meteorite strikes house.

810 AD – Upper Saxony: Charlemagne's horse startled by meteor, throws him to the ground.

1000 AD – Alberta, Canada: The date of this meteor strike is estimated:

> What local hunters in Whitecourt thought for years was a sinkhole is actually the crater left behind by a meteor that fell to earth 1,000 years ago and is now attracting international attention from researchers.... The crater is 36-metres wide and six-metres deep, which is small as far as most craters go.... Herd thinks the meteor came from the asteroid belt and measured one-metre across. However, researchers have so far found 74 different pieces of the original meteor — which is called a meteorite once it hits the ground — scattered around the crater, some up to 70 metres away.[40]

1064 – Chang-chou, China: Daytime fireball, meteorite fall; fences burned.

1178 (June) – Cantebury, England:

> In this year, on the Sunday before the Feast of St. John the Baptist, after sunset when the moon had first become visible a marvelous phenomenon was witnessed by some five or more men who were sitting there facing the moon. Now there was a bright new moon, and as usual in that phase its horns were tilted toward the east; and suddenly the upper horn split in two. From the midpoint of the division a flaming torch sprang up, spewing out, over a considerable distance, fire, hot coals, and sparks. Meanwhile the body of the moon which was below writhed, as it were in anxiety, and to put it in the words of those who reported it to me and saw it with their own eyes, the moon

---

[40] http://www.sott.net/articles/show/169469-Canada-Northern-Alberta-meteor-crater-identified

throbbed like a wounded snake. Afterwards it resumed its proper state. This phenomenon was repeated a dozen times or more, the flame assuming various twisting shapes at random and then returning to normal. Then after these transformations the moon from horn to horn, that is along its whole length, took on a blackish appearance. The present writer was given this report by men who saw it with their own eyes, and are prepared to stake their honour on an oath that they have made no addition or falsification in the above narrative. (Gervase of Canterbury)

1321–68 – O-chia District, China: "Iron rain" kills people, animals, damages house.

1347–48 – Europe: The Black Death — not included in John Lewis' calculations — killed about half the population of Western Europe. The effects of this event were possibly global, though the number of deaths worldwide is unknown.

1348 (January) – Carinthia, Austria: Earthquake, 16 cities destroyed, fire fell from heaven; over 40,000 dead. John Lewis does not include this event in his calculations.

1369 – Ho-t'ao, China: "Large star" fell, starts fire, soldiers injured.

1430 (June) – New Zealand: A huge comet struck the ocean less than a hundred miles from the Chinese fleet of Zhou Man. Up to 173 wrecks have been counted as destroyed by this event. The comet incinerated many ships and hurled the blazing wrecks onto New Zealand South Island and the east coasts of Australia, and across the Pacific and Indian Oceans. Chinese and Mayan astronomers describe a large blue comet seen in *Canis Minor* for 26 days in June 1430 — a date compatible with Professor Bryant's (1410 – 1480) evidence. In November 2003 Dallas Abbott and her team announced they had found where the comet crashed — between Campbell Island and New Zealand South Island. Deaths? Probably in the multiple thousands:

> A mega-tsunami struck southeast Asia 700 years ago rivaling the deadly one in 2004, two teams of geologists said after finding sedimentary evidence in coastal marshes. Researchers in Thailand and Indonesia wrote in two articles in *Nature* magazine that the tsunami hit around 1400, long before historical records of earthquakes in the region began.[41]

Which of course, leads us to ask the question: Was the 2004 mega-tsunami the result of a comet/asteroid strike?

_____

[41] http://www.sott.net/articles/show/168328-Mega-tsunami-hit-southeast-Asia-700-years-ago

1490 (February) – Ch'ing-yang, Shansi, China: Stones fell like rain; more than 10,000 killed.

1492 – Ensisheim, Alsace: 280-pound meteorite landed; in the same year Columbus reported "a marvelous branch of fire" that fell into the sea as he crossed the Atlantic.

1511 (September 5) – Cremona, Lombardy, Italy: Monk killed with several birds, a sheep.

1516 (May) – Nantan, China: "During summertime in May of Jiajing 11th year, stars fell from the northwest direction, five to six fold long, waving like snakes and dragons. They were as bright as lightning and disappeared in seconds." Many of them were recovered by local farmers in 1958 when China needed steel for the "Great Leap Forward" advocated by Mao Zedong. They have coarse octahedral structure and contain 92.35% iron and 6.96% nickel, belonging to IIICD classification of Wasson *et al* (1980). Most Nantan meteorites weigh 150 to 1500 kg. Due to the humid conditions, smaller pieces buried in soils of lower valleys have been extensively weathered and oxidized into limonite.

1620 – Punjab, India: "Hot iron" fell, burned grass; made into dagger knife, two sabres.

1631 – Fall of Magedeburg, Germany:

> [A] grand storm-wind picked up, the town was inflamed at all possible places, so that even little aid (rescue) was of help (appreciated). […] Then I saw the whole town of Magdeburg, except dome, monastery and New Market, lying in embers and ashes, which raged only about 3 or 3-1/2 hours, from which I deduced God's strange omnipotence and punishment. (Geoffrey Mortimer, "Style and Fictionalisation in Eyewitness Personal Accounts of the Thirty Years War," *German Life and Letters*, 54:2)

A "second sun" was seen on and around May 29, 1630; and on May 20, 1631, one year later, Magdeburg fell as described above. The standard historical description of the Fall of Magdeburg goes pretty much as follows:

> The fall of Magdeburg horrified Europe. The city had been starved and then was bombarded unmercifully. The artillery shelling grew so bad, the town caught on fire. Over 20,000 of the citizens perished in the siege and the cataclysm that ended it. The city itself was burned to the ground. The cruel and pointless devastation marked a new low, an act abhorred by a generation well accustomed to horrors.

1639 – China: Large stone fell in market; tens killed; tens of houses destroyed.

1648 – Malacca: Two sailors reported killed on board ship en route from Japan to Sicily.

1654 – Milano, Italy: Monk reported killed by meteorite.

1661 (August 9) – China: Meteorite smashes through roof; no injuries.

1670 (November 7) – China: Meteorite falls, breaks roof beam of house

1761 – Chamblan, France: House struck and burned by meteorite.

1780 – New England, U.S./Eastern Ontario, Canada:

In the midst of the Revolutionary War, darkness descends on New England at midday. Many people think Judgment Day is at hand. It will be remembered as New England's Dark Day.

Diaries of the preceding days mention smoky air and a red sun at morning and evening. Around noon this day, an early darkness fell. Birds sang their evening songs, farm animals returned to their roosts and barns, and humans were bewildered.

Some went to church, many sought the solace of the tavern, and more than a few nearer the edges of the darkened area commented on the strange beauty of the preternatural half-light. One person noted that clean silver had the color of brass.

It was darkest in northeastern Massachusetts, southern New Hampshire and southwestern Maine, but it got dusky through most of New England and as far away as New York. At Morristown, New Jersey, Gen. George Washington noted it in his diary.

In the darkest area, people had to take their midday meals by candlelight. A Massachusetts resident noted, "In some places, the darkness was so great that persons could not see to read common print in the open air." In New Hampshire, wrote one person, "A sheet of white paper held within a few inches of the eyes was equally invisible with the blackest velvet."

At Hartford, Col. Abraham Davenport opposed adjourning the Connecticut legislature, thus: "The day of judgment is either approaching, or it is not. If it is not, there is no cause of an adjournment; if it is, I choose to be found doing my duty."

When it was time for night to fall, the full moon failed to bring light. Even areas that had seen a pale sun in the day could see no moon at all. No moon, no stars: It was the darkest night anyone had seen. Some people could not sleep and waited through the long hours to see if the sun would ever rise again. They witnessed its return the morning of May 20. Many observed the anniversary a year later as a day of fasting and prayer.

Professor Samuel Williams of Harvard gathered reports from throughout the affected areas to seek an explanation. A town farther north had reported "a black scum like ashes" on rainwater collected in tubs. A Boston observer noted the air smelled like a "malt-house or coal-kiln." Williams noted that rain in Cambridge fell "thick and dark and sooty" and tasted and smelled like the "black ash of burnt leaves."

As if from a forest fire to the north? Without railroad or telegraph, people would not know: No news could come sooner than delivered on horseback, assuming the wildfire was even near any European settlements in the vast wilderness.

But we know today that the darkness had moved southwest at about 25 mph. And we know that forest fires in Canada in 1881, 1950 and 2002 each cast a pall of smoke over the northeastern United States.

A definitive answer came in 2007. In the *International Journal of Wildland Fire*, Erin R. McMurry of the University of Missouri forestry department and co-authors combined written accounts with fire-scar evidence from Algonquin Provincial Park in eastern Ontario to document a massive wildfire in the spring of 1780 as the "likely source of the infamous Dark Day of 1780."[42]

Sounds like an impact event.

1790 (July 24) – Barbotan and Agen, Gascony, France: Meteorite crushes cottage, kills farmer and some cattle.

1794 (June 16) – Siena, Italy: Child's hat hit; child uninjured.

1798 (December 19) – Benares, India: Building struck by meteorite

1801 (October 30) – Suffolk, England: "Dwelling-house of Mr. Woodrosse, miller, near Horringermill, Suffolk, was set on fire by a meteor, and entirely consumed, together with a stable adjoining."

1803 (July) – East Norton, England: White Bull public house struck, chimney knocked down, grass burned, flight of object nearly horizontal.

1803 (December 13) – Massing, Czech Republic: Building struck by meteorite.

1810 (July) – Shahabad, India: Great stone fell, five villages burned, several killed.

1811–12 – Eastern North America: Submitted by a reader, from *The Comet Book: A Guide For the Return of Halley's Comet* by Robert D. Chapman and John C Brandt, published in 1984:

In December, 1811, a series of earthquakes began that rocked over three hundred thousand square miles of eastern North America. The great New Madrid Earthquake was just one of the many events of late 1811 and 1812 that followed quickly on the heels of the Great Comet of 1811. First observed from America in late 1811, the comet was described as bright and slightly smaller than the full moon, including its tail. Newspapers in the young republic picked up on the comet and

---

[42] http://www.wired.com/science/discoveries/news/2008/05/dayintech_0519

predicted that it was an omen of evil times. Sure enough, a string of disasters, nature and otherwise, followed.

Though the epicenter of the earthquakes was near a small town on the Mississippi River — New Madrid, Missouri — the numerous shocks were felt as far away as New York and Florida. According to one source, Richmond, Virginia and Boston were shaken so violently by the December 16 shocks that church bells rang. It is said that in some places the current in the Mississippi River flowed backward.

The great naturalist John James Audubon was in Kentucky when the first tremor struck. Initially he thought the roar was a tornado and headed for shelter. Audubon and others reported strange darkenings and brightenings in the sky.

Jared Brookes, a native of Louisville, Kentucky kept accurate records of all the shocks that he experienced. Between December 1811 and May 1812, he tabulated over two thousand separate tremors. Based on the historical evidence, Charles Richter (inventor of the Richter scale used to measure the strength of earthquakes) had estimated that there were at least three severe shocks that exceeded magnitude eight on his scale. The New Madrid Earthquakes thus stand as one of the most severe, if not the most severe, series of quakes recorded in U.S. history.

The New Madrid Earthquakes were not the only disaster to take place in 1811 and 1812. On December 26, 1811, a fire broke out in the new theater in Richmond, which was packed with people. Governor George Smith and almost eighty others perished. The incidents leading to the War of 1812 were moving inexorably forward: the Battle of Tippecanoe; the Guerriere incident in which the British impressed an American seaman from an American vessel onto their warship Guerriere. To round out the problems, severe weather plagued the young republic. All told, 1812 was not a good year.

An interesting contrast to the disasters of 1811-1812 comes from the world of wine. The year 1811 produced a particularly good vintage. In honor of the great comet, the wine was referred to as *vin de la comete* (comet wine). Not everything from the year was bad!

Robert Fritzius' article titled "1811-12 New Madrid Earthquakes: An NEO Connection?"[43] discusses the possibility that the New Madrid Earthquake of 1811 may be related to the Great Comet of that year. Most of the article focuses on the possibility of finding an impact zone related to these events. The evidence seems inconclusive, but based on what I've read, this may be an incorrect approach to understanding the historical data, since comets do have the tenancy to explode in the atmosphere causing much havoc on the ground. In fact, there appears to be eyewitness testimony from people who lived through this event of a

---

[43] http://www.datasync.com/~rsf1/1811.htm

falling charcoal-like substance as well as visible undulations of the earth itself. An eyewitness report:

> ...the earth was observed to be as it were rolling in waves of a few feet in height, with visible depressions between. By and by these waves or swells were seen to burst, throwing up large volumes of water, sand and a species of charcoal....[44]

This evidence may imply an overhead explosion of sorts, which could have also set off some of the earthquakes in this region.

Another point that should be made is *the reference to human conflict* during this time period. Obviously the War of 1812 between the American republic and Britain stands out as one example. Another would be Napoleon's march into Russia and his eventual defeat during an abnormally cold winter. Fritzius draws attention to the following statement by John Kezys ("Napoleon's Comets," *Orbit*, February 1996):

> As Napoleon marched into Russia with an army of seven hundred thousand strong, the Great Comet (of 1811) developed a tail one hundred million miles long. Following initial victories Napoleon overextended himself. After the invasion of Moscow he ran short of supplies and the winter proved unforgiving. Hundreds of thousands died while the comet performed frightening acrobatics by splitting in two.

This may go along with Victor Clube's analysis of the association between comets and turbulence within human civilization.

1823 (November 10) – Waseda, Japan: Meteorite strikes house.

1825 (January 16) – Oriang, India: Man reported killed, woman injured by meteorite fall.

1827 (February 27) – Mhow, India: Man struck on arm, tree broken by meteorite.

1835 (November 14) – Belley, France: Fireball sets fire to barn.

1836–11 (December) – Macaé, Brazil: Several homes damaged, several oxen killed by meteorite.

1841 – Chiloe Archipel, Chile: Fire caused by meteorite fall.

1845–6 – Ch'ang-shou, Szechwan, China: Stone meteorite damages more than 100 tombs.

1847 (July 14) – Braunau, Bohemia: A 37-lb iron smashes through roof of house.

1850 (October 17) – Szu-mao, China: Meteorite falls through roof of house.

---

[44] http://www.showme.net/~fkeller/quake/lib/eyewitness1.htm

1858 (December 9 – Ausson, France: Building hit by meteorite.

1860 (May 1) – New Concord, Ohio, U.S: Colt struck and killed by meteorite.

1868 (August 8) – Pillistfer, Estonia: Building struck.

1869 (January 1) – Hessle, Sweden: Man missed by few meters.

1870 (January 23) – Nedagolla, India: Man stunned by meteorite. (Don't know if this means the man was "amazed" or if he was hit and physically knocked senseless.)

1871 (October 8) – Great Chicago Fire, U.S. See "Comet Biela and Mrs. O'Leary's Cow" (SOTT.net). (Another item that John Lewis has not entered into his calculations.)

1872 – Banbury,U.K: Fireball fells trees, wall

1874 (June 30) – Chin-kuei Shan, Ming-tung Li, China: Thunderstorm; huge stone fell, crushed cottage, killed child.

1876 (February 16) – Judesegeri, India: Water tank struck by meteorite.

1877 (January 3) – Warrenton, Missouri, U.S: Man missed by few meters.

1877 (January 21) – De Cewsville, Ontario, Canada: Man missed by few meters.

1879 (January 14) – Newtown, Indiana, U.S: Leonidas Grover reported killed in bed by meteorite (possible hoax in *Paducah Daily News*).

1879 (January 31): Dun-le-Poelier, France – Farmer reported killed by meteorite.

1879 (November 12) – Huan-hsiang, China: Rain of stones; many houses damaged; sulfur smell.

1881 (November 19) – Grosliebenthal, Russia: Man reported injured by meteorite.

1887 (March 19) – Barque J.P.A., North Atlantic: Fireball "fell into water very close alongside."

1893 (November 22) – Zabrodii, Russia: Building struck by meteorite.

1897 (March 11) – New Martinsville, West Virginia, U.S: A man was reportedly struck, a horse killed, and walls pierced.

1906 (November 4) – Diep River, South Africa: Building struck.

1907 (September 5) – Hsin-p'ai Wei, Weng-Li: Stone fell; whole family crushed to death.

1907 (December 7) – Bellefontaine, Ohio, U.S: Meteorite starts fire, destroys house.

1908 (June 30) – Tunguska valley, Siberia: Two reportedly killed, many injured by Tunguska blast.

1909 (May 29) – Shepard, Texas, U.S: Meteor drops through house.

1910 (April 27) – Mexico: Giant meteor bursts, falls in mountains, starts forest fire.

1911 (June 16) – Kilbourn, Wisconsin, U.S: Meteorite struck barn.

1911 (June 28) – Nakhla, Egypt: Dog struck and killed by meteorite.

1912 (July 19) – Holbrook, Arizona, U.S: Building struck; 14,000 stones fell; man missed by a few meters.

1914 (January 9) – Western France: Meteor explosions break windows.

1914 (November 22) – Batavia, New York, U.S: Meteorites damage farm.

1916 (January 18) – Baxter, Missouri, U.S: Building struck.

1917 (December 3) – Strathmore, Scotland: Building struck.

1918 (June 30) – Richardton, North Dakota, U.S: Building struck.

1921 (July 15) – Berkshire Hills, Massachusetts, U.S: Meteor starts fire in Berkshires.

1921 (December 21) – Beirut, Syria: Building hit.

1922 (February 2) – Baldwyn, Mississippi: Man missed by 3 meters.

1922 (April 24) – Barnegat, New Jersey, U.S: Rocked buildings, shattered windows, clouds of noxious gas, overhead explosion of comet fragment.

1922 (May 30) – Nagai, Japan: Person missed by several meters

1924 (July 6) – Johnstown, Colorado, U.S: Man missed being hit by one meter

1927 (April 28) – Aba, Japan: Girl struck and injured by "dubious" (?) meteorite.

1929 (December 8) – Zvezvan, Yugoslavia: Meteor hits bridal party, kills one.

1930 (August 13) – Brazil: The "Rio Curaca Event" (Brazlilian "Tunguska Event"). Fire and "depopulation"; an ear-piercing "whistling" sound, which might be understood as being a manifestation of the electrophonic phenomena which have been discussed in WGN over the past few years; the sun appearing to be "blood-red" before the explosion. The event occurred at about 8:00 local time, so that the bolide probably came from the sunward side of the earth. If the object were spawning dust and meteoroids — i.e. was cometary in nature — then, since low-inclination eccentric orbits produce radiants close to the sun, it might be that the solar coloration (which, in this explanation, would have been witnessed elsewhere) was due to such dust in the line of sight to the sun. In short, the earth was within the tail of the small comet. There was a fall of fine ash prior to the explosion, which covered the surrounding vegetation with a blanket of white.

1931 (July 10): Malinta, Ohio, U.S: Blast, crater, smell of sulfur, windows broken in farmhouse; four telephone poles snapped, wires down; overhead cometary fragment explosion.

1931 (September 8) – Hagerstown, Maryland, U.S: Meteor crashes through roof.

1932 (August 4) – Sao Christovao, Brazil: Fall destroys warehouse roof.

1932 (August 10) – Archie, Missouri, U.S: Homestead struck, person missed by less than one meter.

1933 (February 24) – Stratford, Texas, U.S: Bright fireball, 4-lb metallic mass falls, grass burned.

1933 (August 8) – Sioux Co., Nebraska, U.S: Man missed by a few meters.

1934 (February 16) – Texas, U.S: Pilot swerves to avoid crash with fireball.

1934 (February 18) – Seville, Spain: House struck, burned.

1934 (September 28) – California, U.S: Pilot escapes fireball shower (one assumes this means he performed evasive maneuvers).

1935 (August 11) – Briggsdale, Colorado, U.S: Man narrowly missed by meteorite.

1935 (December 11) – British Guyana: 21:00 local time; Lat. 2 degrees 10-min North; Long: 59 degrees 10-min West; close to Marudi Mountain. A report from Serge A. Korff of the Bartol Research Foundation, Franklin Institute (Delaware, U.S.) suggested that the region of devastation might be greater than that involved in the Tunguska Event itself. Eye-witness accounts were in accord with a large meteoroid/small asteroid entry, with a body passing overhead accompanied by a terrific roar (presumably electrophonic effects), later concussions, and the sky being lit up like daylight. A local aircraft operator, Art Williams, reported seeing an area of forest more than 20 miles (32 kilometers) in extent which had been destroyed, and he later stated that the shattered jungle was elongated rather than circular, as occurred at Tunguska and would be expected from the air blast caused by an object entering away from the vertical (the most likely entry angle for all cosmic projectiles is 45 degrees).

1936 (March 14) – Red Bank, New Jersey, U.S: Meteorite fell through shed roof.

1936 (April 2) – Yurtuk, USSR: Building struck.

1936 (October 19) – Newfoundland, Canada: Fisherman's boat set on fire by meteorite.

1938 (March 31) – Kasamatsu, Japan: Meteorite pierces roof of ship

1938 (June 16) – Pantar, Phillipines: Several buildings struck.

1938 (June 24) – Chicora, Pennsylvania, U.S: Cow struck and injured.

1938 (September 29) – Benld, Illinois, U.S: Garage and car struck by 4-lb stone.

1941 (July 10) – Black Moshannon Park, Pennsylvania, U.S: Person missed by one meter.

1942 (April 6) – Pollen, Norway: Person missed by one meter.

1940s – Qatar: A crater, believed to have been created by the impact of a falling meteor, found near Dukhan. Sheikh Salman bin Jabor al-Thani, head of the astronomical department at Qatar Scientific Club, said yesterday the Club believed that the meteor had hit Qatar in the 1940s. The Club started a search for evidence three years ago because of stories of a "falling star" told by people of that era. They used Google Earth in the search, and succeeded in locating five craters, which were just visible on the surface.

1946 (May 16) – Santa Ana, Nuevo Leon: Meteorite destroys many houses, injures 28 people.

1946 (November 30) – Colford, Gloucestershire, UK: Telephones knocked out, boy knocked off bicycle.

1947 (February 12) – Sikhote Alin, Vladivostok: An iron meteorite that broke up only about five miles above the Earth rained iron. It produced over 100 craters with the largest being around 85-feet in diameter. The strewnfield covered an area of about one mile by a half mile. There were no fires or similar destruction like that found at Tunguska; shredded trees and broken branches mostly. A total of 23 tons of meteorites were recovered, and it's been estimated that its total mass was around 70 tons when it broke up.

1949 (September 21) – Beddgelert, Wales, U.K: Building struck.

1949 (November 20) – Kochi, Japan: Hot meteoritic stone enters house through window.

1950 (May 23) – Madhipura, India: Building struck.

1950 (September 20) – Murray, Kentucky, U.S: Several buildings struck.

1950 (December 10) – St. Louis, Missouri, U.S: Car struck.

1953 (March 3) – Pecklesheim, FRG: Person missed by several meters.

1954 (January 7) – Dieppe, France: Meteorite-building explosion, smashed windows.

1954 (November 28) – Sylacauga, Alabama, U.S: Mrs. Annie Hodges struck by 4-kg meteorite that crashed through roof, destroyed radio, and left a serious bruise on her hip. It is considered the only documented/verified case of a person being hit by a meteorite. [45]

1955 (January 17) – Kirkland, Washington, U.S: Two irons break through amateur astronomer's observatory dome; one sets a fire.

1956 (February 29) – Centerville, S. Dakota, U.S: Building hit.

1959 (October 13) – Hamlet, Indiana, U.S: Building hit.

1961 (February 23) – Ras Tanura, Saudi Arabia: Loading dock struck.

1961 (September 6) – Bells, Texas, U.S: Meteorite strikes rook of house.

1962 (April 26) – Kiel, FRG: Building hit.

1963 – Massachusetts, U.S: Meteorite fell.

1965 (December 24) – Barwell,U.K: Two buildings struck and a car struck.

1967 (July 11) – Denver, Colorado, U.S: Building struck.

1968 (April 12) – Schenectady, New York, U.S: House hit.

1969 (April 25) – Bovedy, N. Ireland: Building hit.

1969 (August 7) – Andreevka, USSR: Building hit.

1969 (September 16) – Suchy Dul, Czechoslovakia: Building hit.

1969 (September 28) – Murchison, Australia: Building hit.

1971 (April 8) – Wethersfield, Connecticut, U.S: House struck by meteorite.

1971 (August 2) – Havero, Finland: Building hit.

1972 (August 10) – Utah, U.S./Alberta, Canada: The Great Daylight 1972 Fireball (or US19720810). An Earth-grazer meteoroid passed within 57 km of the surface of the Earth at 20:29 UTC on August 10, 1972, or 1.01 Earth radii from the centre of the Earth. It entered the Earth's atmosphere in daylight over Utah, United States (14:30 local time) and passed northwards leaving the atmosphere over Alberta, Canada. It was seen by many people, recorded on film and by space borne sensors. Analysis of its appearance and trajectory showed it was a meteoroid about two to ten meters in diameter, in the Apollo asteroid class, in an orbit that would make a subsequent close approach to Earth in August 1997. In 1994 Zdenek Ceplecha re-analysed the data and suggested the passage would have reduced the meteoroid's mass to about a third or half of its original mass. The meteoroid's 100-second passage through the atmosphere reduced its velocity by about 800 meters per second, and the whole encounter significantly changed its orbital inclination from 15 degrees to 8 degrees.

---

[45] http://www.oberlin.edu/faculty/bsimonso/group9.htm

1973 (March 15) – San Juan Capistrano, California, U.S: Building hit.

1973 (October 27) – Canon City, Colorado, U.S: Building hit.

1974 (August 18) – Naragh, Iran: Building hit.

1977 (January 31) – Louisville, Kentucky, U.S: Three buildings and a car struck.

1979 (June 7) – Cilimus, Indonesia: Meteorite fell in garden.

1979 (September 22) – The Vela Incident (sometimes known as the South Atlantic Flash): The flash was detected on September 22, 1979, at 00:53 GMT, by a US Vela satellite that was specifically developed to detect nuclear explosions. The satellite reported the characteristic double flash (a very fast and very bright flash, then a longer and less-bright one) of an atmospheric nuclear explosion of two to three kilotons, in the Indian Ocean between Bouvet Island and the Prince Edward Islands at 47° S 40° E. Hydrophones operated by the U.S. Navy detected a signal which was consistent with a small nuclear explosion on or slightly under the surface of the water near the Prince Edward Islands. The radio telescope at Arecibo, Puerto Rico, also detected an anomalous traveling ionospheric disturbance at the same time. "There remains uncertainty about whether the South Atlantic Flash in September 1979 recorded by optical sensors on the U.S. Vela satellite was a nuclear detonation and, if so, to whom it belonged."

1981 (June 13) – Salem, Oregon, U.S: Building hit.

1982 (November 8) – Wethersfield, Connecticut, U.S: Pierced roof of house.

1984 (June 15) – Nantong, PRC: Man missed by seven meters.

1984 (June 30) – Aomori, Japan: Building struck.

1984 (August 22) –Tomiya, Japan: Two buildings hit.

1984 (September 30) – Binnigup, Australia: Two sunbathers missed by 5 meters.

1984 (December 5) – Cuneo, Italy: Strong explosion, building flash; windows broken; daytime fireball "bright as Sun."

1984 (December 10) – Claxton, Georgia, U.S: Mailbox destroyed by meteorite.

1985 (January 6) – La Criolla, Argentina: Farmhouse roof pierced, door smashed; 9.5 kg stone misses woman by 2 meters.

1985 (June 26) – Hartford, Connecticut, U.S: A 1,500-pound slab of ice, six-feet long and eight-inches thick flattened a picket fence. The ground shook with the impact. A 13-year-old boy and his friend were standing ten feet away.

1986 (July 29) – Kokubunji, Japan: Several buildings hit.

1988 (March 1) – Trebbin, GDR: Greenhouse struck by meteorite.

1988 (May 18) – Torino, Italy: Building struck.

1989 (June 12) – Opotiki, New Zealand: Building hit.

1989, August 15) – Sixiangkou, PRC: Building hit.

1990 (April 7) – Enschede, Netherlands: House hit by believed fragment of Midas.

1990 (July 2) – Masvingo, Zimbabwe: Person missed by 5 meters.

1991 – Tahara, Japan: Meteorite struck deck of car-transport ship; made crater.

1991 (August 31) – Noblesville, Indiana: Meteorite fall missed two boys by 3.5 meters.

1992 (August 14) – Mbale, Uganda: Forty-eight stones fall; roofs damaged, boy struck on head.

1992 (October 9) – Peerskill, New York, U.S: Car trunk, floor, pierced by meteorite.

1994 (January 18) – Cando, Spain: An explosion that occurred in the village of Cando, Spain, in the morning of January 18, 1994. There were no casualties in this incident, which has been described as being like a small Tunguska Event. Witnesses claim to have seen a fireball in the sky lasting for almost one minute. A possible explosion site was established when a local resident called the University of Santiago de Compostela to report an unknown gouge in a hillside close to the village. Up to 200 square meters of terrain was missing and trees were found displaced 100 meters down the hill.

1994 (July 16) – Fragments of Comet Shoemaker-Levy begin impacting Jupiter.

1994 (October 20) – Coleman, Michigan: Meteorite penetrated roof of house.

1995 – Neagari, Japan: Meteorite penetrated car trunk.

1996 (November 26) – Honduras: According to the Associated Press, "A meteorite slammed into a sparsely populated area of Honduras last month, terrifying residents and leaving a 165-foot-wide crater, scientists confirmed Sunday. Near San Luis, in the western province of Santa Barbara."

1997 (April 11) – Chambrey, France: Meteorite penetrated roof of car; set fire.

1998 (June 13) – Portales, New Mexico: Meteorite penetrated barn roof.

1998 (July 12) – Kitchener, Ontario, Canada: Meteorite falls one meter from golfer.

2000 (January) –Whitehorse, Yukon, Canada: A 150-tonne meteoroid lit the skies over Whitehorse, and exploded over a lake about

100 kilometres south of the city. The Tagish Lake meteor produced a treasure of information about a rare kind of meteorite.

2000 (January) – Iberian Peninsula: Ice chunks weighing up to 6.6 pounds rained on Spain for ten days, causing extensive damage to cars and an industrial storage facility. At first scientists thought the phenomenon was unique to Spain. During the past three years, however, they've accumulated strong evidence that megacryometeors are falling all around the globe. More than fifty falls have been confirmed, and researchers believe that is a small fraction of the actual number, since others may hit unoccupied areas or melt before discovery. Most megacrymeteor falls occur in January, February and March. Megacryometeors show the telltale onionskin layering seen in hailstones. They also contain the dust particles and air pockets found in hail. But they are formed in cloudless skies, a characteristic that defies research on hail formation.

2001 (July 25 to September 23) – Kerala, India: Red rain sporadically fell, staining clothes with an appearance similar to that of blood. Yellow, green and black rain was also reported. The rains were the result of the atmospheric disintegration of a comet, according to a study conducted at the School of Pure and Applied Physics of the MG University by Dr Godfrey Louis and his student Santosh Kumar. The red rain cells were devoid of DNA, which suggests their extra-terrestrial origin. The findings published in the international journal *Astrophysics and Space Science* state that the cometery fragment contained dense collection of red cells.

2002 (June 6) – Asteroid/comet explosion over the Mediterranean: Estimated at five to ten meters in diameter, it released a burst of energy comparable to the nuclear bomb dropped on Hiroshima, Japan.

2002 (September 24) – Near Bodaibo, Irkutsk, Siberia: Eye-witness accounts of the 1:50 a.m. event reported a large luminous object falling to Earth near Bodiabo in Siberia. Hunters in the region have also reported the existence of a crater surrounded by burnt forest, suggesting that an impact event had occurred. The event was detected by near-by geophones as a moderate-earthquake. The event was also detected by a U.S. anti-missile defense military satellite. Some attempts were made to define the magnitude of the explosion. U.S. military analysts calculated it was between 0.2 and 0.5 kilotons, while Russian physicist Andrey Olkhovatov estimates it at 4-5 kilotons. Information about the event appeared in the mass media and among scientists after only a week. Another report says it occurred on the 25th of September at 10:00 p.m.

2004 (June) – Auckland, New Zealand: Meteor crashes through roof of home, damages sofa. The meteorite was a four billion-year-old 1.3 kg rock. "There was this huge bang and a cloud of dust and debris went through the front room. I thought a car had hit the house." In the only account in New Zealand of a meteorite crashing into a house, the chunk of space rock punched a hole through the roof of the Archers' home, bounced off their couch, ricocheted off the ceiling and back on to the couch before ending up on the floor.

2004 (September 3) – Antarctica: A small asteroid exploded in the stratosphere above Antarctica, depositing sufficient micron-sized dust particles to cause "local cooling, and much speculation as to the possible effects on the ozone layer."

2004 (December 26) – Southeast Asia: An undersea earthquake occurred at 00:58:53 UTC with an epicentre off the west coast of Sumatra, Indonesia. It has been said that the earthquake was caused by subduction and triggered a series of devastating tsunamis; however, a comet/asteroid strike cannot be ruled out. Along the coasts of most landmasses bordering the Indian Ocean, more than 225,000 people in eleven countries were killed. Coastal communities were inundated with waves up to 30 meters (100 feet) high. It was one of the deadliest natural disasters in history. Indonesia, Sri Lanka, India and Thailand were hardest hit. See: "Mega-tsunami hit southeast Asia 700 years ago"[46] and compare the events of 1430 with those above.

2006 (February 1) – Calgary, Alberta, Canada: On February 1, twenty people reported seeing a fireball, an exceptionally bright meteor, streak across the sky just before 7 a.m., lasting for several seconds before breaking up into fragments. It was estimated that remnants of the meteorite landed about 400 km south of Calgary somewhere in Montana, about two minutes after it appeared as a ball of fire.

2006 (February 1) – Bangladesh: The Dhaka *Daily Star* newspaper published the following report: "A 'meteor' from outer space fell with a big bang on a field in the Singpara village of Sadar Upazila yesterday afternoon, creating panic and curiosity among people. No one was reported hurt. Superintendent of Police Khandker Golam Farooq rushed to the spot and asked his companions and villagers to dig the earth near the house of one Fazlur Rahman from where smoke was still emitting. To their amazement they found a lead-like black material three feet below the earth. Hot and weighing 2.5 kg, the triangular material

---

[46] http://www.sott.net/articles/show/168328-Mega-tsunami-hit-southeast-Asia-700-years-ago

looked like a mortar shell, witnesses said. The meteor was kept in custody of the Thakurgaon Police Station."

2006 (February 17 and 20) – Scotland: The U.K.'s *Daily Record* reported that "The hunt is on for the crash sites of two meteors near Stirling Castle. Scientists have been spurred into action by reports of spectacular 'balls of fire' falling in the area. If discovered, they would be the first meteorites confirmed to have hit north of the border for almost 100 years. The incidents, reported by several witnesses, were on the evenings of Friday, February 17 and the following Monday, February 20. […] John Faithfull, curator of mineralogy and petrology at Glasgow University's Hunterian Museum, said yesterday: 'Although meteorite falls are rare everywhere, Scotland seems to have escaped remarkably lightly. There have only been four meteorites recovered from Scotland, compared with more than eighteen from England and Wales. Statistically, we are overdue another one.'"

2006 (April 12) – Australia: A Perth astronomer reported that a spectacular light show in the sky was a meteor. Sightings were made as far south as Albany and inland through the Wheat Belt. It lit up the countryside for hundreds of kilometers around the south-west of Western Australia. Witnesses say the sky lit up about 9:00 p.m. AEDT, and the light was followed by a thundering sound that shook buildings.

2006 (May 4) – Texas, U.S: Astronomers said a large meteor shower crossed straight over El Paso just before 9:45 p.m. on May 4. One meteor was so large that it cast an orange glow against the mountain. "The animals were going wild, the horses were bucking and dogs were barking and howling and then, all of a sudden right above my house, there was a big bright light and then just 'Bang!' And it lit up the five acres that are around us, and then I covered my eyes like this because it was bright and when it got past I saw there was a tail and it just went 'shhhh' toward the Hueco Mountains."

2006 (June 2) – Minnesota/Wisconsin/North Dakota, U.S and Canada: A fireball was spotted estimated to be some twenty miles above the Earth's surface. A sonic boom was heard in the Lake of the Woods area of Minnesota, so there may be some pieces of the meteor that survived the fall.

2006 (June 19) – Pennsylvania, U.S: Residents of the Tuscarawas Valley who heard a deafening boom about 12:40 a.m. on Monday the 19th and stepped outside likely saw what one person described as "a marvelous fireball with red streaks in the sky." It probably was a meteor falling through the atmosphere. Numerous callers reported a large red fireball. Several said their homes shook. New Philadelphia police said they received reports from several callers who witnessed the fireball or heard the boom. One woman described it as "a blue light that lit up the sky and went down." Police in Dover said multiple callers reported they heard a loud bang and something rattled their windows. Air Traffic Command in Washington, D.C. confirmed that Cleveland's control center was checking into a meteor shower that occurred within its air space.

2006 (July 10) – South Africa: An ice ball that landed in Douglasdale, South Africa, might be one of the first "megacryometeors" recorded in Africa. The ice ball, which landed on the pavement in suburban Douglasdal, was about the size of a microwave oven. The impact of the ice ball's fall created a small crater on the pavement, which was covered with pieces of broken ice. Despite sharing many chemical characteristics with hail, ice balls are formed under clear-sky conditions. Ice balls have been recorded since the 19th century. They have the potential to damage people, buildings and cars, but no injuries were reported as a result of this one.

2006 (July 14) – Norway: At 10:20 a.m. a bus driver from Ås, south of Oslo, was sitting in the outhouse at his holiday cabin near Rygge on the 14th of July when he heard an enormous blast. Right after that, some particles from a meteor that exploded over the Oslo area rained down just outside. He said he didn't think too much about the surprising blast at first, dismissing it as probably coming from an exercise at a nearby military air station at Rygge. But he said the blast and the rumbling it caused was terrible. He was just hooking the door when he heard a new noise, a whistling sort of sound, followed by a new bang on some aluminum plates lying near the outhouse. Sure enough, it was particles from a meteor that exploded somewhere over the Oslo Fjord area on Friday morning. Astronomers confirm Martinsen's remarkable discovery of meteorite particles on his property. "This is Norway's 14th meteorite, but we've never heard about a meteorite landing so close to a person before." A family from Moss, south of Oslo, came home from their summer holidays to find a meteorite in their garden. It's another remnant of the meteor that exploded over the Oslo Fjord area on the 14th of July. Astronomers in Norway are calling the discovery of meteorites around southeast Norway "incredible," and urge local residents to keep looking for more. "Two branches on our plum tree were broken. I lifted them up and there lay this stone." It had made a hole measuring about seven centimeters in his lawn.

2006 (September 12) – New Zealand: A small piece of rock found in a paddock in New Zealand may be a piece of the meteorite that streaked across the sky there on Tuesday the 12th, panicking residents who flooded emergency hotlines. A farmer found a 10 by 5 centimeter piece of "almost weightless" rock in his field today near the town of Dunsandel, south of Christchurch. It was sent to New Zealand's National Radiation Laboratory for analysis. The meteorite tore across the sky over the northern half of the South Island in the afternoon, leaving a bright, burning trail behind it, and causing a sonic boom that rattled houses and shook the ground. It then apparently erupted into a fireball, sending forth a thick puff of smoke. People were sent running from the homes and offices when they heard the boom, fearing buildings could collapse. The sonic boom was registered on earthquake-detecting equipment. The boom meant the meteorite was probably travelling "very low." It was probably about the size of a basketball as it shredded through the sky and became a "terminal fireball" at a speed of about 40,000 kph. "If this had happened at night, it would have lit up the whole countryside."

2006 (October 10) – Bonn, Germany: A fire that destroyed a cottage near Bonn and injured a 77-year-old man was probably caused by a meteor, and witnesses saw an arc of blazing light in the sky. Burkhard Rick, a spokesman for the police in Siegburg east of Bonn, said the fire gutted the cottage and badly burnt the man's hands and face in the incident on October 10.

2006 (November 17) – The Moon: NASA reports that meteoroids are smashing into the Moon a lot more often than anyone expected. That's the tentative conclusion of Bill Cooke, head of NASA's Meteoroid Environment Office, after his team observed two Leonids hitting the Moon on November 17, 2006. "We've now seen 11 and possibly 12 lunar impacts since we started monitoring the Moon one year ago," says Cooke. "That's *about four times more hits than our computer models predicted.*" [This author's emphasis]

2007 (January) – Tampa, Florida, U.S: A 200-pound chunk of ice streaked through the clear Florida sky and landed in the back seat of a really nice red Ford Mustang. The car was totaled.

2007 (January 4) – New Jersey, U.S : Authorities were trying to identify a mysterious metallic object that crashed through the roof of a house in eastern New Jersey. Nobody was injured when the golf-ball sized object, weighing nearly as much as a can of soup, struck the home and embedded itself in a wall Tuesday night. Approximately 20 to 50 rock-like objects fall every day over the entire planet, said Carlton Pryor, a professor of astronomy at Rutgers University. "It's not all that uncommon to have rocks rain down from heaven," said Pryor, who had not seen the object that struck the Monmouth County home. "These are usually rocky or a mixture of rock and metal."

2007 (January 10) – Russia: A meteorite fell in January in the Altai Territory in southern Siberia and searchers found an extraterrestrial substance which could be meteorite fragments. "We have collected about 50 samples, and vitreous threads (traces of comet substance) were discovered in the first of them using a microscope." Local motorists and residents witnessed the impact of a fiery ball, which eventually ended in a loud sound resembling an explosion.

2007 (January 24) – Virginia, U.S.: Giles County residents were a little shaken after a tremor-like event, others say they heard a loud "thunder-like" sound. Virginia Tech researchers say they received several calls about a meteor sighting the same time of the tremors. The bizarre incident took place around 8:00 p.m. Researchers say the seismic station in Giles County did get a very short but intense seismic signal.

2007 (January 31) – Turkey: Police were inundated with calls from scores of people from Didim to Bodrum after they heard a big bang and a flash of light across the skies. The flashing green, yellow and red lights were from a meteorite which crashed through the earth's atmosphere and landed in Yesilkent. A startled man revealed that the rock had smashed a hole in the ground at the Green Park Complex, at Yesilkent, narrowly missing him by ten meters. Police reported that people from Bodrum, Milas and Didim had heard a bang and seen the flashing light across the skies at about 5:30 p.m.

2007 (February 4) – Midwestern U.S.: Scores of people all over the Midwest and Upper Midwestern United States reported seeing flames and fiery explosions in the sky Sunday night. From southeastern Wisconsin to as far as Des Moines, Iowa and St. Louis, people reported seeing balls of fire, possibly meteors, streaking across the sky on Sunday night. "We had a pilot reporting seeing a meteor." Reports came from residents in central Missouri, Illinois, Kansas, Wisconsin and Minnesota.

2007 (February 15) – Ohio, U.S: Something happened at around 9 p.m. that a lot of people heard. But nobody seems to have any idea what it was. "It" was a loud bang, something loud enough to be heard all over the county, and strong enough to make small objects move in houses. Rumors range from an earthquake to a meteor strike, a sonic boom or something ice-related. At least one scientist believes the meteor could be the answer. There's no evidence to suggest an earthquake could have caused the bang, especially not over the range specified. One man said he saw a meteor with a relatively long trail, with red, green and gold coloration. It was headed east to west and lasted about three seconds; after it faded, the sonic boom washed over him. "I saw it first. It was the most eerie, cool, scary, wonderful thing. You just see this dragon tail going across the sky. All of a sudden, everything goes boom."

2007 (February 22) – Rajasthan, India: Three people were killed and four injured in a mysterious blast in a village in India's northern Rajasthan state Thursday that villagers claim was caused by a meteorite, news reports said. Residents of Banchola village in the Bundi district, about 200 kilometers south of Rajasthan capital Jaipur, said the victims were sitting with some iron scrap in an open field when an "object" fell from the sky and hit them.

2007 (February 23) – Panama: Panamanian geologists found a meteorite at Rio Hato, a coastal town west of the capital Panama City. The meteorite fell onto Rio Hato's beach. The landing was witnessed by a security guard, who described it as a ball of fire crashing down from the sky onto the sand. The 4.2 kg red object, measuring 20 centimeters in diameter, was to be X-rayed for more details. The meteorite shows burn marks on its exterior, and appears to be mainly carbon-based, in contrast to most meteorites, which mainly contain iron.

2007 (March 15) – Ontario, Canada: What Richard Yip-Chuck saw fall into a farmer's field Sunday evening looked like a long, white ball with orange sparks shooting off the back. The Holland Landing resident was driving along Highway 7 with his wife and sons when they saw what looked like a fireball plummet to earth.

2007 (March 29) – New Zealand: Flaming debris of a possible meteor almost hit a plane. The pilots of a Chilean passenger jet reported seeing flaming debris fall past their aircraft as it approached the airport at Auckland, New Zealand. The captain "made visual contact with incandescent fragments several kilometers away." The pilots reported the near-miss to air traffic controllers, reportedly saying the noise of the debris breaking the sound barrier could be heard above the roar of his aircraft's engines.

2007 (May 10) – Spain: A fireball was spotted across central Spain. Scientists think some fragments may have fallen to earth in the Ciudad Real area. The fireball fell across the center of the country with sightings in Cuenca, Toledo, Ciudad Real and Valladolid. Scientists believe it was a meteorite and say it is quite a normal phenomenon, possibly a fragment from a comet which fell from earth orbit.

2007 (May 14) – Vermont, U.S: Recorded as a 2.1 temblor on the Richter scale, a quake hit Hubbardton, Vermont at 4:10 a.m. One resident, who said he was wide awake at 4 a.m., said he not only felt the earthquake, he saw what caused it. He said he saw something in the sky to the northeast of Lake Hortonia. He believes he saw a meteorite and that's what triggered the earthquake. "It was like a streak of fire. I've heard meteorites hit before and that was what it sounded like. It was no earthquake, it was a meteor."

2007 (May 26) – Woburn, Massachusetts, U.S: A meteorite reportedly punched a hole through a warehouse roof.

2007 (June 7) – Norway: A large meteorite struck in northern Norway, landing with an impact an astronomer compared to the atomic bomb used at Hiroshima. The meteorite appeared as a ball of fire just after 2 a.m., visible across several hundred miles in the sunlit summer sky above the Arctic Circle. "I saw a brilliant flash of light in the sky, and this became a light with a tail of smoke. I heard the bang seven minutes later. It sounded like when you set off a solid charge of dynamite a kilometer away." The meteor struck a mountainside in Reisadalen. The country's leading astronomer said he expects the meteor to prove to be the largest to hit Norway in modern times, even bigger than the 198-pound Alta meteorite of 1904. "If the meteorite was as large as it seems to have been, we can compare it to the Hiroshima bomb. Of course the meteorite is not radioactive, but in explosive force we may be able to compare it to the bomb."

2007 (June 10) – Sri Lanka: The strange objects that lit the night skies on June 10 have now been confirmed as meteors. "This is the first time that meteors of such magnitude have fallen in Sri Lanka. The shockwaves and vibrations have been heard throughout the country, from Galle to Puttalam." A Senior Consultant believes that two large meteoroids entered the atmosphere, the larger one splitting into two and the smaller one into about 25 fragments. The loud explosions were some of the particles exploding, probably about 50 to 100 kilometers above the ground. In Kovinna, Andiambalama, at 9.05 p.m. on the 10th, a woman had noticed something unusual in the western sky. A bright light, almost as large as the full moon, appeared to be moving towards her in a wide arc. Alarmed by thoughts of terrorist air attacks, she called out to her neighbour. Together they watched fearfully as the glowing object drew closer, landed on the roof and vanished completely. A few minutes later the air vibrated with a loud explosion. The next day they discovered that parts of the asbestos sheets on the roof were charred and cracked. A few pieces of rock and sand were scattered around the damaged area. Similar incidents were reported around the country that night. Several people in areas such as Puttalam, Maho and Bingiriya also noted the appearance of the bright light in the sky, as well as the loud explosion. In Kimbulapitiya a woman watched a flaming object land on a house and heard the booming sounds soon afterwards. In Campbell Place, Dehiwala, the roofs of two buildings were damaged, and a loud noise was heard.

2007 (July 6) –Colombia, South America: An incoming object broke apart in the lower atmosphere with a trio of ferocious explosions that shattered windows and shook the ground violently. Moments later, stones rained from the sky and pelted homes in the poor barrios surrounding the city. Some smashed through the roofs of homes. Recovered objects were chondritic (rocky) meteorite.

2007 (July 26) – Iowa, U.S: A Dubuque woman said she was lucky to be alive after a 50 pound chunk of white ice crashed through the roof of her home, landing about 15 feet away from where she was standing. She said it sounded like a bomb exploded when the massive ball of ice hit her roof. Other large chunks of ice fell from the sky in this northeast Iowa city, tearing through nearby trees. Dubuque had clear skies at the time the ice fell.

2007 (August 1) – India: Hotipur (Sangrur) village near Khanauri hit the headlines when a meteorite fell in the, leaving many villagers baffled. The police took possession of the 8-centimeter meteorite and handed it over to a three-member team of the Geological Survey of India. Curious villagers queued up in the fields to see the "heavenly object," while the farmer, who was the only witness to the fall of the "fireball," said, "I got scared of the big fireball that was coming my way.... I ran for cover as I felt that it will fall on me." [May be a hoax.]

2007 (August 11) – California, U.S: Representatives with the Sonora Police Department and both the Tuolumne and Calaveras County Sheriff's Departments say they fielded numerous calls early in the morning in regards to a "loud boom," and "structures shaking." There were several calls from residents who reported seeing "a blue light," just before the "loud boom." The incident reportedly occurred at 12:09 a.m. The Police Department notes that it also received a call from a resident in Tuolumne, in which a female reported seeing what she thought was fireworks, and then something spiraling over her house. Early indication from the law enforcement agencies is that the loud boom was somehow the result of a meteor shower.

2007 (September 15) – Peruvian Highlands: A meteorite's impact sent debris flying up to 820 feet away, with some material landing on the roof of the nearest home 390 feet from the crater. Nearby residents who visited the impact crater complained of headaches and nausea.

2007 (October 3) – Minnesota, U.S: People across the Twin Cities reported seeing a "metallic" object or "flaming ball" falling from the sky. Broadcasters and emergency dispatchers got hundreds of calls from people who saw the object traveling from the northeast to the southwest. Residents of Lyon County in far southwestern Minnesota reported a loud boom that might have been connected with the sightings in the Twin Cities. A man who lives near the town of Amiret says it shook his house and sounded like a sonic boom from an F-14 breaking the sound barrier at close range. Coincidentally, at the same time, drivers in the Twin Cities metro were dodging debris in the middle of Interstate 94. Some drivers said the debris fell from the sky shortly after 2:00 p.m. Wednesday.

2008 (January 31) – Didim, Turkey: Police were inundated with calls from scores of people from Didim to Bodrum after they heard a big bang and a flash of light across the skies. A startled Abdullah Arıtürk revealed that the rock had smashed a hole in the ground at the Green Park Complex, at Yeşilkent, narrowly missing him by ten meters.

2008 (February 19) – Northwest U.S: An apparent meteor streaked through the sky over the Pacific Northwest, drawing reports of bright lights and sonic booms in parts of Washington, Oregon and Idaho. Although a witness reported seeing the object strike the Earth in a remote part of Adams County, in southeast Washington, it still has not been found. People in Washington, Oregon, Idaho, Montana and British Columbia reported seeing the bright fireball streaking across the sky about 5:30 a.m. At least one person said the object exploded on impact in eastern Washington and another report from southeastern Washington said someone felt tremors from the blast.

2008 (March 5) – Ontario, Canada: The Physics and Astronomy Department at the University of Western Ontario has a network of all-sky cameras in Southern Ontario that scan the sky monitoring for meteors. Associate Professor Peter Brown, who specializes in the study of meteors and meteorites, says that Wednesday evening (March 5) at 10:59 p.m. EST these cameras captured video of a large fireball. The department also received a number of calls and emails from people who actually saw the light.

2008 (March 8) – Turkey: A resident of Yaka said he heard a loud roaring noise at around 11:20 a.m. sounding as if "a plane had crashed." "We were amazed to find such a small stone after that thunderous sound. It was black and about 40 centimeters in diameter, weighing three kilograms at most," another said, adding that the meteorite opened a small crater in the ground and created a cloud of dust.

2008 (March 10) – Sudbury, Canada: Great balls of fire were seen falling from the sky. While most sightings were reported around 1:30 p.m. near Sudbury, Hagar, Highway 69 North and North Bay, Wayne

Lachance spotted something in the sky earlier in the morning. Lachance was driving home to Massey after a night shift at Vale Inco Ltd. when something caught his eye around 7:30 a.m. "I thought it was a real bright star," he said. "It was getting brighter and coming down with sparks." Lachance arrived home and looked outside his bedroom window to see "spirals of smoke" falling.

2008 (March 13) – The Moon: Meteorite videotaped hitting the Moon.

2008 (April 6) – Argentina: A space rock reportedly crashed somewhere in Entre Rios Province, some 260 miles northwest of Buenos Aires. Milton Blumhagen, a witness and astronomy buff said: "For three or four seconds I saw an object in flames, changing color until it turned blue when it approached the ground." A fire department source said the impact was felt for miles around. No damage was reported.

2008 (April 15, 16, 18) – Illinois, U.S: Maybe there was a comet-fragment impact or two (or three) over a period of several nights; perhaps a couple of overhead explosions and then, later, a ground impact. That would explain booms, earthquake and lights in the sky spread out over three days. Read the following stories and judge for yourself:

"Damage Control: Mysterious booms, lights over Indiana were just F-16s"[47]

A sonic boom and fireballs and flaming debris that Kokomo-area residents reported seeing in the sky Wednesday night prompted Howard County's police agencies to conduct a two-hour search for what many residents thought was a crashed aircraft.

As it turned out, the fireballs were flares fired by F-16s that are part of the 122nd Fighter Wing, an Indiana Air National Guard unit based at Fort Wayne International Airport. [...] Staff Sgt. Jeff Lowry, with Indiana National Guard's headquarters in Indianapolis, said the jets taking part in the training are not supposed to exceed the speed of sound, which is about 760 mph, because supersonic speeds produce sonic booms.

He said the 122nd's commander, Col. Jeff Soldner, will investigate why at least one jet reached supersonic speeds Wednesday night over Howard and Tipton counties, and also on Tuesday night over the Logansport area, shaking the ground below. [...] He said F-16 training often involves the aircraft dropping flares from more than 10,000 feet

[47] http://www.sott.net/articles/show/154005-Damage-Control-Mysterious-booms-lights-over-Indiana-were-just-F-16s

above the ground, a technique that can allow the jets to evade heat-seeking missiles in combat. [...]

Logansport Police Chief A.J. Rozzi said he heard a loud sonic boom on Tuesday night, and then heard the sound of a jet high overheard. He said residents also reported seeing fire streaks in the sky. He said it is common for the 122nd to conduct missions in the area and believes F-16 training almost certainly explains the sights and sounds.

"They've been doing that training for quite a while. I don't know what maneuvers they're actually doing, but they do shoot out streaks of light," he said.

"5.4 earthquake rocks Illinois; felt 350 miles away"[48]

A 5.4 earthquake that appeared to rival the strongest recorded in the region rocked people awake up to 350 miles away early Friday, surprising residents unaccustomed to such a powerful Midwest temblor.

The quake just before 4:37 a.m. was centered 6 miles from West Salem, Illinois, and 66 miles from Evansville, Indiana. It was felt in such distant cities as Chicago, Cincinnati and Milwaukee, 350 miles north of the epicenter, but there were no early reports of injuries or significant damage. [....]

"You could hear a roaring sound and the whole motel shook, waking up the guests," Vibha Ambelal, manager of the Super 8 Motel in Mount Carmel, Illinois, near the epicenter, said in a telephone interview.

"4.5 Magnitude Earthquake Hits Illinois, Continuing Series"[49]

A 4.5-magnitude tremor struck southern Illinois on Monday continuing the series of aftershocks initiated by the 5.2 earthquake which hit the region Friday morning, the U.S. Geological Survey (USGS) informed.

This was the 18th earthquake in that series and its epicenter was approximately six miles below ground and about 37 miles (60 km) north-northwest of Evansville, Indiana, or about 131 miles (211 km) east of St. Louis, the USGS revealed. [...]

The 18 aftershock earthquakes which followed Friday's tremor haven't measured more than 3.9 on the Richter scale, but the first one was the biggest to shake the region, called the Illinois basin-Ozark dome, in over 40 years.

2008 (April 16) – Argentina: The Entre Ríos Astronomy Society in Argentina announced that on Wednesday, April 16th, 2008, at approximately 19:30 hours, they observed a highly luminous object that

---

[48] http://www.sott.net/articles/show/153997-5-4-earthquake-rocks-Illinois-felt-350-miles-away

[49] http://www.sott.net/articles/show/154193-4-5-Magnitude-Earthquake-Hits-Illinois-Continuing-Series

had all the characteristics of a bolide. This object was sighted from Paraná, Oro Verde and San Benito. According to witnesses, the bolide was intensely bright, with colours fluctuating between green, yellow and red. It followed *a roughly north-east trajectory towards the south-west*, with an angle of 75 degrees. *One observer has stated that the bolide exploded before disappearing.* It is not possible to discount the idea that this meteor relates to a similar fall which occurred the previous week over central Entre Ríos province, and which was observed across a wide part of Argentina. *The AEA also received over the past few days many reports of sightings of very luminous objects in different parts from the country*, e.g. from Mar del Plata, Tucumán, Zárate, Concordia, Ituzaingó (Prov. de Corrientes), etc.

2008 (April 17) – Argentina: [This may be the same event as reported on April 16th above.] A fireball fell somewhere in or nearby Entre Rios, 260 miles northwest of Buenos Aires.[50] Mariano Peter from the Entrerriana Astronomy Association said there were reports from four witnesses. One of them described "a strong light that passed at a high speed through the sky and at a low altitude, going towards the south and then fell in the distance." Another witness said, "It was very bright and it changed color between green and red." The first fireball was reported in Entre Rios on April 6th, 2008 [see above]. A witness said: "For three or four seconds I saw an object in flames, changing color until it turned blue when it approached the ground." A fire department source said *the impact was felt for miles around.* The next day a fragment of the space rock was recovered.

And now: "Smoke chokes Argentina's capital"[51]

> Smoke blanketed the Argentine capital Friday as *brush fires apparently set deliberately* consumed thousands of acres in the provinces of Buenos Aires and Entre Ríos.
>
> The smoke, from about 300 fires, is blamed for at least two fatal traffic accidents this week that left eight people dead. Sections of major highways and the Buenos Aires port, among the busiest in the world, have been closed. Incoming flights to the city's domestic airport, Jorge Newbery Airpark, have been diverted.
>
> *The Argentine government has blamed farmers looking to clear their land for crops and grazing* for the fires, which are estimated to cover 173,000 acres (70,000 hectares).

50

http://www.jornadaonline.com/LeerNoticia.asp?Tabla=Noticias&Seccion=Nacional&id=11866
[51] http://www.sott.net/articles/show/154066-Smoke-chokes-Argentina-s-capital

"This is the largest fire of this kind that we've ever seen," Argentine Interior Minister Florencio Randazzo said Thursday. Randazzo called the situation a "disaster."

As of Friday morning, little progress had been made extinguishing the blazes.

2008 (April 20) – Russia: "Another overhead explosion? Two killed, 300 left homeless in Russian Far East fires"[52]

Two people have died and 325 people including 18 children have been left homeless by fires that ripped through the Amur Region in Russia's Far East, local emergency services said.

The fires began on Sunday evening and continued until Monday morning in seven districts of the region. *Locals had set light to dry grass to free land for farming and other purposes, and the flames were spread by high winds*, a police source told RIA Novosti.

A total of 104 houses have been destroyed. One of those who died in the fires was a disabled man who was unable to leave his home. A total of 50 rescuers have been involved in the firefighting operation. People injured in the fires will receive 20,000 rubles ($900) in compensation, local authorities said.

Over 11,000 hectares have been destroyed in an estimated *59 forest fires* currently burning in Russia's Far East, the Natural Resources Ministry said.

Curious how this report is similar to what happened several days ago in Argentina. And again, farmers all decided to set fires on the same day, and burning grass and high winds are blamed for the vast damage and considerable destruction. We wonder what kind of excuse authorities will invent when this kind of event happens in a non-agricultural area.

---

[52] http://www.sott.net/articles/show/154226-Another-overhead-explosion-Two-killed-300-left-homeless-in-Russian-Far-East-fires

Appendix B

03-11-95
Q: (L) At one point we were told that time was an illusion that came into being at the "time" of the "Fall" in Eden, and this was said in such a way that I inferred that there were other illusions put into place at that time....
A: Time is an illusion that works for you because of your altered DNA state.
Q: (L) Okay, what other illusions?
A: Monotheism, the belief in one separate, all powerful entity.
Q: (T) Is separate the key word in regard to Monotheism?
A: Yes
Q: (L) What is another one of the illusions?
A: The need for physical aggrandizement.
Q: (L) What is another of the illusions?
A: Linear focus.
Q: (L) Anything else at this time?
A: Unidimensionality. [...]
Q: (L) Were these illusions programmed into us genetically through our DNA?
A: Close. [...]
Q: (L) Can you tell us a little bit about how these illusions are enforced on us, how they are perceived by us?
A: If someone opens a door, and behind it you see a pot of gold, do you worry whether there is a poisonous snake behind the door hidden from view, before you reach for the pot of gold?
Q: (L) What does the gold represent?
A: Temptation to limitation.
Q: (L) What does the door represent?
A: Opening for limitation. [...] What is snake? [...]
Q: (L) Who was the snake?
A: Result of giving into temptation without caution, i.e. leaping before looking. [...]
Q: (L) So what you are saying to us is that the story of the temptation in Eden was the story of Humankind being led into this reality as a result of

being tempted. So, the eating of the fruit of the Tree of Knowledge of Good and Evil was....

A: Giving into temptation. [...] Free will could not be abridged if you had not obliged.

Q: (T) What were we before the "Fall"?

A: 3rd density STO. [...]

Q: (T) We are STS at this point because of what happened then?

A: Yes.

Q: (T) Okay, now, we were STO at that time. [...] Was this after the battle that had transpired? In other words, were we, as a 3rd density race, literally on our own at that point, as opposed to before?

A: Was battle.

Q: (L) The battle was in us?

A: Through you. [...]

Q: (T) Okay, we were STO at that point. You have said before that on this density we have the choice of being STS or STO.

A: Oh Terry, the battle is always there, it's "when" you choose that counts! [...]

Q: (T) This must tie into why the Lizards and other aliens keep telling people that they have given their consent for abduction and so forth. We were STO and now we are STS. [...]

A: Yes, continue.

Q: (T) We are working with the analogy. The gold was an illusion. The gold was not what we perceived it to be. It was a temptation that was given to us as STO beings on 3rd density. The door was opened by the Lizards.

A: No temptation, it was always there. Remember Dorothy and the Ruby Slippers? [...]

Q: (T) It's always there.... (J) It's there now....

A: Yes, think of the Ruby Slippers. What did Glenda tell Dorothy???

Q: (J) You can always go home. (L) You have always had the power to go home....

A: Yes.

Q: (L) So, we always have the power to return to being STO? Even in 3rd density?

A: Yes. [...] "When" you went for the gold, you said "Hello" to the Lizards and all that that implies.

Q: (T) The door was always there and always open. I was just trying to work with the analogy. So, the concept is that, as STO beings we had the choice of either going for the gold or not. By going for the gold, we became STS beings because going for the gold was STS.

A: Yes.

Q: (T) And, in doing so, we ended up aligning ourselves with the 4th density Lizard beings...

A: Yes.

Q: (T) Because they are 4th density beings and they have a lot more abilities than we at 3rd density....

A: You used to be aligned with 4th density STO.

Q: (T) And we were 3rd density STO. But, by going for the gold, we aligned ourselves with 4th density STS.

A: Yes.

08-28-99

Q: I have this book, this Marcia Schafer thing — *Confessions of an Intergalactic Anthropologist* — and [...] one thing she says: "The snake is associated with the sign of wisdom and higher learning, and is often regarded quite highly in mystical circles." She had an interaction with a rattlesnake, for which she felt sympathy, and she also has sympathetic interactions with Lizzies. I would like to have a comment on the idea of the snake as a "sign of wisdom and higher learning." Does this, in fact, represent what the snake symbolizes?

A: Snake is/was reported in context of the viewpoint of the observer. [...] Maybe the observer was just "blown away" by the experience. [...] If you were living in the desert, or jungle, about 7,000 years ago as you measure time, would you not be impressed if these Reptoid "dudes" came down from the heavens in silvery objects and demonstrated techno-wonders from thousands of years in the future, and taught you calculus, geometry and astrophysics to boot?!?

Q: Is that, in fact, what happened?

A: Yup. [...]

Q: As I understand it, or as I am trying to figure it out from the literature, prior to the "Fall in Eden," mankind lived in a 4th density state. Is that correct?

A: Semi/sort of. [...] 4th density in another realm, such as time/space continuum, etc.

Q: Okay, so this realm changed, as a part of the cycle; various choices were made. The human race went through the door after the "gold," so to speak, and became aligned with the Lizzies after the "female energy" consorted with the wrong side, so to speak. This is what you have said. This resulted in a number of effects: the breaking up of the DNA, the burning off of the first ten factors of DNA, the separation of the hemispheres of the brain....

A: Only reason for this: You play in the dirt, you're gonna get dirty. [...]

Q: Was there any understanding, or realization of any kind, that increased physicality could be like Osiris lured into his own coffin by Set? That they would then slam the lid shut and nail him in?

A: Obviously, such understanding was lacking.

Q: Sounds like a pretty naive bunch! Does the lack of this understanding reflect a lack of knowledge?

A: Of course. But more, it is desire getting in the way of.... [...] These events took place 309,000 years ago, as you measure it. This is when the first prototype of what you call "modern man" was created. The controllers had the bodies ready, they just needed the right soul-matrix to agree to "jump in."

Q: So, prior to this time, this prior Edenic state....

A: Was more like 4th density.

Q: But that implies that there was some level of physicality. Was there physicality in the sense of bodies that look like present-day humans?

A: Not quite. Cannot answer because it is too complex for you to understand.

Q: Does this mean that the... bodies we possibly would move into as 4th density beings, assuming that one does, would also be too complex for us to understand? You are saying that this "sort of 4th density" pre-Fall state, in terms of the physical bodies, is too complex to understand. If going back to 4th density is anything like coming from 4th density, does that mean that what we would go back to is something that is too complex to understand? This variability of physicality that you have described?

A: Yes.

04-08-00

Q: Now, I have this book entitled *Arktos*. He says something here that echoes a remark you once made. He says: "It is a very remarkable thing that enlightenment seems to have come from the North against the common prejudice that the Earth was enlightened as it was populated from South to North. The Scythians are one of the most ancient nations; the Chinese descend from them. The Atlanteans themselves, more ancient than the Egyptians, descend from them." You said that the civilizing influence came from the North to the South. Of course, all the standard texts claim that civilization came from South to North, starting in Mesopotamia. Now, getting....

A: Okay, just a minute here. Thinking Mesopotamia is the beginning is like thinking that the beginning starts at the 12th chapter.

Q: I know that! The problem is: finding artifacts. I've been searching and digging, and I find a little bit here and there, but my God! Either nothing survived....

A: Artifacts have a limited shelf life! Specimens survive by sheer luck. [...]

Q: ...[D]uring the time Neanderthal man was on the Earth, did he live alongside modern man?

A: Yes. Except modern-type man was different then.

Q: In what ways?

A: DNA and psycho/electrical frequencies.

Q: Does this mean that their physical appearance was different from what we consider to be modern man?

A: Radiance.

Q: What do you mean "radiance?"

A: You find out!

Q: Oh, that's interesting. Well, there are legends that the Northern people had "light" in their veins. Very ancient belief. Is this what you are referring to?

A: Maybe.

04-15-00

Q: (L) So, in effect, we *are* the new Neanderthals on the eve of extinction. You have said that those who transition into 4th density in the body will go through some kind of rejuvenation process or body regeneration or something. Does that mean that these present "Neanderthal"-type bodies that we presently occupy will morph into something more in line with the new model? Is it genetically encoded into some of them to do so?

A: Something like that.

Q: (L) So, that's why they have been following certain bloodlines for generation after generation; they are tinkering with the DNA and arming genetic time-bombs that are waiting to go off. (A) What is interesting is how do those who are trying to get these people, to abduct them, how do they spot them? How do they get the information? By following the bloodline, or by some kind of monitor you can detect from a long distance — and they can note that "here is somebody of interest" or "here is somebody dangerous" or "let's abduct this one" or whatever? How do they select? Do they search the genealogies or is it some kind of remote sensing?

A: Now this is interesting, Arkadiusz, as it involves the atomic "signature" of the cellular structure of the individual. In concert with this is the etheric-body reading and the frequency-resonance vibration. All these are interconnected, and can be read from a distance using remote viewing technology/methodology.

Q: (L) Can it be done in a pure mechanical way without using psychic means?

A: At another level of understanding, the two are blended into one.

Q: (T) Computerized psychic remote viewing, maybe. Like artificial intelligence. Maybe a mind connected to a computer?

A: That is close, yes.

07-22-00

Q: (L) I had a call from Vincent Bridges, who informed me that the *Wave* series was really creating a stir. It seems that he has had a connection to this Dr. Hammond of the Greenbaum lecture fame, and also had a number of exchanges with Andrija Puharich, and it is Vincent's contention that the UFO phenomenon, the alien abduction phenomenon, and the many and varied other things we talk about and study and discuss, are a product of super-advanced technological, *human-controlled* mind-programming projects using the technology of Puharich and Tesla. Yes, it is supposed to be so advanced that they can not only read minds and can control minds, but that it is, in the end, merely human-engineered programming. Is he, even in part, correct?

A: Well, there are elements of the phenomenon which may be connected to human, 3rd density STS engineering, *but by and large, this is not the case.*

Q: (L) He also said that it was his opinion, that the center of the web of all of this mind programming conspiracy is in Tyler, Texas. Is that correct?

A: The what?!?

Q: (L) Well, what about the center of the human branch of the programming conspiracy?

A: We feel that Vincent needs to recharge his batteries a bit.

Q: (L) He also said that the area we are living is the center of a particular programming experiment, something like Nazi/Black Magick cultists or something like that.

A: Better not to get too carried away. Remember, the root of all "negative" energies directed at 3rd density STS subjects, coming from 4th density, is essentially the same. [...] Suggest a review of the transcripts relating to the situation in Nazi Germany for better understanding here. [...] The concept of a "master race" put forward by the Nazis was merely a 4th density STS effort to create a physical vehicle with the correct frequency-resonance vibration for 4th density STS souls to occupy in 3rd density. It was also a "trial run" for planned events in what you perceive to be your future.

Q: (L) You mean, with a strong STS frequency so they can have a "vehicle" in 3rd density, so to speak?

A: Correct. Frequency-resonance vibration! Very important.

Q: (L) So, that is why they are programming and experimenting? And all these folks running around who some think are "programmed," could be individuals who are raising their nastiness levels high enough to accommodate the truly negative STS 4th density — sort of like walk-ins or something, only not nice ones?

A: You do not have very many of those present yet, but that was, and still is, the plan of some of the 4th density STS types.

08-05-00

Q: Okay, last session you brought up the subject of Frequency Resonance Vibration. You suggested that there are certain STS forces who are developing or creating or managing physical bodies that they are trying to increase the frequency so that they will have bodies that are wired so that they can manifest directly into 3rd density, since that seems to be the real barrier that prevents an all-out invasion, the fact that we are in 3rd density and they are in 4th. Now, I assumed that the same function could be true for STO individuals. It seems that many individuals who have come into this time period from the future, coming back into the past via the incarnational cycle so as not to violate free will, have carefully selected bodies with particular DNA, which they are, little by little, activating so that their 4th density selves, or higher, can manifest in this reality. Is it possible for those energies to manifest into such bodies which have been awakened or tuned in 3rd density?

A: STO tends to do the process within the natural flow of things. STS seeks to alter creation processes to fit their ends.

12-21-96

Q: (A) Which part of a human extends into 4th density?

A: That which is effected by pituitary gland.

Q: (L) And what is that?

A: Psychic.

Q: (A) Are there some particular DNA sequences that facilitate transmission between densities?

A: Addition of strands.

Q: (L) How do you get added strands?

A: You don't get, you receive.

Q: (L) Where are they received from?

A: Interaction with upcoming wave, if vibration is aligned.

Q: (L) How do you know if this is happening?

A: Psychophysiological changes manifest.

05-03-97

Q: Reading through the session of May 23, last year, when Tom was also here, and the issue of his being in O'Brien was addressed, you asked who had begged him to stay there. Then there was a remark about an EM vector. The way I understood it is that a person can be an EM vector. Is that possible?

A: Vector means focuser of direction.

Q: Could that mean that EM waves can be vectored by a human being simply by their presence? I also noticed that several of us have been involved with persons and relationships that seem designed to confuse, defuse, and otherwise distort our learning, as well as drain our energy. Basically, keeping us so stressed that we cannot fulfill our potential. Is there some significance to this observation?

A: That is elementary, my dear Knight!

Q: One of the things I have learned is that these individuals seem to attach via some sort of psychic hook that enters through our reactions of pity. Can you comment on the nature of pity?

A: Pity those who pity.

Q: But the ones who are being pitied, who generate sensations of pity, do not really pity anybody but themselves.

A: Yes…?

Q: Then is it true as my son said, when you give pity, when you send love and light to those in darkness, or those who complain and want to be "saved" without effort on their own part, when you are kind in the face of abuse and manipulation, that you essentially are giving power to their further disintegration, or contraction into self-ishness? That you are powering their descent into STS?

A: You know the answer!

Q: Yes. I have seen it over and over again. Were the individuals in our lives selected for the extremely subtle nature of their abilities to evoke pity, or were we programmed to respond to pity so that we were blind to something that was obvious to other people?

A: Neither. You were selected to interact with those who would trigger a hypnotic response, that would ultimately lead to a drain of energy.

Q: (L) What is the purpose of this draining of energy?

A: What do you think?

Q: (T) So you can't concentrate or do anything. You can't get anywhere with anything.

A: Or, at least not the important things. […]

Q: (T) Is it the area or the person?

A: Both. One is wrapped within the other.

Q: (L) Why is it that it seems to be one of the primary things about us that prevents us from acting against such situations, is our fear of hurting another person? […] Why are we so afraid of hurting someone's feelings if they are hurting us?

A: Not correct concept. You do not need to "act against them," you need to act in favor of your destiny.

Q: But, when you do that, these persons make you so completely miserable that there seems to be no other choice but a parting of the ways.

A: Yes, but that is not "acting against." Quite the contrary. In fact, remember, it takes two to tango, and if you are both tangoing when the dance hall bursts into flames, you both get burned!!!

Q: Why is it that when one tries to extricate from such a "tango," why is there such violent resistance to letting you go when it is obvious, clearly obvious, that they do not have any feeling for you as a human being?

A: It is not "they." We are talking about conduits of attack.

07-18-98

Q: (T) So, the complete UFT is known to someone here on the planet?

A: Yes.

Q: (T) And they are not making it available…

A: Oh no, because "The Truth Will Set You Free!" [...] You may access hyperspatial truths with UFT [Unified Field Theory].

Q: (A) Is it a good time for me to know more on that, to work on that? Can you give me a pointer so that I can discover it for myself?

A: Back up to where you were in "69."

Q: (A) I was reading books by Lichnerowicz on UFT....

A: Yes. Check the notes.

Q: (L) Well, we have a *real* problem with these notes and papers and things because of the fact that the bags that they were packed in have disappeared!

A: Gee, we wonder why?!? [...] Even without notes, the lonely young man walks down the concrete walk with the clumsily arranged light poles, contemplating the truth, the *real* truth. You were in an alpha state, a crossroads, wondering "Where do I go from here?" and "Why are all these things being pushed onto me?" Go back to then, Arkady. You know you are really a "Russian" at heart!

Q: (A) The question is whether such activity or knowing such things will lead to other densities? Is it just for satisfaction, or is there real value in knowing more in this direction?

A: Well, the Unified Field Theory unlocks the door completely to the higher densities.

Q: (T) I think there is more to it than this HAARP. UFT is a major step....

A: Grids. [...] The planet has been enshrouded with EM grid.

Q: (T) Are these the ley lines?

A: No.

Q: (L) Are they artificially generated?

A: Contoured.

Q: (L) They are artificially contoured. What is the result of this shrouding?

A: Manipulated for use by $3^{rd}/4^{th}$ Consortium.

Q: (A) What kind of EM grid? (L) The natural EM grid is being contoured....

A: Like a gently waving geometric "blanket."

Q: (T) Is it on the surface of the planet, through the planet, or where?

A: Above.

Q: (J) Do microwave towers factor into this?

A: Indirectly discovered by same principal. [As in person?]

Q: (T) The gravity waves, whether they exist or not, are a controversy, yet they are part of the UFT, and someone already knows how it works. Therefore, it is only controversy to those who don't know what the answer is, and it is not a controversy to those who know. They know what it is and how to measure it and how to use it.

A: Of course.

Q: (A) Some power is used to sustain this grid. What is it?

A: Land and space-based generators.

Q: (T) What can it be used for?

A: Multiple uses. [...] You are dancing on the 3rd density ballroom floor. Alice likes to go through the looking glass at the Crystal Palace. Atlantean reincarnation surge brings on the urge to have a repeat performance.

Q: (T) The Atlanteans who have reincarnated are getting ready to do the same thing they did before with the crystals. So, this is an Atlantean-type thing that is being done now? Different equipment, but the same type of thing?

A: All lessons must be learned before you can move onto bigger and better things. [...]

Q: (A) Now, how did we come to this grid from UFT?

A: Grid construction represents application of....

Q: (L) Somehow we went from the increased gravity of the sun, to UFT, to the grid....

A: UFT explains the "increased" gravity of Sol. But is there not something in UFT about increase/decrease???

Q: (A) There is no reason for it to increase or decrease... but this is Einstein's theory, which we were told is incorrect... we were told that there is some interaction between gravity and EM wave, and this is what UFT is about.... If we use other dimensions which we are supposed to use in this UFT, going with Kaluza-Klein, then the very concept of mass is something which is not so clear, and mass can be variable....

A: Yes, variability of physicality.

Q: (T) 4th density. (A) We were told earlier that this UFT opens the door to other densities....

A: Yes.

Q: (A) Can we have a UFT which unifies EM and gravity and does not include the concept of other densities? In other words, can we put in a textbook all about the gravity and electromagnetics, and a student could learn all of this and still know nothing about other densities?

A: No. Other densities become apparent when....

Q: (A) So, it means that Einstein and Von Neumann knew about these other densities?

A: Yes, oh yes!!!

Q: (T) Just a thought: Having UFT, and being able to manipulate different fields within it, creates different effects. So as we understand it in the apparent present state of science, we have to spin something in space in order to create gravity. But with the UFT, one small offshoot is that one could create real gravity without spinning anything. So the problem of weightlessness is really already solved....

A: Elementary my dear Terry, elementary.

Q: (T) So this whole thing with the space station and all the trouble they are having readapting to gravity when they come back, is all a game....

A: When you "let the cat out of the bag," you create an entire feline "nation."

Appendix C

During the period that Bridges was working on me to invite him to come and do hypnosis on me, part of his spiel to convince me that he was on the "same side," was a long series of suggestions that there was a dangerous Masonic/Black Magic Cult in my own town, and the threads all led back to Tyler, Texas. Here is what he wrote:

Date sent: Mon, 26 Feb 2001 02:00:35 -0500 Subject:

Re: (Fwd) (Para)Psychological triggers — The story...

From: Vincent Bridges

Dear Laura,

Well, I hope you know your group very well because outing this kind of information will have repercussions. As I have said all along, you are too close to the truth, on many levels. We seem to be in a race too get the info out and the sources protected before, literally, all hell breaks loose. Someone like you asking the right questions could be very dangerous to the Program. But, as my granny used to say, in for a penny, in for a pound, and if we all stand together, at least we will present a large enough target. (Note grim humor. . .) I read your missive while watching the X-Files, and once again, Chris Carter makes me wonder what the hell is going on. A new character, Laura Reas, is a specialist in ritual crime; she smokes but is apologetic about it, believes that there is something to the UFO business, but also believes in a human agency involved. She looks like a younger and thinner version of you and says fricking a lot. The universe with a sense of humor, or what? Carter used to do this regularly while we were working on AMET, truly freaky! I was interesting to see what the Cs [Cassiopaeans] had to say, but truly, they were not very helpful, to me another point in their favor. One thing that makes the Cs so unique are exchanges such as this one:

> Q: (L) What was the source or cause of the complete loss of my voice for several weeks?
>
> A: Q plasma bacterial infection. Ask Wu.
>
> Q: (L) You mean Dr. Vu?
>
> A: Yes. Pronounced Wu in southern-oriental inflective tone. [Yes, it *is* Wu, and I *did* ask. This is a definition of an "unknown" or

"unidentifiable" bacteria which the medical profession labels "quelle" to indicate its unknown origin.]

So the Cs not only know current medical jargon, unknown to you, but also can tell Cantonese from Mandarin, also unknown to you. Any other channeler in the business would have a field day with a hit this correct, it would be the source of their ad campaign! Now there's this:

Q: (L) Is this something that happens in altered states or in sleep states?

A: Not happens, happened.

Q: (L) Something that happened in the past?

A: Laura, you need to consult a powerful, practiced, effective hypnotherapist to unlock these questions for you.

From the same session and dated 04-24-96. Now you had shared this before, but somehow I didn't catch the date. When I read it tonight, something niggled at me and I went and checked my appointment book for 1996. I was still practicing then, and practically every moment of every day is recorded. On Wednesday, April 24th, 1996 I was in Tampa, the book says University of Tampa, but all I remember is a large and funky 1890s hotel deep in downtown, doing a seminar with my partner Dr. Jean Templeton for a group of therapists and psychs on early childhood abuse and our light and sound entrainment procedures for treating it. How far is that from New Port Richey? That night, a lady from Clearwater came up afterward with her therapist and told us the story of the Masonic Temple group in of all places, New Port Richey. My scribbled notes say "Blacke mer lodge, R(oyal)A(rch) Temple, 228 Front St. New Port Richey, Fla." Now this is too strange for words. I probably could have helped you recover the memories, but I didn't know then what I know now. So maybe there is a method here. Now this is also curious:

A: Yes. And other.

Q: (L) Can you tell me what ages?

A: 2,4,7,10,17,22,44

Q: (L) When I was 2, I was abducted?

A: Yes.

Q: (L) And when I was 4? I was kidnapped and programmed?

A: Yes.

Q: (L) And when I was 7 I was abducted?

A: Yes, and....

Q: (L) I was programmed then, too?

A: Yes.

Q: (L) When I was 10.... Abducted?

A: Yes...

Q: (L) Well, I am just going to have to think about all that.

You didn't ask about 44, why? You turned 44 in '96 so what did that mean? Any idea? And then there is this one:

A: O'Brien is "lyin'"

Q: What is it about O'Brien?

A: Discover. Why is Tom there, of all places?!?

Q: Is there something in that area, some frequency from the earth, some electromagnetics or something, that can tend to....

A: Maybe, maybe, maybe, maybe....

Q: It keeps a person quiescent and in the dark?

A: Stalling frequency.... And by the way, can anyone come up with a purpose for the existence of Camp Blanding? Well??... First, some blockbuster stuff for the Knighted ones.... Look upon a detailed map, and reflect, remember lonely journeys from long ago, and begin to unlock shattering mysteries which will lead to revelations opening the door to the greatest learning burst yet!!

So who is Tom and why was he at Camp Blanding? And then:

Q: (L) No. Who did this work, can I have that? Can I know that?

A: Consortium.

Q: (L) Is there any possibility, to some extent, that I have overcome this influence at the the present time?

A: No. Was partial, then aborted, leaving fragments of trigger response programs that have been in remission.

Q: (L) Why was it aborted?

A: Because STO forces intervened.

Q: (L) And when was this?

A: Mid "fifties."

Q: (L) So it was when I was three or four years old.

This is very curious, and we've already discussed your loose cannon status. This sounds like it was planned by the STO forces, so that you would have only so much programming, and no more. And then, the clincher:

Q: (TM) Or any two people who have had the programming?

A: The programming is mainly intended to produce erratic behavior, for the purpose of "spooking" the population so that they will welcome, and even demand, a totalitarian government.

Q: (L) So the programming is designed to... in other words, when the people are just being erratic....

A: Think of the persons who have inexplicably entered various public and private domains, and shot large numbers of people.... Now, you have "met" some of these Greenbaum subjects....

Q: (L) Let me say this. If this is what we're saying, well, what I'm saying is, that... is this Greenbaum programming something that goes along the line of what I've just described a part; a part, I don't think that's all of it....

A: In part.

Q: (L) Is there also the implanted triggers to activate at a certain point in future time, to create a mass chaos, in the public domain?

A: Better to discover that one on your own.

Q: (L) Ok, that's another one that's dangerous to know right now.... (V) Was the person I met last week, [Name deleted], was he Greenbaumed? Has he been Greenbaumed. (L) He was bizarre, wasn't he?

A: Now, some history... as you know, the CIA and NSA and other agencies are the children of Nazi Gestapo... the SS, which was experiment influenced by Antareans, who were practicing for the eventual reintroduction of the Nephalim on to 3rd and/or 4th density earth. And the contact with the "Antareans" was initiated by the Thule Society, which groomed its dupe subject, Adolph Hitler, to be the all time mind programmed figurehead. Now, in modern times, you have seen, but so far, on a lesser scale: Oswald, Ruby, Demorenschildt, Sirhan Sirhan, James Earl Ray, Arthur Bremer, Farakahan, Menendez, Bundy, Ramirez, Dahmer, etc....

Q: (L) Is there any particular individual who is currently being programmed to take a more prominent position in terms of this....

A: Later... you must know that Oswald was programmed to be the "patsy." So that he would say many contradictory things. Demorenschildt was both a programmer and programmed. Ruby was hypnotically programmed to shoot Oswald. With an audio prompt, that being the sound of a car horn.

Q: (L) The question has been brought up, is there some way or means that one can distinguish or discern a victim of Greenbaum or other mind programming by some clues?

A: *Not until it is too late.*

Now a disincarnate source that talks intelligently about the JFK assassination is truly unique. I have a good deal of evidence that everyone on the Cs list is actually part of the Program, even the two that don't fit, Farakahan and Bundy. Who else knows about the audio prompt that triggered Ruby? He didn't even remember going into the garage, just the sound of a car horn and the impulse to shoot. You won't find it in the literature unless you look through the transcripts of Ruby's conversation with Earl Warren, and Warren clearly didn't want to hear it. So how do the Cs know this? If you know simply as much as the Cs have told you about the JFK business, you can figure the whole thing out, and it ain't what Oliver Stone thinks it is. Who programmed Oswald? The New Orleans, Tampa, east-Texas axis at work! Who programmed DeMorenschildt and who was he working for? Answer that and you are one jump away from Dr. Greenbaum himself! You can get burned being this close to the fire. Jean Cocteau, Pope John XXIII, and John Kennedy all died that same year. Why? Find that link and the whole ball game unravels before your eyes. Look, back in 1995 and 96, when all of this started becoming very clear to me, I made a decision that has kept me out of all of this. It's better to know the truth and act accordingly, than it is to convince anyone else that you do indeed know that truth. Less unwanted attention that way. But you have been very public and are becoming more so every day. I think all of this makes the matter of your mastery of the mirror work even more important. Your lack of success and the lingering programming seem to be connected. J*** certainly expected you to respond to his triggers and we can't be completely sure you didn't, but as soon as J*** left the group, S*** popped up with two clues of the most extreme significance. It's almost as if J*** couldn't be allowed to get his hands on that info. Again, why? If they know what we know, or even more, why is it important to keep something like iron gall ink and oak galls from them? *The secret is in plain sight*, but no one can see it unless it is first pointed out by some one who can already see it. If everyone sees it at the same time that They do, then it can't be kept secret. Knowledge protects, *ignorance engenders*. Well, it's late, and I'm talking in riddles. LVX, V. [This author's emphases]

Aside from noting the psychopathic clue of the next to the last sentence — which could, indeed, merely be a typo, or it could be a clear "Freudian slip" — the reader is probably quite capable of identifying the "winding up" that was the objective of this missive. And, the truth is, to some extent it worked. I was about 65 percent convinced that I really needed Vincent to come right away and save me from myself and that dreadful programming, which he, and only he, could do. We notice that Vincent is very astute in many ways. He noted things about the Cassiopaean material that were truly insightful. The only thing was, everything was subsumed to his agenda — to get inside my head.

But there was a glitch: the story about the Masonic Lodge. Since it is a small town, and since I had lived there most of my life, there isn't much that goes on that isn't known by those who are "old-timers." I can guaran-damn-tee that there is no Black Magic mumbo-jumbo going on in the Lodge down the street. Another glitch was his assumption that the Cassiopaeans were talking about Chinese when discussing the pronunciation of my doctor's name. They were not: he is Vietnamese. And that bugged me. But it was too small to comment on, so I let it slide under the rug. The Masonic Lodge, however, I could not let slide. I wrote back.

From: Laura Knight-Jadczyk

To: Vincent Bridges

Subject: Re: (Fwd) (Para)Psychological triggers — The story...

Date sent: Mon, 26 Feb 2001 09:00:32 -0500

" told us the story of the Masonic Temple group in of all places, New Port Richey. My scribbled notes say "Blacke mer lodge, R(oyal)A(rch) Temple, 228 Front St. New Port Richey, Fla."

I'll have to have a look. Never heard of Front Street. I know that there is a Masonic Temple about 4 blocks away on Illinois avenue right across from the old High School I attended (though it wasn't there then) and the cross street is Harrison which is also the side street by our house here, though it zags a half a block before running by the school. I had a look at the map and there is no Front Street in NPR. There is a Front Dr. in Beacon Woods, a fancy schmancy subdivision with major golf courses and so forth, but it is considered to be in Hudson, about 10 miles north of here. They have a country club there, but I will have to ask H**, who lives there, about any Masonic Temple. Also, there are no 3-digit numbers at all in the entire county. All building numbers are based on plats and have four digits at the very least. So, I will have to investigate this further. My uncle and cousin are both Masons, and my aunt is former Grand Holy Pooh Bah Matron or something of the Eastern Star, so I will ask them about the local lodges. They come down here occasionally to attend lodge shindigs in Clearwater so they know who is who. My aunt gets highly upset when anybody suggests that the Masons are up to anything at all because, from their experience and perspective, it's just a glorified social club. I will note that my grandmother's uncle was a 33rd degree and an architect who probably designed the old hotel you stayed in. He was instrumental in designing many of the "historical" buildings of old Tampa, including the semi- famous Tampa Theater.

As is always the case with these things, the story morphs to adjust to what the victim "feeds back." They suddenly "know already" what you have told them and were just "checking" or something.

Date sent: Mon, 26 Feb 2001 11:58:25 -0500

Subject: Re: (Fwd) (Para)Psychological triggers — The story...

From: Vincent Bridges

Laura wrote:

" " Also, there are no 3 digit numbers at all in the entire county. All building numbers are based on plats and have four digits at the very least. So, I will have to investigate this further.

OK, I wanted you to confirm it, but that is exactly what I found. No such address. When I called her therapist to check, I discovered that she had completely disappeared, she had even given a false name and address to her therapist. I chalked it up to the general level of weirdness, but it seemed rather strange, given that about a year later I discovered that there was a Masonic Temple in New Port Richey and that it did have a rather sordid history of ritual-abuse claims against it. But nothing about anyone similar to the lady who approached us. All I can surmise is that someone wanted us to get in the car the next morning and go looking around NPR. Would we have found you, I wonder? Strangely enough, that was my last seminar with Jean. She went off to Masters and Johnson that summer and became famous for our technique. I manned the store while she was gone and then that fall I decided I had had enough. I started pulling back on clients, keeping only those that were willing to do the work.

By early '97 I was officially retired. I kept my **Tyler Texas** family because I had two sisters and a mother with serious problems and a history of involvement with the Program with going back to the 1930s. Their family connections, including William Bailey III, an oil man who was a friend of Prescott Bush's and the first person J. Edgar Hoover called after he was notified of the assassination, led me to the core of the whole JFK thing. That's how I know that the Cs interpretation is correct. They were just not telling you the whole story. So, while it's fine for me to dig the story out of other people's unconsciousness, it's too dangerous just to spill the whole thing out to you through the Ouiji board. Sheesh!

As I may have mentioned before, some of the truly strange events in a truly bizarre life have happened in your part of the world. In 1971, at Fla. Pres. College, now I think it is Eckard University, out at St. Pete Beach, a group of us experienced repeated UFO sightings, and some other rather strange experiences, as well as the most overwhelming sense of evil and imminent destruction that I have ever felt. Then again, a decade later in the aftermath of the Greensboro debacle, Reagan's election and the YIPie fiasco at the inauguration, I was conned into coming to Clearwater for a vacation that resulted in my being arrested for burglary. After a week in the Pinelas County jail, my family managed to get the charges dropped. I guess I had more resources than they figured, because the rest of my friends from that period are either dead or still in jail on various charges. So, I'm just a little reluctant to come to your area. Not rational, but there it is.

Now, I've sought guidance on the issue of what and how I might be able to help you unlock some of this. Since I'll only get one shot, of perhaps a few days length, you need to be absolutely ready to do the work. It sounds facetious to say I'm waiting on a sign from you, but there it is. When the timing is right, we'll both know it and I'll already be on the way. Ultimately, you are right that 10,000 copies of The Wave circulating the globe will be your best protection. But only if you know how to use that protection. If you're still hooked into the victim programming, you'll get eaten alive. More than a race, it is sort of like an orchestra surging toward that cymbal clash of climax and needing to arrive at the same point at the same time. More later, gotta go now and sell some books. Been chatting with M***, and have some questions, but later on that. - V.

Bibliography

Abell, George O., Morrison, D., and Wolff, S.C. *Exploration of the Universe.* Third edition. Saunders College Pub., 1975.

Ashe, Geoffrey. *The Ancient Wisdom.* Macmillan, 1977.

Attar, Farid Ud-Din. *Conference of the Birds.* New York: Random House, 1974.

Baillie, Mike. *Exodus to Arthur: Catastrophic Encounters with Comets.* London: Batsford, 1999.

———. *New Light on the Black Death The Cosmic Connection.* Stroud, Gloucestershire: Tempus, 2006.

Bamford, James. *Inside America's Most Secret Agency, The Puzzle Palace.* Houghton Mifflin Company, 1982.

Carroll, Lee and Jan Tober. *The Indigo Children: The New Kids Have Arrived.* Carlsbad, CA: Hay House Inc., 1999.

Castaneda, Carlos. *The Active Side of Infinity.* HarperCollins, 1998.

Chapman, C.R. and D. Morrison. *Cosmic Catastrophes.* New York: Plenum Press, 1989.

Cleckley, Hervey. *The Mask of Sanity: An attempt to reinterpret the so-called psychopathic personality.* St. Louis: The C. V. Mosby Company, 1941.

Clube, Victor and Bill Napier. *The Cosmic Serpent.* Londong: Faber and Faber, 1982.

———. *The Cosmic Winter.* Oxford: Blackwell, 1990.

Cremo, Michael A., Richard L. Thompson. *Forbidden Archaeology: The Hidden History of the Human Race.* San Diego: Bhaktivedanta Institute, 1993.

Eliade, Mircea. *Shamanism: Archaic Techniques of Ecstasy.* New York: Oxford University Press, 1964.

Elkins, Don, Carla Rueckert and Jim McCarty. *The Law of One: Book Three.* Schiffer Pub, 1982.

Festinger, Leon, et al. *When Prophecy Fails.* University of Minnesota Press, 1956.

Firestone, Richard, Allen West, and Simon Warwick-Smith. *The Cycle of Cosmic Catastrophes: Flood, Fire, and Famine in the History of Civilization.* Inner Tradition, 2006.

Freke, Timothy and Peter Gandy. *Jesus and The Lost Goddess.* New York: Harmony Books, 2001.

Fujii, Yoshiyuki and Okitsugu Watanabe. "Microparticle Concentration and Electrical Conductivity of A 700 m Ice Core from Mizuho Station Antarctic." *Annals of Glaciology* 1-, 1988, 38-42.

Gehrels, T. (ed.). *Hazards due to Comets and Asteroids*. Tucson: Univ. Arizon Press, 1994.

Godwin, Joscelyn. *Arktos: The Polar Myth in Science, Symbolism, and Nazi Survival*. Kempton: Adventures Unlimited Press, 1996.

Godwin, Joscelyn, Christian Chanel, and John P. Deveney. *The Hermetic Brotherhood of Luxor*. York Beach: Samuel Weiser, 1995.

Grieve, R.A.F. "Impact cratering on the Earth." *Scientific American* (1990), v. 262, 66-73.

Guggenbuhl-Craig, Adolph and James Hillman. *The Emptied Soul: On the Nature of the Psychopath*. Spring Publications, 1980.

Hare, Robert. *Without Conscience: The Disturbing World of Psychopaths Among Us*. New York: Guildford, 1999.

Harpending, H.C., and J. Sobus. "Sociopathy as an adaptation." *Ethology & Sociobiology*. 8: 63S-72S (1987).

Hedsel, Mark and David Ovason (ed.). *The Zelator: A Modern Initiate Explores the Ancient Mysteries*. London: Century Books, 1998.

Hildebrand, A.R. "The Cretaceous/Tertiary boundary impact (or the dinosaurs didn't have a chance)." *Journal of the Royal Astronomical Society of Canada* (1993), v. 87, 77-118.

Horne, John B. "Pyschopathic Personality: Where are the good guys now?" http://personal.cfw.com/~write/sociopth.html (accessed February 13, 2007).

Jacobs, David. *The Threat*. New York: Simon & Schuster, 1998.

Jessup, Morris K. *The Case for the UFO*. New York: Citadel Press, 1955.

Jung, Carl G. *Flying Saucers: A Modern Myth of Things Seen in the Sky*. London: Routledge, 1958.

Klein, Naomi. *The Shock Doctrine: The Rise of Disaster Capitalism*. New York: Metropolitan Books/Henry Holt, 2007

Legrand, Michel R. and Robert J. Delmas. "Soluble Impurities in Four Antarctic Ice Cores Over the Last 30,000 Years." *Annals of Glaciology* 10, 1988, pp 116-120.

Levy, Steven. *The Unicorn's Secret: Murder in the Age of Aquarius*. New York: Simon & Schuster, Inc., 1988.

Lorgen, Eve. *The Love Bite: Alien Interference in Human Love Relationships*. Bonsall, CA: ELogos & HHC Press, 2001.

McCafferty, Patrick and Mike Baillie. *The Celtic Gods: Comets in Irish Mythology*. Stroud, Gloucestershire: Tempus, 2005.

Mealey, L. "The sociobiology of sociopathy an integrated evolutionary model." *Behavioral & Brain Sciences*, 18, 523-599 (1995).

Meloy, J. Reid. *Violent Attachments*. Northvale, NJ: Jason Aronson, 1992.

Nasar, Sylvia. *A Beautiful Mind*. New York: Simon & Schuster, 1998.

Osborn, Nancy. *The Demon Syndrome*. New York: Bantam Books, 1983.

Ouspensky, P. D. *In Search of the Miraculous*. New York: Harcourt, Brace, Jovanovich, 1949.

Patterson, William Patrick. *Ladies of the Rope: Gurdjieff's Special Left Bank Women's Group*. Fairfax, CA: Arete, 1999.

Peckham, Morse. "Towards a Theory of Romanticism," *PMLA* 66 (1951): 5-23.

——. *Explanation and Power: The Control of Human Behavior*. New York: Seabury Press, 1986.

Peiser, Benny. "Comets and Disaster in the Bronze Age." *The Journal of the Council for British Archaeology* 30, December 1997, pp. 6-7.

Pennick, Nigel. *Secret Games of the Gods*. Maine: Samuel Weiser, Inc., 1992.

Pitchford, I. "The Origins of Violence: Is Psychopathy an Adaptation?" *Human Nature Review*. 1: 28-36, 2001, http://humannature.com/nibbs/01/psychopathy.html

Reich, Wilhelm. *Ether, God, and Devil*. NY: Orgone Institute Press, 1949.

Romer, Paul, Ph.D. University of Chicago, 1983, cited at: http://www.econ.uiuc.edu/~seppala/econ102/

Shaeffer, Claude. *Stratigraphic Comparée et Chronologie de l'Asie Occidentale*. London: Oxford University Press, 1948.

Schafer, Marcia. *Confessions of an Intergalactic Anthropologist*. Phoenix: Cosmic Destiny Press, 1999.

Schechner Genuth, Sara J. *Comets, Popular Culture and the Birth of Modern Cosmology*. Princeton University Press, 1997.

Schellhorn, G. Cope, Ph.D. "Evidence of Cyclical Earth Changes." *When Men are Gods*. Inner Light – Global Communications, 1991.

Scott, Ernest. *The People of the Secret*. London: Octagon Press, 1983.

Strieber, Whitley. *Communion: A True Story*. New York: Avon, 1995.

——. *Transformation: The Breakthrough*. New York: Avon, 1997.

Topper, Michael. "Precis on The Good and The Evil," *Thunderbird Journal*.

Turner, Elton. "Alien Behavior: Concept or Precept?" *Contact Forum*, September/October, 1994: http://www.karlaturner.org/articles/alien_behavior.html

Velikovsky, Immanuel. *Worlds in Collision*. New York: Dell, 1965.

Von Neumann, John, and Oskar Morgenstern. *Theory of Games and Economic Behavior*. Princeton: Princeton University Press, 1953.

Watkins, Leslie with David Ambrose, and Christopher Milles. *Alternative 3*. London: Sphere Books, 1978.

Online Einhorn Resources

Baker, Russ. "Ira's tour de France: How Ira Einhorn hijacked the French conscience." *Inquirer, Sunday Magazine*. [On Einhorn's flight to France] http://www.russbaker.com/Ira%27s%20tour%20de%20France.htm

DeMaio, Don. "The Hippie Killer." [Reminiscences of a friend of Einhorn's] http://web.archive.org/web/20011205000051/http://whitecanary.com/personal/columns/hippie_killer.htm

Einhorn, Ira. "Ira's Defense Statement." http://web.archive.org/web/20010611042412/http://www.ion.com.au/%7Ephil/IraLetter.html

Furia, Edward. "Subject: NBC 'Unicorn' Miniseries, which Knowingly Perpetuates Lie That Accused Murderer, Ira Einhorn, Was Founder of Earth Day." [Letter to John Katzman, producer of an NBC promo perpetuating Einhorn's myth] http://www.edfdad.addr.com/NBCkatzman4unicorn.html

Gorightly, Adam. "PKD, the Unicorn, and Soviet Psychotronics." [On Einhorn's correspondence with sci-fi author Philip K. Dick] http://www.alphane.com/moon/PalmTree/unicorn.htm

Lopez, Steve. "The Search for the Unicorn." [*TIME*'s coverage of the Einhorn case]
http://www.time.com/time/nation/article/0,8599,168382,00.html

Mind Control Corner. "Synchronicity Conspiracy." *Journal of Possible Paradigms* 7: 1998. http://www.elfis.net/elfol7/e7mkcie.htm
["Damage control" speculating that Einhorn was set-up as a "Manchurian Candidate" in order to silence him]

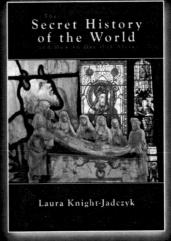